NMR
Spectroscopy
of
Biological
Solids

NMR
Spectroscopy
of
Biological
Solids

Edited by A. Ramamoorthy

CRC Press
Taylor & Francis Group
Boca Raton London New York

CRC Press is an imprint of the
Taylor & Francis Group, an **informa** business
A TAYLOR & FRANCIS BOOK

CRC Press
Taylor & Francis Group
6000 Broken Sound Parkway NW, Suite 300
Boca Raton, FL 33487-2742

First issued in paperback 2019

ISBN-13: 978-1-57444-496-4 (hbk)
ISBN-13: 978-0-367-39208-6 (pbk)

Library of Congress Card Number 2005050723

Library of Congress Cataloging-in-Publication Data

NMR spectroscopy of biological solids / edited by Ayyalusamy Ramamoorthy.
     p. cm.
  Includes bibliographical references and index.
  ISBN 1-57444-496-4
  1. Nuclear magnetic resonance spectroscopy. 2. Biomolecules--Analysis. 3. Solids. I. Ramamoorthy, Ayyalusamy.

QP519.9.N83N697 2005
570'.28--dc22   2005050723

Visit the Taylor & Francis Web site at
http://www.taylorandfrancis.com

and the CRC Press Web site at
http://www.crcpress.com

Preface

A plethora of techniques developed in the last decade has dramatically advanced solid-state nuclear magnetic resonance (SSNMR) spectroscopy as a unique tool for amazing applications into such diverse realms as analytical chemistry, bioengineering, material sciences, and structural genomics. Recent studies have shown that SSNMR is one of the most valuable techniques providing high-resolution details on the structural and functional aspects of many important biological solids (viruses, fibril-forming molecules such as amyloidogenic peptides, silk, collagen, cell walls, and molecules embedded in the cell membrane, such as lipids, cholesterol, proteins, and carbohydrates). The increasingly broad spectrum of developments and applications continues to glorify the intrinsic beauty and wealth of SSNMR. As a result, the number of research laboratories making significant contributions to solving problems related to biological solids has also increased considerably in the last decade. Therefore, I decided that publishing a book comprising all the recent developments in SSNMR and its applications to biological systems would benefit the scientific community. The goal of this book was not to provide an exhaustive compendium of information relevant to SSNMR but rather to provide a critical selection of SSNMR methods with demonstrated utility in the investigation of structure and function of biomolecules in solid-state.

Because there are no intrinsic limitations to the size and tumbling rate of molecules, SSNMR is applicable to all non-isotropic systems such as single crystalline, polycrystalline, amorphous, and semi-solids. While NMR parameters of these systems are tensorial in nature, they can be reduced to scalar parameters under magic-angle spinning (MAS), which provides solution-like spectra, but without any chemical modification. This unique advantage has been utilized thoroughly to measure accurately anisotropic interactions such as chemical shift anisotropy (CSA) and dipolar and quadrupole couplings. These anisotropic interactions contain a wealth of information about molecules: for example, chemical bonding, molecular shape and geometry, local and global molecular dynamics, folding and intermolecular interactions. In addition, the ability to measure both isotropic (chemical shift and scalar couplings) and anisotropic (mainly coherent homo- and heteronuclear dipolar couplings) from the same sample provided avenues to determine high-resolution structures of biological molecules in a non-soluble state. As a result, it is possible to determine three-dimensional structures of proteins in a microcrystalline state using SSNMR experiments under MAS. This topic is elaborately discussed in several chapters of this book. These techniques can also be extended to obtain information about the three-dimensional organization of molecules in nano (and pico) crystalline states, which could ultimately yield atomistic-resolution pictures of molecules in solid-state.

Another powerful and rapidly growing dimension of NMR spectroscopy is the study of aligned systems. In these systems, high-resolution spectral lines are rendered by the macroscopic alignment of molecules while the NMR parameters are still anisotropic. The art of preparing aligned molecules provide systems with various degrees of alignment defined by order parameters that vary from 0 (isotropic and completely unordered as in liquids) to 1 (rigid and completely ordered as in frozen solids). The last few years have witnessed tremendous progress in the development of multidimensional SSNMR methods to study aligned molecules. As discussed in several chapters of this book, significant strides have been made with respect to studying the structure and topology of membrane proteins using static SSNMR experiments on mechanically or magnetically aligned lipid bilayers. Such studies are important not only because they help us to understand basic issues related to protein folding in membranes but also to understand protein-induced damages to the cell membrane. Thus, these studies provide basic information about the mechanisms of action of a variety of biologically active molecules such as antimicrobial peptides, fusion peptides, toxins, ion channels, metal transporters, and signaling and receptor proteins.

It is important to note that the above-mentioned SSNMR studies are increasingly aimed at solving biological problems that have so far been considered nearly impossible. For example, the biogenesis of biological membranes is one of the most significant unresolved problems of biochemistry and cell biology. It should be apparent from recent biophysical studies that membrane proteins yield their structural secrets grudgingly. Therefore, it has been of considerable importance to develop a variety of biophysical techniques capable of solving, initially, the structure of membrane proteins and the host lipid bilayer within the membrane in as much detail as possible and, ultimately, of obtaining the response of each of these structures to a perturbation.

For the foreseeable future, progress in determining high-resolution structures of membrane proteins will depend greatly on SSNMR techniques. Poor sensitivity, resolution, and sample stability are some of major issues in this field and are addressed in this book. This volume also covers the theoretical and computational analysis of NMR experiments and experimental data.

I hope that *NMR Spectroscopy of Biological Solids* provides an answer for students and researchers with an interest in biological solids. I thank all the authors who contributed to this book, despite many other pressing obligations, for their understanding and patience during the prolonged birth of this project.

The cover picture of the book is a gift from Dr. Marassi (Burnham Institute) and was adapted from *Protein Science* (12, 403, 2003).

A. Ramamoorthy

Contributors

Sergii Afonin
IFIA
Forschungszentrum Karlsruhe
Karlsruhe, Germany

Marc Baldus
Department for NMR-Based Structural
  Biology, Solid-State NMR
Max Planck Institute for Biophysical
  Chemistry
Göttingen, Germany

Marina Berditchevskaia
Institute of Organic Chemistry
University of Karlsruhe
Karlsruhe, Germany

Peter V. Bower
University of Washington
Seattle, Washington, U.S.A.

Anna A. De Angelis
Department of Chemistry and
  Biochemistry
University of California, San Diego
La Jolla, California, U.S.A.

Gary P. Drobny
Department of Chemistry
University of Washington
Seattle, Washington, U.S.A.

Ulrich H.N. Dürr
Institute of Organic Chemistry
University of Karlsruhe
Karlsruhe, Germany

Carla M. Franzin
The Burnham Institute
La Jolla, California, U.S.A.

Ralf W. Glaser
Institute of Biochemistry and
  Biophysics
University of Jena
Jena, Germany

Stephan Grage
IFIA
Forschungszentrum Karlsruhe
Karlsruhe, Germany

Gerard S. Harbison
Department of Chemistry
University of Nebraska at Lincoln
Lincoln, Nebraska, U.S.A.

Mei Hong
Department of Chemistry
Iowa State University
Ames, Iowa, U.S.A.

Tortny Karlsson
University of Washington
Seattle, Washington, U.S.A.

Joanna R. Long
University of Washington
Seattle, Washington, U.S.A.

Gary A. Lorigan
Department of Chemistry and
  Biochemistry
Miami University
Oxford, Ohio, U.S.A.

Elizabeth A. Louie
Department of Chemistry
University of Washington
Seattle, Washington, U.S.A.

Francesca M. Marassi
The Burnham Institute
La Jolla, California, U.S.A.

Alexander A. Nevzorov
Department of Chemistry and
  Biochemistry
University of California, San Diego
La Jolla, California, U.S.A.

Neils Chr. Nielsen
INANO
Department of Chemistry
University of Aarhus
Aarhus, Denmark

Stanley J. Opella
Department of Chemistry and
  Biochemistry
University of California, San Diego
La Jolla, California, U.S.A.

Nathan A. Oyler
University of Washington
Seattle, Washington

Sang Ho Park
Department of Chemistry and
  Biochemistry
University of California, San Diego
La Jolla, California, U.S.A.

John Persons
Department of Chemistry
University of Nebraska at Lincoln
Lincoln, Nebraska, U.S.A.

Tatyana Polenova
Department of Chemistry and
  Biochemistry
Brown Laboratories
University of Delaware
Newark, Delaware, U.S.A.

Jennifer M. Popham
University of Washington
Seattle, Washington, U.S.A.

Ayyalusamy Ramamoorthy
Biophysics Research Division and
  Department of Chemistry
University of Michigan
Ann Arbor, Michigan, U.S.A.

Chad M. Rienstra
Department of Chemistry, Biophysics,
  and Biochemistry
University of Illinois at Urbana-
  Champaign
Urbana, Illinois, U.S.A.

Paolo Rossi
Center for Advanced Biotechnology and
  Medicine
Rutgers University
Piscataway, New Jersey, U.S.A.

Kay Saalwächter
University of Freiburg
Institute for Macromolecular
  Chemistry
Freiburg, Germany

Carsten Sachse
Institute of Biochemistry and
  Biophysics
University of Jena
Jena, Germany

Wendy J. Shaw
University of Washington
Seattle, Washington, U.S.A.

Patrick S. Stayton
Department of Bioengineering
University of Washington
Seattle, Washington, U.S.A.

Erik Strandberg
IFIA
Forschungszentrum Karlsruhe
Karlsruhe, Germany

Pierre Tremouilhac IFIA
Forschungszentrum Karlsruhe
Karlsruhe, Germany

Anne S. Ulrich
Lehstuhl Biochemie
Institut für Organische Chemie
Karlsruhe, Germany

Parvesh Wadhwani
IFIA
Forschungszentrum Karlsruhe
Karlsruhe, Germany

Sungsool Wi
Virginia Tech Department of Chemistry
Blacksburg, Virginia, U.S.A.

Alan Wong
Department of Chemistry
Queen's University
Kingston, Ontario
Canada

Gang Wu
Department of Chemistry
Queen's University
Kingston, Ontario
Canada

Jinghua Yu
The Burnham Institute
La Jolla, California, U.S.A.

Table of Contents

1 Magic-Angle Spinning Recoupling Techniques for Distance Determinations among Spin-1/2 Nuclei in Solid Peptides and Proteins

Chad M. Rienstra

CONTENTS

ABSTRACT Magic-angle spinning (MAS) recoupling techniques for determining long-range (>4 Å) distances among spin-1/2 nuclei are reviewed. These MAS experiments were originally designed to examine individual dipolar interactions between spin pairs to yield a few high-precision distances between site-specifically labeled nuclei. In recent years, many new pulse sequences and analysis methods have been devised for the purpose of examining spin clusters of several spins and even uniformly $^{13}C,^{15}N$-labeled samples with dozens to hundreds of resolved resonances. Such experiments sometimes require a tradeoff of measurement precision (data quality) to examine a larger number of distances per sample (data quantity). Here we consider both heteronuclear and homonuclear sequences, and discuss — from a qualitative and pragmatic perspective — the requirements and relative merits of various techniques in terms of available samples and instrumentation. Approaches to measure many semiquantitative distances in multidimensional spectra are emphasized, as these techniques will likely play a prominent role in total structure determination of solid proteins by magic-angle spinning nuclear magnetic resonance methods.

KEY WORDS: *distances, magic-angle spinning, peptides, proteins, recoupling, solid-state NMR, structure determination*

1.1 INTRODUCTION

More than any other property, strong (>4 kHz) dipolar couplings among nuclear spins differentiate solid-state from solution nuclear magnetic resonance (NMR). Without methods and instrumentation sufficient to decouple these interactions, solid-state NMR (SSNMR) spectra cannot be acquired with satisfactory sensitivity or interpreted in terms of structural properties. Pioneering studies in the first decades of SSNMR research focused on methods for removing the effects of dipolar couplings to obtain high-resolution spectra, and harnessing their power for purposes of sensitivity enhancement. Andrew et al.[1] and Lowe[2] showed that magic-angle spinning (MAS) attenuates the broadening effects of dipolar couplings and chemical shift anisotropy (CSA) at low magnetic fields. Haeberlen and Waugh employed coherent averaging[3] of the spin Hamiltonian to develop homonuclear decoupling sequences[4] that produce spectra possessing line widths of less than a part-per-million (ppm) despite the presence of >20 kHz ^1H-^1H dipolar couplings. Pines, Gibby, and Waugh recognized that the favorable Boltzmann polarization of ^1H spins could be exploited by Hartmann–Hahn[5] cross polarization[6,7] to improve the detection sensitivity for isotopically rare, lower–gyromagnetic ratio spins such as ^{15}N, ^{13}C, and ^{31}P. Schaefer and Stejskal combined MAS, cross polarization (CP), and high-power decoupling to acquire high-resolution spectra of organic solids.[8] All of these developments addressed issues unique to strong dipolar couplings, which especially at low magnetic field dominate the internal Hamiltonian, relative to chemical shift and indirect (scalar) coupling terms.

   Once it was established that strong coupling effects could be attenuated to acquire high-resolution spectra, attention became more directly focused on the goal of measuring weak dipolar couplings to extract structural information. For purposes of this review, we define weak dipolar couplings as those less than 200 Hz, to differentiate them from the moderate (200 Hz to 4 kHz) couplings and strong (>4 kHz) couplings, both of which must be removed by MAS to achieve high resolution. Moderate couplings are often used to establish correlations among ^{15}N and ^{13}C nuclei for assignment purposes (as reviewed recently by Baldus in Ref. 9 and elsewhere in this volume). Strong interactions, such as those involving directly bonded ^{1}H or ^{19}F spins, must usually be decoupled by high-power radio frequency (rf) irradiation or very fast MAS to obtain high-resolution spectra. Weak dipolar couplings arise from pairs of ^{15}N, ^{13}C, ^{31}P, ^{19}F, and ^{1}H nuclei at least ~2.5–3.0 Å apart; this corresponds to molecular geometries in which the precise distance depends on at least one torsion angle, or more generally on the overall tertiary fold of the molecule, whereas the magnitudes of strong couplings depend only on bond lengths and angles, which we presume in peptides and proteins to be generally well known. Measurements of these strong ^{1}H-^{13}C, ^{1}H-^{15}N, ^{13}C-^{15}N, and ^{13}C-^{13}C couplings can be valuable in at least two ways: first, the time-averaged coupling depends on the amplitude and timescale of molecular motions (most typically small amplitude librations), and so the measured values can be interpreted in terms of molecular order parameters; second, the trajectory of correlated dipolar evolution under two coupling tensors (e.g., ^{1}H^N-^{15}N and ^{1}Hα-^{13}Cα) depends on the relative orientations of the tensors, which in turn can be used to constrain torsion angles (in this example, ϕ), as reviewed by Hong elsewhere in this volume. However, in this chapter, we focus on weak couplings that can be directly interpreted in terms of internuclear distances, to be used subsequently for structure determination.

Weak dipolar couplings can be observed experimentally under MAS conditions only if pulse sequences selectively recover (recouple) a particular desired interaction while suppressing competing couplings.[10,11] This task can be simplified by considering both sample and experiment pulse sequence design parameters together, with appropriate consideration of the relative magnitudes of dipolar couplings. Table 1.1 presents a matrix of spin-1/2 nuclei that are commonly present in chemically synthesized peptides or drug molecules, or biosynthetically produced

TABLE 1.1
Accessible Range of Internuclear Distance Measurements (in Å) among the Spin-1/2 Nuclei Most Commonly Encountered in Peptides, Proteins, and Nucleic Acids, Assuming a 50-Hz (20-Hz) Coupling Detection Threshold

	^{15}N	^{13}C	^{31}P	^{19}F	^{1}H
^{15}N	2.9 (4.0)	3.9 (5.3)	4.6 (6.3)	6.1 (8.3)	6.2 (8.5)
^{13}C	3.9 (5.3)	5.3 (7.2)	6.3 (8.5)	8.3 (11.2)	8.5 (11.5)
^{31}P	4.6 (6.3)	6.3 (8.5)	7.3 (9.9)	9.7 (13.2)	9.9 (13.4)
^{19}F	6.1 (8.3)	8.3 (11.2)	9.7 (13.2)	12.9 (17.5)	13.1 (17.8)
^{1}H	6.2 (8.5)	8.5 (11.5)	9.9 (13.4)	13.1 (17.8)	13.4 (18.2)

proteins and nucleic acids. The matrix illustrates the potential range of distance determinations, assuming experiments that are able to measure 50 or 20 Hz couplings, respectively.

These values are calculated from the standard equation for the dipolar coupling constant (neglecting orientation dependence)

$$b_{IS} = \frac{-\mu_0}{4\pi} \frac{\gamma_I \gamma_S}{r_{IS}^3} , \qquad (1.1)$$

where the γ_I and γ_S are the gyromagnetic ratios of the nuclei, and r is the internuclear distance. The other fundamental constants have a net effect resulting in a coupling of 120.1 kHz for two ^{1}H spins separated by 1.00 Å; this value serves as a reference point for calculating other dipolar couplings but is much larger than any value ever observed in standard organic solids such as peptides and proteins. For example, the homonuclear coupling between methylene protons would be ~21.3 kHz, assuming a 1.78-Å distance; directly bonded ^{1}H-^{15}N and ^{1}H-^{13}C pairs have couplings on the order of 10.5 and 22.7 kHz, respectively, assuming 1.05- and 1.10-Å (effective) bond lengths.

Two points implicit to this table and equation are worth emphasizing. First, the tremendous potential for constraining molecular geometry is immediately evident, even if 50-Hz couplings could be measured with ±30% precision, which translates into about a ±10% error in the distance determination. In fact, such experimental uncertainties can be achieved with modern spectrometers and pulse sequences if sample relaxation properties and sensitivity are favorable. In less fortunate circumstances, the error in distance increases to ±20–30% or more, whereas the error may be as little as ±2% if the spectrometer is well optimized and the sample is rigid organic molecule (such as a small microcrystalline peptide). Likewise, measurements of 50-Hz couplings (i.e., ~20-ms dephasing trajectories in the time domain) are routine, and favorable circumstances may allow measurement of couplings as weak as 20 Hz. This would correspond to 5.4 Å ^{13}C-^{13}C, 8.5 Å ^{1}H-^{15}N, and 18.2 Å ^{1}H-^{1}H distances — a range much greater than can give rise to observable nuclear Overhauser effects (NOEs) in solution.[12] So why then is it not a trivial problem to determine complete protein structures in the solid state?

The answer to this question lies within the second implication of Table 1.1: weak dipolar couplings are often obscured by stronger couplings that have the same symmetry under sample spinning and rf manipulations. For example, the phenomenal prospect of measuring a 20-Hz coupling between ^{1}H spins separated by 18.2 Å is made practically impossible by the fact that a typical organic solid would have hundreds of other protons within the sphere of radius 18.2 Å. The competing couplings presented by these "spectator" spins complicate measurements of all structurally interesting dipolar couplings. The effect is least severe when naturally rare spin pairs can be uniquely labeled. For example, if two ^{13}C nuclei are synthetically placed in close proximity (<5 Å) within a sample that elsewhere contains only naturally abundant (1.1%) ^{13}C, dipolar dephasing or polarization transfer trajectories often agree extremely well with theory. However, three issues must be carefully considered. First, one must realize that the molecule is not always enriched to 100% ^{13}C at the site of interest and that all other carbon sites have natural abundance (i.e., 1.1% of ^{13}C). These factors contribute statistically to the experimentally observed dephasing or polarization transfer and must be considered for very precise work. Second, intermolecular couplings may be comparable to the intramolecular coupling of interest, especially when the latter is >4 Å or when small peptides are being examined; thus, it is necessary to prepare samples diluted in natural-abundance material (potentially increasing the magnitude of the corrections noted above) or to explicitly consider intermolecular interactions in the theoretical modeling of the spin dynamics. Third, the discussion presumes that sufficiently high-power ^1H pulsed magnetic (B_1) fields can be applied to attenuate the heteronuclear couplings. It is this last issue that defines the required instrumentation for dipolar recoupling experiments and most often compromises the precision of experiments by reducing echo lifetimes or complicating multiple quantum relaxation effects.

Therefore, the spectroscopist's ability to measure structurally informative distances depends on a combination of molecular environment and instrumental capabilities, which together define the most suitable type of pulse sequence for the desired application. The strong relationship between sample and experiment design motivates the organization of this chapter (Figure 1.1).

First, in the spin-pair limit (Section 1.2), the sample is engineered to have only two sites isotopically labeled. Historically, this has been the most tractable problem, and therefore its solutions are the most intellectually mature and commonly used in the SSNMR community. Spin-pair distances can be probed by either narrow- or broad-band pulse sequences, depending on the details of the problem. Classes of

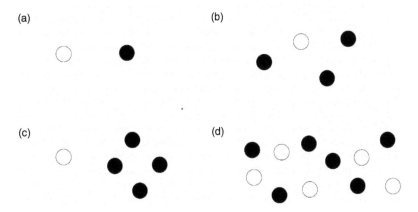

FIGURE 1.1 Schematic of different isotopic labeling limits discussed in this review. Open circles denote I spins; filled circles denote S spins. (a) The spin-pair limit. (b) Weakly coupled clusters of S spins. For some experiments, such as I-dephased, S-observe REDOR in this case, the behavior can be adequately described as the sum of several commuting spin pair interactions. (c) The spin cluster limit. Strong couplings among the S spins must explicitly be considered to interpret dipolar recoupling data. (d) The uniformly labeled limit. Multiple spin clusters exist, requiring site resolution through two- or three-dimensional experiments.

pulses sequences developed for this purpose include the REDOR[13] and transferred echo double resonance (TEDOR)[14] family of heteronuclear recoupling experiments, rotational resonance (R^2)[15,16] and related zero quantum (ZQ) homonuclear recoupling sequences (including radio-frequency driven recoupling (RFDR)),[17] and double-quantum (DQ) homonuclear recoupling experiments including homonuclear rotary resonance (HORROR),[18] its adiabatic variant dipolar recoupling enhanced by amplitude modulation (DREAM),[19,20] windowless sequences such as dipolar recoupling with a windowless sequence (DRAWS), melding of spin locking and DRAMA (MELODRAMA),[24] C7,[25] permutationally offset compensated C7 (POST-C7),[26] combined MLEV refocusing and C7 (CMR7),[27] supercycled POST C5 (SPC5),[28] and more recent R-symmetry sequences.[29] Most typically, a single distance per sample is measured this way although it is possible to prepare samples that have several isolated spin pairs. Given the ability to determine only one or a few distances from each sample, experimental precision is at a premium, and so through the years a great emphasis in this area of research has been placed on developing sequences for improved stability, bandwidth, and tolerance to experimental imperfections to enable applications to heteronuclear and homonuclear spin systems with varying chemical shift and CSA values. As a result, near-quantitative agreement between experiments and exact numerical simulations of the spin dynamics can be achieved. Most internuclear distances so far determined by MAS NMR have used these approaches, and they form the foundation for more complex experiments and sample conditions.

Second, in the small cluster limit (Section 1.3), sets of several spins may be in close physical proximity and therefore strongly coupled, but only a small fraction of the entire molecule is labeled; for example, one $^{13}C_3$-Ala and one ^{15}N-Gly residue may be incorporated into a peptide. To measure distances from the ^{15}N of the Gly to the ^{13}C, signals of the Ala would then require techniques to recouple the ^{15}N-^{13}C interaction selectively while suppressing the effects of the strong ^{13}C-^{13}C couplings within the Ala residue. The advantage of this approach is that it is often simpler and more cost effective to introduce such sets of ^{15}N and ^{13}C labels by amino acid (either by chemical synthesis or biosynthetic expression), and having multiple sites labeled in principle allows several distances to be determined in a single experiment. The disadvantage is that applying pulse sequences, designed for spin pairs, to spin clusters can produce substantially more complicated dephasing or polarization transfer trajectories.[30,31] Especially when several ^{13}C nuclei are strongly coupled to each other, dipolar truncation effects[28,30,32] obscure the structurally informative weak couplings in homonuclear experiments. This issue can be addressed in the following ways: samples can be designed to have several weak couplings of interest, but no strong couplings under a particular experiment (e.g., I-dephased, S-observe REDOR in the case of Figure 1.1b); the complicated multispin dynamics can be directly interpreted with advanced computational and theoretical tools; the data can be simplified by exploiting spectral selectivity, usually based on chemical shift differences among the labeled spins; or longitudinal spin order terms in the recoupled Hamiltonian can be used to avoid truncation effects.[33,34] Through combinations of these four strategies, increasingly sophisticated solutions to the spin cluster problem have been developed in recent years.

Third, in the uniformly (randomly) labeled limit (Section 1.4), spin density is constant throughout the molecule, so that all sites of a given type (e.g., Gly residues) are assumed to have the same probability of being ^{13}C or ^{15}N labeled. However, this probability does not necessarily have to be 100% — a point that will be considered in more detail. This final case presents the greatest spectroscopic challenge because not only are spin cluster dynamics present but also large numbers of signals (typically many dozens in small peptides or hundreds in larger proteins) must be resolved in two- or three-dimensional spectra. As such, site-specificity must be derived from sequential chemical shift assignment methodologies. Subsequently, site-resolved measurements of distances must come from spectra that are at minimum two-dimensional (two dimensions of chemical shifts), and more likely three-dimensional (two dimensions of chemical shifts and a dipolar dimension, or three dimensions of chemical shifts). Thus, both major spectroscopic problems — the spin dynamics of tightly coupled clusters and the site-specific assignment problem — must be addressed simultaneously. However, along with these challenges comes a huge potential payoff, as hundreds of distances could be measured from a few experiments on a single sample. This approach therefore has the greatest potential to improve the throughput of global protein structure determination by SSNMR. Experiments of this type are in early stages of development by many investigators, and it is certain to be a highly active area in the coming years.

In the concluding section, we discuss how current techniques might be combined to yield — with a few highly labeled samples — both global structural information and high-precision local measurements of direct relevance to function.

1.2 DISTANCE MEASUREMENTS BETWEEN SPIN PAIRS

1.2.1 SAMPLE DESIGN AND EXPERIMENT GOALS

Designing experiments and suitably labeled samples for spin-pair measurements requires sufficient background information (from mutagenesis, homology modeling, crystallography, previous NMR experiments, etc.) to construct specific structural hypotheses. Given the range of distance measurement techniques summarized above, some foresight about distances that are likely to be observed is the first prerequisite to successful application of spin-pair methods. The second requirement is the ability to prepare sufficient quantities of the sample with labels at the desired positions; in the case of small peptide studies, this is usually achieved by solid-phase peptide synthesis (SPPS).[35] For larger proteins, synthesis is possible but can be cost prohibitive and require special expertise.[36,37] Alternatively, if unique residues (typically Cys or Trp) exist in a protein sequence, incorporation of singly ^{15}N- or ^{13}C-labeled amino acids to the bacterial growth media may allow unique labeling by residue position. In general, substantial investments of time and isotope costs are required to prepare appropriately labeled samples, and these issues can restrict the available range of problems. Furthermore, scrambling of isotopes during biosynthesis (see, e.g., references 38 and 39) can complicate the data analysis.

In those cases in which appropriately labeled samples are available, there are two principal advantages of the spin-pair approach (relative to the cluster and uniform labeling limits discussed below). First, site-specific ^{13}C or ^{15}N signals are easy to identify and resolve in one-dimensional spectra, so that two- or three-dimensional chemical shift correlation spectra are not required, although suppression or subtraction of natural abundance background signals may require additional control experiments. Relative to two- or three-dimensional methods, one-dimensional spectra offer a sensitivity advantage for the one particular site of interest. Spectral sensitivity in a two-dimensional experiment is typically three to five times lower, because polarization transfer efficiencies are typically 30–60%, and another factor of two is lost as a result of hypercomplex sampling of the indirect chemical shift dimensions.

The second advantage of measuring couplings between spin pairs is that the spin dynamics are relatively simple in comparison with the multispin effects discussed further in Section 1.3. In the isolated spin-pair limit, the dephasing of one signal or polarization transfer from one signal to the other is dictated primarily by the dipolar coupling between the nuclei (presuming the use of a well-designed multiple pulse sequence). This approximation is especially well justified in the case of the REDOR experiment, used to measure heteronuclear distances. For high-precision homonuclear distance measurements, more detailed analysis of relaxation effects, pulse errors, and chemical shift differences is required. Over the years, these issues have been demonstrated to be tractable in many contexts.

1.2.2   HETERONUCLEAR RECOUPLING BY ROTATIONAL ECHO DOUBLE RESONANCE

1.2.2.1   The REDOR Concept

The rotational echo double resonance (REDOR) effect forms the basis of most heteronuclear distance measurement techniques. REDOR[13] and its closely related, coherence transfer technique (TEDOR)[14,40] are to date the only techniques established to provide quantitative measurements of structurally informative, long-range (>4 Å) heteronuclear distances. (Other schemes, such as double CP (DCP)[41] and its adiabatic variants,[42,43] are useful for establishing correlations between nearby nuclei but have not yet been demonstrated to be reliable for quantitative distance determinations.) In its simplest implementation, REDOR consists of a rotor-synchronized train of pulses on S (observe) and I (dephasing) spins. A Hahn echo is observed on the S spin, giving rising to the S_0 signal intensity; with pulses applied to the dephasing spin, the S spin echo is attenuated. The REDOR effect is typically reported as $\Delta S = (S_0 - S)/S_0$ as a function of time and can be subsequently analyzed in terms of the universal REDOR dephasing curve,[13] analytical REDOR transform techniques,[44] or numerical simulations.[45] Thus, the heteronuclear spin Hamiltonian is theoretically straightforward to interpret and apply to spin pair interactions.

Despite the elegant simplicity of its theory, REDOR has only in the last several years become broadly and routinely available to the SSNMR community. REDOR experiments require probes that can handle high power (at least ~80 kHz) ^1H decoupling and pulses with power levels of ~30–40 kHz on at least two other channels (typically ^{13}C and ^{15}N) for echo periods of 20 ms or more. Such probes must also demonstrate excellent interchannel isolation and have the long-term stability to obtain accurate ΔS data. It is therefore no coincidence that the majority of early developments and applications of REDOR, TEDOR, and DCP occurred in the Schaefer laboratory, where novel transmission-line probes were developed by McKay.[46] Dissemination of transmission line probe technology to several academic laboratories in the mid-1990s,[47] as well as improved commercial triple resonance probes, led to an increased rate of activity in REDOR research among a wider cross section of the SSNMR community. As REDOR has been applied to an increasing variety of problems, modifications to the original sequence have been devised to address spin clusters, as will be discussed in Section 1.3. First, in this section, we survey some recent examples of REDOR applications to accurate spin-pair measurements in peptides and proteins. The intent of this section is not to provide an exhaustive review of REDOR applications but to highlight strategies that foreshadow the discussion of uniformly labeled proteins.

1.2.2.2 Applications of REDOR to Spin Pairs

REDOR can be used to determine small peptide backbone structure by a series of measurements, as exemplified by Nishimura et al., who determined the structure of the pentapeptide Leu-enkephalin dihydrate (Tyr-Gly-Gly-Phe-Leu) by measuring ^{15}N-^{13}C distances in six samples labeled at various backbone (^{15}N and ^{13}CO) sites (Figure 1.2).[48] The authors implemented a strategy of measuring several four-bond distances and one five-bond distance, as illustrated in Figure 1.2. Each distance depends on at least two backbone torsion angles. For example, distance IV was

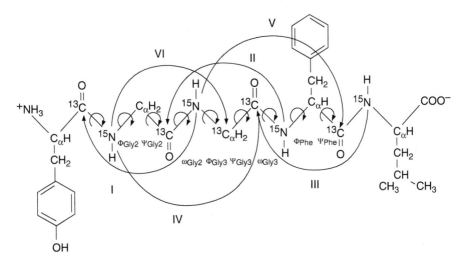

FIGURE 1.2 Schematic of six site-specifically $^{13}C,^{15}N$-labeled Leu-enkephalin (Tyr-Gly-Gly-Phe-Leu) samples used to determine backbone conformation with REDOR experiments. Each distance constrains at least two degrees of freedom (torsion angles) along the backbone. Reprinted from Ref. 48. Copyright 1998, American Chemical Society.

determined to be 5.10 ± 0.10 Å, and distance II to be 3.79 ± 0.10 Å, constraining the ϕ and ψ values of Gly2 and Gly3. This approach reduced the possible conformations of each residue to four, which when combined with chemical shift information yielded a unique backbone structure. One conclusion from this study was that a minimum of $2(N - 2)$ samples would need to be made to determine the (backbone) structure of a peptide of N residues.

A recent example of ^{15}N-dephased ^{13}C-observe REDOR to a larger protein was applied to the α-spectrin SH3 domain, with labels selectively positioned at Ala-^{13}Cβ (residues 11, 55, and 56) and the indole ^{15}N of Trp (41 and 42), along with the side-chain ^{15}N of Gln (16 and 50) and Asn (35, 38, and 47).[49] The Ala signals were clearly resolved in the one-dimensional ^{13}C spectrum, and a significant dephasing effect could be attributed to the ~4-Å distance to the indole ^{15}N of Trp42. One complication arose in this context of residue-specific isotopic labels for REDOR: it was necessary to determine independently the percentage of isotopic incorporation at each site to interpret the REDOR dephasing data accurately. Petkova et al. applied REDOR in a similar manner by labeling Trp indole ^{15}N signals in bacteriorhodopsin.[50] The interactions of Trp residues were examined, especially Trp182, which is close to the retinal; the ^{15}N chemical shifts of the Trp182 were assigned, and the distance from [20-^{13}C]retinal to [indole-^{15}N]Trp182 were determined not to change (within statistical limits) from the light adapted to M_O states.

The theoretical range of REDOR with ^{19}F dephasing is especially beneficial. An early application of ^{19}F-dephased ^{13}C-detected REDOR reported on an ~8-Å distance in a nine-residue fragment of the antibiotic peptide emerimicin.[51] This synthetic peptide contained a flouroacetyl group at its N-terminal Phe residue, and a ^{13}CO-^{15}N pair across a peptide ([^{13}CO]-methylalanine-[^{15}N]-Val) bond later in the sequence. TEDOR was used to select the uniquely labeled site, and ^{19}F dephasing applied to ascertain the distance in excellent agreement with the crystal structure (Figure 1.3).[51] Similarly, a study of rat cellular retinol binding protein, labeled with 6-^{19}F-Trp and 2-^{13}C-Trp, was used to constrain inter-Trp distances in the range of 7–11 Å,[52] and interactions between subunits of Trp synthase (a 143-kDa complex) were determined by 4-^{19}F-Phe to 4-^{13}C-Trp labeling and REDOR experiments, yielding three spin-pair interactions within <6 Å.[52] Weakly coupled clusters of spins were also examined to describe the molecular geometry of vancomycin binding sites in *Staphylococcus aureus*, using ^{13}C labels on the protein and ^{19}F labels on the antibiotic.[53] Finally, ^{19}F REDOR has been used to measure 12 to 12-5 Å ^{31}P-^{19}F distances in DNA labeled with phosphorothioates, in which the S^{31}PO$_3$ signal is shifted ~60 ppm downfield from the O^{31}PO$_3$ signal,[54] and to measure a ligand-induced conformational change between two α-helices in the serine bacterial chemoreceptor.[55] Notably, such experiments require probes that can decouple ^{1}H and ^{19}F simultaneously while allowing ^{31}P, ^{13}C, or ^{15}N observation.[56] Such probes have recently become commercially available from the major instrumentation vendors.

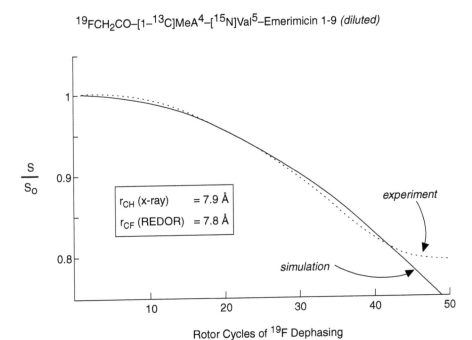

^{19}FCH$_2$CO–[1–^{13}C]MeA^4–[^{15}N]Val^5–Emerimicin 1-9 *(diluted)*

r$_{CH}$ (x-ray)   = 7.9 Å

r$_{CF}$ (REDOR) = 7.8 Å

experiment

simulation

$\dfrac{S}{S_0}$

Rotor Cycles of ^{19}F Dephasing

FIGURE 1.3 REDOR curve for a ^{13}C signal selected by a ^{15}N-^{13}C TEDOR period and dephased by a ^{19}F. The ^{13}C-^{19}F distance was determined to be 7.8 Å. Reprinted from Ref. 51. Copyright 1992, American Chemical Society.

1.2.3 HOMONUCLEAR RECOUPLING EXPERIMENTS

For spin-pair measurements, homonuclear experiments have primarily been applied to ^{13}C, and less often to ^{31}P and ^{19}F. Homonuclear couplings between ^{15}N spins in peptides and proteins are generally too weak to be structurally informative, although some exceptions have been demonstrated.[57] Although possible in special circumstances,[58] it is in general difficult to prepare protein samples with unique ^{1}H-^{1}H spin pairs; therefore, ^{1}H-^{1}H measurements require techniques for addressing clusters of spins, as discussed below. In contrast to the heteronuclear (REDOR) family of pulse sequences, homonuclear experiments may be less instrumentally demanding but more difficult to interpret uniquely in terms of an internuclear distance, because of the greater complexity of the homonuclear dipolar Hamiltonian under MAS, in the regime in which chemical shifts, MAS rates, and applied rf fields have the same order of magnitude. For most ^{13}C measurements, the required instrumentation has been commercially available for some time. However, the requirements for probe power (short pulses or spin locks on the ^{13}C channel and high-power decoupling of ^{1}H) vary greatly among the many pulse sequences; the exact amplitude and type of ^{1}H decoupling can have implications for the relaxation rates, which (unlike REDOR) can have nontrivial effects on the interpretation of internuclear distances.

1.2.3.1 The Rotational Resonance Family (Zero Quantum Recoupling)

1.2.3.1.1   Rotational Resonance (R^2)

Rotational resonance (R^2) occurs when the chemical shift difference between spins is equal to one or two times the MAS rate ($\Delta\omega = n\omega_r$, where $n = 1, 2$). Under this circumstance, the energy of MAS compensates for the (small) difference in Zeeman energy, permitting efficient longitudinal zero quantum (ZQ) polarization transfer. Although the R^2 effect was observed very early in the history of NMR,[59] its utility for measurement of distances was not realized until much later.[15] Relative to REDOR, the R^2 method is experimentally simple to implement but somewhat more difficult to interpret precisely.[16] Early applications of rotational resonance included ^{31}P-^{31}P measurements in a 50-kDa enzyme-inhibitor complex, which identified the bound species as a phosphonate;[60] studies of bacteriorhodopsin, which confirmed the presence of a 6-s-trans-retinoic acid by measurements between the ^{13}C-8 and ^{13}C-18 of the retinal[61] and determined the structure of the retinal-protein linkage in the dark-adapted state by measurements of retinal ^{13}C-14 to Lys216 $^{13}C\varepsilon$, a distance that varied from 3.0 ± 0.2 Å in bR555 to 4.1 ± 0.3 Å (syn C=N bond) in the bR568;[62] and applications to pair-wise labeled amyloid samples, permitting low-resolution structural models to be developed for pancreatic amyloid (residues 20–29 of the human islet amyloid polypeptide)[63] and the C-terminal fragment of β-amyloid (residues 34–42).[64] Based in part on the latter studies, the absolute precision of the R^2 method was determined to be limited by chemical shift dispersion, as well as relaxation effects.

1.2.3.1.2   Modifications to R^2

Several modifications of the original R^2 experiment have been proposed to address its dependence on relaxation and chemical shift dispersion. Costa et al. developed "rotational resonance tickling" (R^2T), using a ramped field through the R^2 condition to minimize the dependence on ZQ relaxation parameters and to enable highly precise (±0.2 Å) measurements of 4.3- and 5.05-Å distances in model compounds; at this range, the corrections for natural abundance background became significant.[65] The effect of differential relaxation was analyzed further by Levitt and coworkers[66] to understand subtle features of the line shapes that could accurately be interpreted in terms of internuclear distances.[67] Practical issues relevant to applying R^2 to larger biomolecules have also been systematically investigated,[68] and highly precise measurements have been demonstrated in large membrane proteins.[69] Most recently, Costa and Griffin have shown that measurements of R^2 as a function of spinning frequency provide accurate constraints on both ZQ relaxation and the dipolar couplings.[70] Applications of R^2 to uniformly labeled peptides will be discussed further below. Tilted frame versions of R^2 (R^2TR)[71,72] permit application to spin pairs whose chemical shifts are not separated by a convenient multiple of the spinning rate.

1.2.3.1.3   RFDR

The limitations of R^2 for purposes of correlating spins with arbitrary chemical shifts differences was recognized early in the work of Bennett.[17] The simplest picture of

RFDR is as a rotor-synchronized train of pulses that scales the first-order effect of isotropic chemical shift differences; initially, the motivation was to acquire broadband chemical shift correlation spectra from U-^{13}C-labeled samples (an application for which RFDR remains popular).[73–75] To use RFDR for quantitative distance determinations, as with R^2, the effect of ZQ relaxation must be explicitly considered. Full numerical density matrix simulations were necessary to account for finite pulse and relaxation effects; nevertheless, the precision was ±20 Hz, limited by natural-abundance background signals.[76] Constant time methods were developed[77] and applied to determine the antibody-dependent conformation of the third variable loop of the HIV-1 envelope protein gp120.[78] Further modifications of RFDR include band-selective,[79,80] compound,[81] finite pulse,[82,83] and adiabatic inversion[84] variations. RFDR has primarily been applied to ^{13}C-^{13}C measurements, although one recent example of ^{19}F-^{19}F distance measurements in the range a 5–12 Å has been presented.[85] As with REDOR, the success of ^{19}F-based homonuclear recoupling techniques depends strongly on spectrometer and probe instrumentation.

1.2.3.1.4 Proton-Driven Spin Diffusion

Proton-driven (longitudinal) spin diffusion (PDSD) is the simplest homonuclear mixing scheme to implement experimentally, but it is theoretically more difficult to describe.[86–89] Thus, it is less accurate for purposes of extracting internuclear distances. Semiquantitative (±0.6–1.0 Å) distances were determined in bacteriorhodopsin by RFDR and spin diffusion.[90] The accuracy of a distance determination from this method is limited by how well the ZQ line shape can be described. Applying ^{1}H-^{13}C recoupling during the spin diffusion period effectively broadens the ZQ line shape, enabling more reliable correlation spectroscopy (by a decreased dependence of polarization transfer on the proximity to R^2 conditions).[91,92] This experiment, named DARR (dipolar-assisted R^2) has also been applied to the membrane protein rhodopsin, labeled (in the spin-pair limit) with 4-^{13}C-Tyr and 8,19-^{13}C-retinal.[93] Rhodopsin has 18 Tyr residues, two of which are less than 5.5 Å from the retinal. The two-dimensional DARR spectrum of 150 nmol (6 mg) of this labeled rhodopsin sample provided two sets of cross peaks: one corresponding to the 8-^{13}C to 19-^{13}C retinal interaction, which confirmed chemical shift assignments and provided an internal calibration of the buildup curve (a known 2.5-Å distance, dependent only on bond lengths and angles). The 4-^{13}C-Tyr to 19-^{13}C-retinal peak was assigned to Tyr268, known to be 4.8 Å from 19-^{13}C. A two-dimensional experiment at a longer mixing time (1.5 s) showed an additional peak, assigned to Tyr191 (5.5 Å from the retinal in the crystal structure). These especially long mixing times (100–500 ms) are experimentally possible because of the modest decoupling field requirement of DARR; with high-power decoupling, dielectric heating of conductive protein samples might be prohibitive, although a new, commercially available probe design addresses this issue.[94]

1.2.3.2 The Dipolar Recoupling at the Magic-Angle Family (Double Quantum Recoupling)

The original DRAMA experiment illustrated principles for DQ dipolar recoupling.[21,22] Although the original experiment was too sensitive to chemical shift offsets

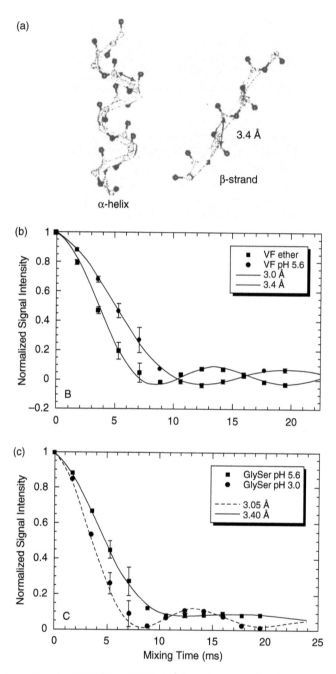

FIGURE 1.4 (a) Expected $^{13}CO[i]$ to $^{13}CO[i+1]$ distances from α-helix and β-strand peptides. (b) DRAWS dephasing data for ^{13}CO-Val_{18}-^{13}CO-Phe_{19} labeled β-amyloid(10–35) in ether-precipitated and fibrillar forms. (c) DRAWS dephasing data for ^{13}CO-Gly_{25}-^{13}CO-Ser_{26} labeled β-amyloid(10–35). Reprinted from Ref. 100. Copyright 2000, American Chemical Society.

and B_1 inhomogeneities to be useful for quantitative internuclear distance measurements, DRAMA provided the intellectual foundation for development of rotating frame experiments, including MELODRAMA[24,95] and DRAWS.[30,31,96,97] DRAWS has become the most fully developed experimental DQ scheme for accurate distance determinations; for example, it has been applied to studies of β-amyloid[98–100] to determine intramolecular $^{13}CO[i]$ to $^{13}CO[i+1]$ distances that depend on secondary structure (Figure 1.4). In a helix, this distance is ~3.0 Å; in a β-sheet, it is ~3.4 Å. The differences can be easily resolved, in this case demonstrating that the ethyl ether–precipitated form of β-amyloid(10–35) is helical, whereas the fibrillar form is a β-sheet.[100] Further, it was possible to determine that the supramolecular structure of the fibril was parallel and in register. DRAWS also has been applied to examine peptide structure on surfaces[101,102] to examine the extended backbone conformation of a peptide that inhibits hydroxyapatite growth.[101,103] Such applications to biomineralization problems are uniquely accessible to SSNMR recoupling techniques.

Pulse sequences designed from symmetry principles, including supercycled versions of $R14_2^6$, have also been used for distance determinations, especially in the context of ^{13}C sites with large CSA tensor magnitudes.[104] Although this initial application focused on precise bond length determinations, more recent developments include the determination of a longer distance with very high precision (3.82 ± 0.04 Å) with the SR26 sequence.[105] Several of these windowless recoupling pulse sequences have been compared numerically[106] and experimentally.[107] The continued development of robust DQ recoupling experiments will be critical for homonuclear distance measurements in spin pairs and likely will find applications in spin clusters.

Two additional DQ homonuclear recoupling experiments are worth mentioning. The HORROR experiment of Nielsen and coworkers[18] exploits the condition in which the sum of effective fields on two ^{13}C spins is equal to the MAS rate, providing a high scaling factor for rapid polarization transfer. The original HORROR experiment suffered from strong dependence on isotopic chemical shifts and rf field homogeneity. The DREAM experiment, consisting of adiabatic passage of the effective Hamiltonian through the HORROR condition, has been effectively used to overcome this shortcoming.[19,20] HORROR and its field-ramped variants may also be useful for band-selective measurements, as discussed further below.

1.2.4 CONCLUDING REMARKS ABOUT SPIN-PAIR MEASUREMENTS

Many examples of informative spin pair measurements in the literature illustrate the power and flexibility of such techniques, which are particularly well suited to directed studies of function, drug binding, and structure refinement. The risk of this approach is a null result: the lack of observable dephasing within the experimental signal-to-noise ratio. To distinguish among the possible reasons for the null result (longer distance than expected, motional averaging, etc.), additional samples must be prepared. Each new sample then would require expression or synthesis, purification, analytical characterization (by enzymatic assays, electrophoresis, mass spectrometry, optical spectroscopy, etc.), successful transfer into the NMR rotor in a functionally relevant state, and spectral validation of sample quality by NMR control experiments. These practical concerns add greatly to the total time investment required for this methodology, beyond the spectroscopic measurement time itself. If larger numbers of distances are required, as is the case in total structure determination, more efficient approaches to measuring them with fewer samples are highly desirable.

1.3 THE SPIN CLUSTER LIMIT: A FEW TO A DOZEN SPINS

1.3.1 HETERONUCLEAR EXPERIMENTS

Early efforts to measure multiple heteronuclear (I-S) couplings in a single, three-dimensional REDOR experiment demonstrated good accuracy,[108] but the range was limited to ~3 Å because of limitations on the ^{13}C echo (from scalar and residual dipolar ^{13}C-^{13}C couplings). Schaefer increased the echo lifetimes by employing a multiple pulse decoupling scheme on the ^{13}C channel during the REDOR period (multiple-pulse decoupled REDOR),[109] enabling measurement of a 7.7–8.1-Å distance between a CF_3 group and a ^{13}C of p-trifluoromethylphenyl[1,2-$^{13}C_2$]acetate. To interpret this experiment, it was necessary to calibrate the multiple pulse scaling factor separately for each type of spin cluster and for the number of pulses in the echo train.

Several authors have considered how to interpret REDOR data in $I_n S$ spin systems more precisely. The REDOR transform[110] enables a direct frequency-domain interpretation, but Fyfe and Lewis concluded that precise I_1-S and I_2-S distance determinations would be imprecise in this context, because of the effects of relative orientations of the I-S dipole vectors.[111] One approach to reducing the effect of multiple spin interactions is to replace the pulse in the middle of the I spin dephasing period with a θ pulse (~40°), reducing the probability that a given S spin would be affected by all I spins (i.e., θ-REDOR).[112] This simplifies interpretation at the cost of sensitivity (the overall dephasing effect is also scaled as θ varies from π). This approach was improved by combining the S_0 (i.e., $\theta = 0$), S (or $S\pi$, where $\theta = \pi$), and $S_{\pi/2}$ experiments in linear combinations to extract a signal function with higher sensitivity.[113] If both traditional and θ-REDOR data sets are available, the angle between the I_1-S and I_2-S vectors can be determined.[114] These effects can also be analyzed by correlated-tensor decomposition.[115] Thus far, analysis of systems beyond three spins in this manner has not been demonstrated to be tractable. For distance determinations in larger spin clusters, therefore, spectral simplification is required.

Frequency selectivity in the REDOR experiment can be achieved by an approach analogous to DANTE, dubbed FDR,[116] or with selective Gaussian pulses.[117] A single selective pulse applied to the desired S spin (^{13}C) decouples it from other S spins; in a IS_n spin system this results in near-ideal REDOR dephasing curves if a separate experiment is performed for each S spin of interest. Likewise, in an $I_m S_n$ spin system, $n*m$ separate dephasing trajectories can be acquired by selective inversion pulses on both I and S channels (Figure 1.5).[118] This frequency selective REDOR (FS-REDOR) experiment provides nearly ideal spin-pair behavior in cases in which the one-dimensional ^{13}C and ^{15}N spectra are resolved, or as long as the other spins within the excitation bandwidth of the selective pulse are not strongly coupled (e.g., the entire ^{13}CO region, from ~160 to ~190 ppm, can be inverted in the ^{15}N-dephased,

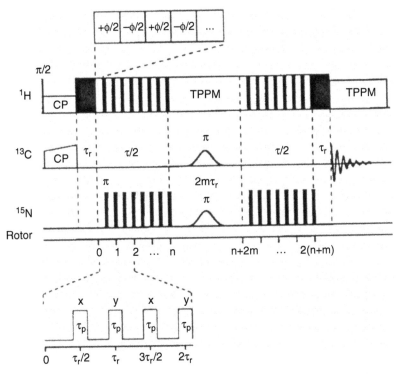

FIGURE 1.5 Frequency selective REDOR pulse sequence. The REDOR effect between a individual ^{13}C and ^{15}N spins is measured by employing narrow-band Gaussian inversion pulses. Reprinted from Ref. 118. Copyright 2001, American Chemical Society

^{13}CO-observed experiment, thus reducing experimental effort). In the extensively studied tripeptide formyl-Met-Leu-Phe-OH, several distances in the range of 4–6 Å were determined in this manner, with typical uncertainties of ±0.2 to ±0.4 Å. Examples of the quality of the dephasing data are shown in Figure 1.6, and the corresponding distances in crystal structure are in Figure 1.7.[119] Combined with torsion angle restraints,[120] these distances enabled the determination of this tripeptide structure to high resolution.[121]

The problem of suppressing homonuclear dipolar coupling while recoupling heteronuclear dipolar interactions can be addressed by symmetry. Gross and Griffin developed a tilted n-fold symmetric ^1H-X (X = ^{15}N, ^{13}C, etc.) recoupling scheme using frequency-switched Lee–Goldburg[122] homonuclear (^1H-^1H or ^{13}C-^{13}C) decoupling within a C-symmetric experiment.[25] Hohwy and coworkers further advanced this approach with a transverse Mansfield–Rhim–Elleman–Vaughan (T-MREV) scheme,[123] and generalized R-class symmetries have been applied successfully as well.[124] These solutions address strongly coupled ^1H spin systems; applications to suppress ^{13}C-^{13}C couplings in ^{15}N-^{13}C REDOR experiments have also been demonstrated.[125–127] The most advanced application to date of this strategy, involving protons, was developed by Schmidt-Rohr and Hong, who combined the approaches of MPDR and frequency-selective REDOR (FS-REDOR) to achieve homonuclear

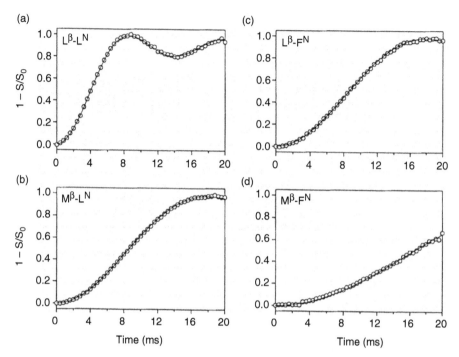

FIGURE 1.6 Frequency selective REDOR data from different pairs of spins in a U-^{13}C,^{15}N-labeled tripeptide, formyl-Met-Leu-Phe-OH. The distances determined were (a) Leu-N to Leu-Cβ, 2.46 ± 0.02 Å; (b) Leu-N to Met-Cβ, 3.12 ± 0.03 Å; (c) Phe-N to Leu-Cβ, 3.24 ± 0.12 Å; and (d) Phe-N to Met-Cβ, 4.12 ± 0.15 Å. Distances as long as 5.7 ± 0.7 Å were determined. Reprinted from Ref. 118. Copyright 2001, American Chemical Society

^{1}H-^{1}H decoupling and site-specificity in the same experiment, opening up the use of ^{1}H in REDOR experiments on solid peptides.[128] This scheme (Figure 1.8) uses two types of multiple-pulse decoupling, first to suppress homonuclear ^{1}H-^{1}H couplings while a (scaled) ^{1}H-^{13}C dipolar interaction is recovered, and then to suppress all couplings and shifts on the ^{1}H nuclei during the selective ^{13}C pulse. This experiment has the same limitations inherent to the FS-REDOR experiment; namely, that the ^{1}H spin will evolve under the influence of couplings to all ^{13}C nuclei within the selective pulse bandwidth. Data interpretation may also be complicated by intermolecular effects; however, in the model compound studied, it was possible to account with high accuracy for the statistical distribution of 3.1, 4.5, and 6.0 Å ^{1}H-^{13}C distances (Figure 1.9). The technique has been demonstrated to determine distances that constrain torsion angles in small peptides[129] and distributions of distances in the range of 3.3–4.3 Å in the elastin mimetic, (Val-Pro-Gly-Val-Gly)$_3$.[130]

1.3.2 HOMONUCLEAR EXPERIMENTS

Principles similar to the FS-REDOR experiment can be adapted to homonuclear spin systems. Individual R^2 conditions can be matched, and an initial condition

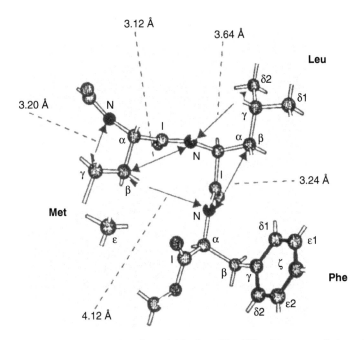

FIGURE 1.7 X-ray structure of N-formyl-Met-Leu-Phe-OH with some of the solid-state nuclear magnetic resonance–measured distances labeled. Reprinted from Ref. 118. Copyright 2001, American Chemical Society.

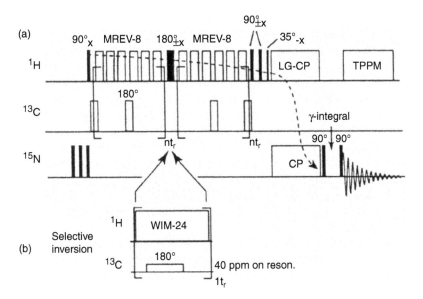

FIGURE 1.8 Pulse sequence for ^{15}N-detected ^{13}C-^1H REDOR. MREV-8 suppresses ^1H-^1H couplings during the ^1H-^{13}C REDOR period, enabling precise measurements of the scaled the heteronuclear coupling. Reprinted from Ref. 128. Copyright 2003, American Chemical Society.

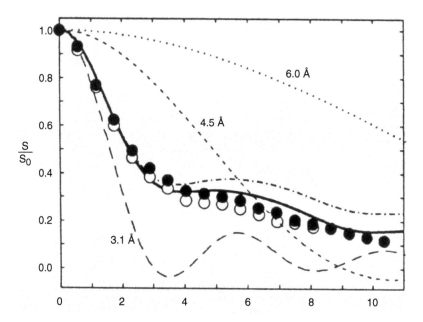

FIGURE 1.9 Experimental dephasing for an intermolecular ^{13}CO-$^1H^N$ interaction in ^{15}N-labeled N-t-BOC-Gly. The open circles represent data from the standard experiment of Figure 1.8a; the filled circles are data from the experiment with selective ^{13}C inversion, minimizing the dephasing caused by natural abundance aliphatic sites. The solid line is a simulation based on the crystal structure, including intermolecular interactions. The dotted lines represent REDOR curves for single distances, as labeled. Reprinted from Ref. 128. Copyright 2003, American Chemical Society.

created by selective inversion of one ^{13}C spin, yielding distances of up to 4.5 Å with 10% accuracy (Figure 1.10). Williamson, Meier, and coworkers were able to demonstrate excellent agreement between experiment, theory, and x-ray structure distances in small molecules.[131] Similar approaches were demonstrated to be applicable to β-amyloid(11-25) fibrils.[132] Constant-time R^2 width measurements may improve accuracy.[133] All of these approaches suffer from the fact that R^2 conditions from "spectator" spins may cause line broadening of the desired observation spin; for example, the measurement of $^{13}CO[i]$ to $^{13}C\alpha[i+1]$ may be structurally informative but would require acquiring data (in general) near the $^{13}CO[i]$-$^{13}C\alpha[i]$ R^2 condition. The line broadening contributions from R^2 in this context can be avoided by applying the R^2TR experiment, as demonstrated in a series of measurements on a U-^{13}C,^{15}N-labeled dipeptide.[134,135]

As with REDOR experiments, attempts to apply standard implementations of homonuclear recoupling experiments result in complicated trajectories, which can be analyzed in terms of distances out to ~3.5 Å.[30,31] Band-selective homonuclear DQ recoupling sequences[18,19,136,137] reduce the effective size of the spin cluster, thereby increasing polarization transfer efficiency and simplifying interpretation. Although not sufficient to extract structurally informative distances from U-^{13}C-labeled peptides, this approach foreshadows the three-dimensional experiments discussed in the following section.

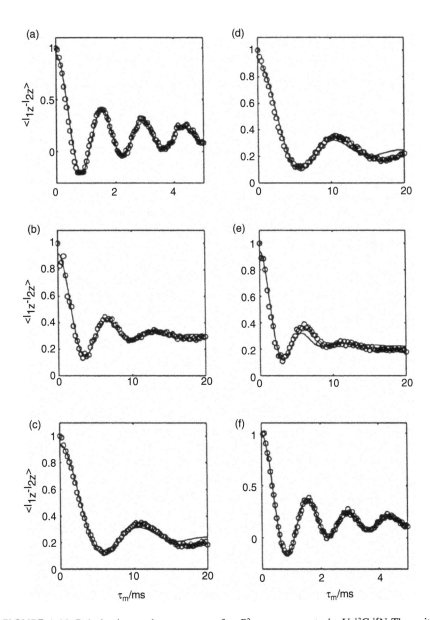

FIGURE 1.10 Polarization-exchange curves for R^2 measurements in U-^{13}C,^{15}N-Thr, with nuclear magnetic resonance (with error bar) and x-ray distances noted in parentheses. (a) ^{13}CO-^{13}Cα [1.556(4)-Å solid-state nuclear magnetic resonance, 1.54 Å x-ray]; (b) ^{13}CO-^{13}Cβ [2.47(3) Å, 2.55 Å]; (c) ^{13}CO-^{13}Cγ [2.93(3) Å, 3.09 Å]; (d) ^{13}Cγ-^{13}CO [2.91(4) Å, 3.09 Å]; (e) ^{13}Cγ-^{13}Cα [2.37(4) Å, 2.55 Å]; (f) ^{13}Cγ-^{13}Cβ [1.554(6) Å, 1.52 Å]. Reprinted from Ref. 131. Copyright 2003, American Chemical Society.

1.4   THE UNIFORMLY LABELED LIMIT: DOZENS TO HUNDREDS OF SPINS

1.4.1   SAMPLE PREPARATION AND CHEMICAL SHIFT ASSIGNMENTS

The leap to U-^{13}C,^{15}N labeling requires site-specific assignments as a starting point. As such, the major advances in recent years by the Oschkinat, McDermott, Zilm, and Baldus groups[73–75,138–140] have presented the foundation on which structure-determination strategies will be based. Both sample preparation methodologies and pulse sequences appear to be generally applicable, and many direct analogies to the solution of NMR structure determination process are evident. It is therefore reasonable to expect that *de novo* assignments will become more common in the near future, as commercial instrumentation and pulse programs are now available for this purpose. Along with assignments comes semiquantitative secondary structure analysis via secondary chemical shift indices and empirical databases (e.g., TALOS).[141] For example, Böckmann and colleagues assigned the regulatory protein Crh in microcrystalline form with a variety of highly resolved two-dimensional experiments, from which informative backbone dihedral predictions could be made.[139] However, to determine a global fold, a protein requires a significant number of experimentally determined, long-range internuclear distances. To this end, we now consider techniques aimed at encoding distance information in the dozens to hundreds of resolved cross peaks in two- and three-dimensional spectra of U-^{13}C,^{15}N-labeled proteins.

1.4.2   UNIFORM AND 100% LABELING

Fully ^{13}C,^{15}N-labeled samples (typically expressed biosynthetically from ^{13}C glucose and ^{15}N ammonium chloride) may be suitable for certain types of distance measurement strategies, in which strong couplings among ^{1}H spins are emphasized, or selective ^{15}N-^{13}C or ^{13}C-^{13}C Hamiltonians can be engineered. Many of the important advances so far have been based on explicit consideration of what sets of spins are likely to give the most informative distances.[142] Longer is not necessary better. For example, Lange, Luca, and Baldus have described a conceptually simple but especially powerful experiment that probes ^{1}H-^{1}H couplings with site-resolution through a two-dimensional ^{13}C-^{13}C plane.[143] This experiment works despite dipolar truncation effects because the interresidue ^{1}H spins in the hydrophobic core of proteins are significantly closer to each other (as close as ~2–2.5 Å in van der Waals contact) than the interresidue ^{13}C spins. Thus, qualitative or semiquantitative constraints can be leveraged very effectively. For example, the authors observed many peak assignments consistent with long-range distances in ubiquitin. This same strategy was applied to a macroscopically disordered fibril sample by Tycko and coworkers, who readily measured intermolecular contacts, based on molecular modeling that indicates the shortest distance between Hα sites on different strands of an antiparallel β-sheet to be 2.1 Å.[144] Thus they identified intermolecular cross peaks (Figure 1.11) between Val18 and Phe19 in fibrils prepared at pH 7.4, whereas a shift in registry occurred at pH 2.5 (cross peaks between Val18 and Ala21). Beyond the simplicity of the experimental scheme, the very short mixing times minimize variations in

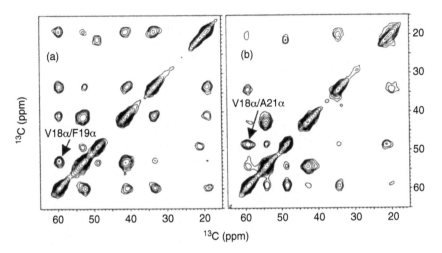

FIGURE 1.11 Aliphatic regions of 2D ^{13}C-^{13}C spectra in which polarization transfer proceeds through a ^{13}C-^1H-^1H-^{13}C pathway. The cross-peak intensities therefore are most intense for those $^1H\alpha$ sites close in space. Fibrils of β-amyloid(16–22) with U-^{13}C,^{15}N labeling of residues 18–21 were prepared at pH 7.4 (a) and pH 2.5 (b). Reprinted from Ref. 144. Copyright 2003, American Chemical Society.

signal decay over the protein (e.g., caused by motion) and avoid lengthy bursts of high-power ^1H decoupling.

Distances between ^{15}N and ^{13}C in U-^{13}C,^{15}N peptides can be measured by a three-dimensional z-filtered and band-selective TEDOR experiment.[33] This experiment bears close resemblance to the three-dimensional REDOR experiment of Michal and Jelinski,[108] but with the critical addition of a z-filter or selective pulses and phase cycling to remove undesired antiphase components. The build-up curves from the three-dimensional z-filtered TEDOR version show quantitative agreement for ^{15}N-^{13}C distances as long as 4.7 Å (Figure 1.12) and semiquantitative estimations of distances as long as 6.1 Å. The band-selective version is an elaboration of the FS-REDOR experiment, designed to avoid the shortcomings of a finite selective pulse bandwidth. A selective pulse is used to select a band of resonances in the ^{13}C spectrum, but the ^{15}N dimension is frequency labeled by evolution of the antiphase coherence generated from the initial REDOR period. These experiments were together used to measure 35 distance restraints (including 10 of greater than 4 Å) in a fibril-forming 11-residue peptide fragment from transthyretin, using a total of three samples, each with consecutive four-residue stretches of U-^{13}C,^{15}N-labeled amino acid residues.[145] The value of REDOR and TEDOR schemes derives from the form of the recoupling Hamiltonian as a longitudinal spin order ($2\,I_z\,S_z$), which avoids direct truncation effects; this point has been discussed in more detail in a recent publication by Nielsen and coworkers, in which it was also demonstrated that the γ-encoded TEDOR experiment gamma-encoded transferred echo (GATE) has the additional advantage of larger theoretically achievable coherence transfer (because of the dependence of only the dipolar coupling phase, not amplitude, on the Euler angle γ defining the orientation of the interaction to the MAS rotor frame of reference).[34]

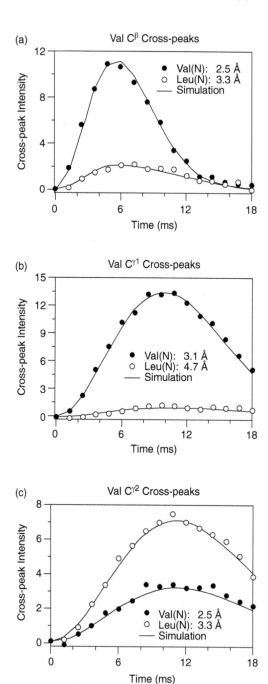

FIGURE 1.12 Cross-peak buildup curves from a three-dimensional z-filtered TEDOR experiment applied to the U-^{13}C,^{15}N-labeled dipeptide N-acetyl-Val-Leu. The distances noted on the figures agree well with the known crystal structure. Reprinted from Ref. 33. Copyright 2002, American Chemical Society.

Two recent developments applicable to ^{13}C-^{13}C distance measurements in U-^{13}C-labeled samples are derived from band-selective pulse sequences. Ladizhandsky and Griffin developed a band-selective three-dimensional R^2 tickling resonance width (R^2TRW) scheme to enable polarization transfer from ^{13}CO to $^{13}CH_3$ signals with site resolution.[146] Distances in the range of 2.9 to 6 Å were measured with about ±0.5-Å precision. The Baldus group found similar success in the R^2TR limit, where the polarization transfer among closely spaced resonance in U-^{13}C His were analyzed out to ~3.6 Å with precision of better than ±0.5 Å. This experiment was also demonstrated in ubiquitin.[147] Although so far the limit near R^2 has been explored the most (revealing CO-Cα and CO-methyl distances), the application of HORROR[18] and/or DREAM[19,20] may be well suited to recoupling resonances with closely spaced chemical shifts (e.g., methyl-methyl and aromatic-aromatic ^{13}C-^{13}C distances are typically within 4–5 Å in the hydrophobic protein core).

1.4.3 SPIN DILUTION

Beyond the problem of overcoming dipolar truncation in U-^{13}C,^{15}N-labeled samples, resolution must be adequate to assign peaks uniquely. It is in this respect that isotopic spin dilution schemes are especially helpful. Protein expression in bacterial grown on 1,3-^{13}C or 2-^{13}C glycerol, originally implemented by LeMaster for solution relaxation studies,[148] was adapted for SSNMR sample preparations by Hong and Jakes to improve resolution.[149] Oschkinat and coworkers then demonstrated this approach to be especially valuable for structure determination in the α-spectrin SH3 domain.[150] In combination with PDSD[86] ^{13}C-^{13}C two-dimensional experiments on a 750-MHz spectrometer, the resulting "checkerboard" isotopic labeling patterns (Figure 1.13a) increase the percentage of cross peaks that arise from long-distance interactions. In comparison with the two-dimensional ^{13}C-^{13}C spectrum from a U-^{13}C-labeled sample (Figure 1.13b), the line widths from the 2-^{13}C (Figure 1.13c) and 1,3-^{13}C (Figure 1.13d) samples are narrower, and a much greater number of long-range interactions are observed. Mixing times of up to 500 ms can be used because no RF pulses are on during the mixing period, avoiding problems caused by sample heating. In total, 286 interresidue ^{13}C-^{13}C constraints (149 long range) were assigned from these spectra. Structure calculations yielded a backbone root mean squared deviation (RMSD) of 1.6 ± 0.3 Å with respect to the average SSNMR structure. This study was subsequently supplemented with three-dimensional versions of the experiments, using an additional ^{15}N dimension to improve resolution, despite the use of a much lower field (400 MHz) instrument.[151] From the three-dimensional data, an additional 374 distance constraints were determined, improving the resolution to 1.1 Å, or 0.7 Å when torsion angle likelihood obtained from shifts and sequence similarity (TALOS) backbone dihedral constraints were also included.

The final strategy useful for distance measurements in highly enriched protein samples is to prepare proteins from ^{15}N,^2H media for bacterial expression and then back-exchange the labile ^1H sites during purification or refolding of the protein. This labeling scheme, combined with water suppression techniques, results in ^1H-detected heteronuclear correlation spectra with especially high sensitivity and resolution.[152] Furthermore, combination with an indirect ^1H chemical shift evolution period

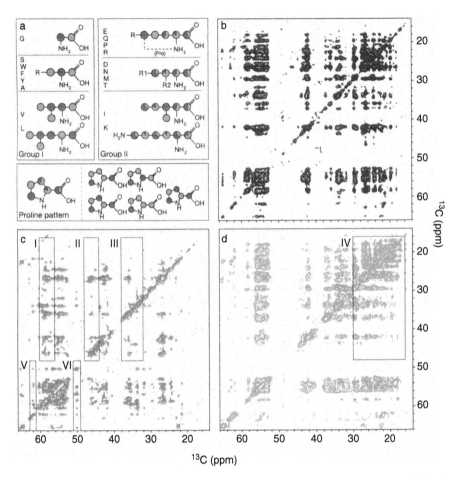

FIGURE 1.13 Isotopic labeling patterns and nuclear magnetic resonance spectra derived from three preparations of the α-spectrin SH3 domain. (a) Schematic of effective ^{13}C enrichment for the indicated residues. The dark gray sites derive a ^{13}C label from 2-^{13}C-glycerol (with ^{13}C carbonate); the light gray sites derive from 1,3-^{13}C-glycerol (with ^{12}C carbonate). Fractional labeling is represented by the pie chart within each site. (b) Two-dimensional ^{13}C-^{13}C spectrum of U-^{13}C,^{15}N SH3. (c) Spectrum of SH3 derived from 2-^{13}C-glycerol. (d) Spectrum of SH3 derived from 1,3-^{13}C-glycerol. Reprinted from Ref. 150. Copyright 2002, Hartmut Oschkinat.

enables the observation of dipolar exchange and NOEs in a three-dimensional experiment.[153] The example shown in Figure 1.14 includes a sequential walk through a portion of the ubiquitin sequence in a β-turn conformation, where strong correlations are observed for ^{1}H spins 2.7 to 3.0 Å apart, as well as weaker peaks corresponding to 4.1 Å distances. Similar distances were observed in ^{15}N-^{15}N two-dimensional experiments based on ^{1}H-^{1}H mixing in a ^{15}N,^{2}H,^{1}H-back-exchanged sample of the α-spectrin SH3 domain.[154]

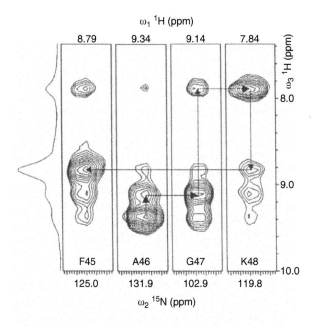

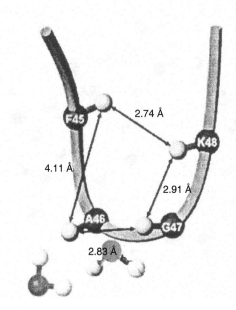

FIGURE 1.14 Strip plots from a three-dimensional experiment correlating $^{15}N[i]$, $^{1}H^{N}[i]$, and $^{1}H^{N}[j]$. Correlations around the β-turn (Phe45 to Lys48) of ubiquitin are illustrated, arising from spins as far as 4.1 Å apart. Reprinted from Ref. 153. Copyright 2003, American Chemical Society.

1.5 CONCLUSIONS AND PROSPECTS FOR HIGH-RESOLUTION PROTEIN STRUCTURE DETERMINATION

With the recent advances in so many aspects of distance measurement techniques, we conclude our discussion by considering the logic of *de novo* SSNMR protein structure determination. Embracing this approach requires accepting a compromise (at least temporarily) of data quality in favor of quantity. Whereas in the spin-pair limit, many recoupling experiments could be implemented with especially high precision, in the uniformly labeled limit, interpretation is clouded by multispin effects and imperfect resolution. Confidence that this paradigm shift will succeed comes not only from early examples in SSNMR (both of a total structure and of quantitative measurements in small peptides) but also from the wide success of solution NMR, in which NOE constraints are classified in a semiquantitative manner. Therefore, the SSNMR community should continue to proceed with confidence to the task at hand.

Applying MAS recoupling methods to determine complete, high-resolution protein structures will likely require several samples and pulse sequences. Presuming that chemical shift assignments are available and predominantly determined from a 100% U-^{13}C,^{15}N sample (derived from ^{13}C-glucose and ^{15}N ammonium chloride), one could apply three-dimensional, band-selective TEDOR[33] and R^2T experiments[146] to the same U-^{13}C,^{15}N samples. Either experiment could be extended to three dimensions, including an additional ^{13}C or ^{15}N chemical shift dimension to increase the resolution and minimize the ambiguity of cross-peak assignments; in this scenario, a single or small number of ^{15}N-^{13}C mixing times might be selected to minimize experiment time, cognizant of the goal to identify long-range interactions semiquantitatively. With the same sample, as described by Lange et al.[143,155] and Tycko and Ishi[144] ^{1}H-^{1}H distance measurements built on resolved ^{13}C-^{15}N or ^{13}C-^{13}C planes would provide another set of NOE-like constraints to provide interdomain contacts at a semiquantitative level of precision. Diluting the ^{13}C,^{15}N sample in naturally abundant material will generally be necessary to avoid confusion of intramolecular with intermolecular cross peaks.

With a second pair of samples derived from 1,3-^{13}C-glycerol and 2-^{13}C-glycerol, two-dimensional ^{13}C-^{13}C and three-dimensional ^{15}N-^{13}C-^{13}C experiments from Castellani et al.[150,151] would provide additional large numbers of semiquantitative ^{13}C-^{13}C distances. Such samples would also result in improved resolution and sensitivity for the aforementioned experiments. For example, removal of one-bond scalar couplings by chemical means avoids signal loss from selective pulses in the three-dimensional TEDOR sequence and provides a sensitivity and resolution benefit in the direct dimension as well. Glycerol-derived samples have especially high resolution in the CO and CH$_3$ spectra regions; therefore, techniques such as R^2TRW are likely to work even better than already demonstrated in the U-^{13}C,^{15}N samples.

The third set of samples would likely be ^{13}C,^{15}N,^{2}H labeled with exchangeable ^{2}H sites (amide, carboxyl)[156] replaced with ^{1}H (and, optionally, with protonated methyl groups from α-keto acids).[157] The techniques developed by Reif et al.[58,154] Zilm,[152,153] and coworkers could then be applied to obtain yet another unique set of distance constraints. Such samples would also be well suited to the selective ^1H-^{13}C or ^1H-^{15}N REDOR experiments of Schmidt-Rohr and Hong,[128] which would benefit from the improved resolution of the ^1H chemical shift dimension.

Together, these sets of experiments would provide approximately 10 or more distance constraints per residue. Presuming that the majority of these peaks can be resolved and assigned in the spectra, more than adequate information would be available to calculate global folds with acceptable resolution. The extent to which the spectra will be resolved, of course, will have a major effect on the viability of this approach. However, this same problem of assignment ambiguity has been considered in the solution NMR community. It is likely that SSNMR data sets will be compatible with these algorithms that consider ambiguous constraints.[158,159] With an initial global fold determined, structure refinement can proceed by incorporation of torsion angle constraints and conformation-dependent chemical shifts[141] (if not already included in initial structure calculations). Furthermore, spectrally selective spin-pair experiments can be revisited to target specific regions of the protein of special functional interest. Such an approach is inefficient if the hypothesis is ill formed, but benefits greatly from initial low-resolution data.

Only a few complete protein structures have thus far been determined exclusively by two-/three-dimensional MAS methods on highly enriched samples,[150,151,160,161] although it appears that several groups are progressing toward this goal in other systems. Many challenges remain to address major fundamental questions: What type and size of proteins will be amenable to analysis? How precisely can protein structures be determined by this approach? Along this path, many more practical issues will also be addressed. With many laboratories now actively developing distance measurement techniques that are directly applicable to uniformly labeled proteins, it is reasonable to be optimistic that the methods will continue to advance at the remarkable pace of the last several years.

ACKNOWLEDGMENTS

The author thanks the University of Illinois Department of Chemistry for start-up funds to support this and other ongoing research in the Rienstra group, and Professor Ann McDermott (Columbia University) for insightful discussions.

REFERENCES

1. Andrew, E.R., Bradbury, A. and Eades, R.G., Nuclear magnetic resonance spectra from a crystal rotated at high speed, *Nature (London)*, 182, 1659, 1958.
2. Lowe, I.J., Free induction decays of rotating solids, *Phys. Rev. Lett.*, 2, 285, 1959.
3. Haeberlen, U. and Waugh, J.S., Coherent averaging effects in magnetic resonance, *Phys. Rev.*, 175, 453, 1968.
4. Waugh, J.S., Huber, L.M. and Haeberlen, U., Approach to high-resolution NMR in solids, *Phys. Rev. Lett.*, 20, 180, 1968.
5. Hartmann, S.R. and Hahn, E.L., Nuclear double resonance in the rotating frame, *Phys. Rev.*, 128, 2042, 1962.

6. Pines, A., Gibby, M.G. and Waugh, J.S., Proton-enhanced nuclear induction spectroscopy, *J. Chem. Phys.*, 56, 1776, 1972.

7. Pines, A., Gibby, M.G. and Waugh, J.S., Proton-enhanced NMR of dilute spins in solids, *J. Chem. Phys.*, 59, 569, 1973.

8. Schaefer, J. and Stejskal, E.O., ^{13}C-NMR of polymers spinning at the magic-angle, *J. Am. Chem. Soc.*, 98, 1031, 1976.

9. Baldus, M., Correlation experiments for assignment and structure elucidation of immobilized polypeptides under magic-angle spinning, *Prog. Nucl. Magn. Reson. Spectrosc.*, 41, 1, 2002.

10. Griffin, R.G., Dipolar recoupling in MAS spectra of biological solids, *Nat. Struct. Biol.*, 5, 508, 1998.

11. Dusold, S. and Sebald, A., Dipolar recoupling under magic-angle spinning conditions, *Ann. Rep. NMR Spectroscopy*, 41, 185, 2000.

12. Overhauser, A.W., Polarization of nuclei in metals, *Phys. Rev.*, 92, 411, 1953.

13. Gullion, T. and Schaefer, J., Rotational-echo double-resonance NMR, *J. Magn. Reson.*, 81, 196, 1989.

14. Hing, A., Vega, S. and Schaefer, J., Transferred-echo double-resonance NMR, *J. Magn. Reson.*, 96, 205, 1992.

15. Raleigh, D.P., Levitt, M.H. and Griffin, R.G., Rotational resonance in solid state NMR, *Chem. Phys. Lett.*, 146, 71, 1988.

16. Levitt, M.H., Raleigh, D.P., Creuzet, F. and Griffin, R.G., Theory and simulations of homonuclear spin pair systems in rotating solids, *J. Chem. Phys.*, 92, 6347, 1990.

17. Bennett, A.E., Ok, J.H., Griffin, R.G. and Vega, S., Chemical shift correlation spectroscopy in rotating solids: radio-frequency dipolar recoupling and longitudinal exchange, *J. Chem. Phys.*, 96, 8624, 1992.

18. Nielsen, N.C., Bildsøe, H., Jakobsen, H.J. and Levitt, M.H., Double-quantum homonuclear rotary resonance: efficient dipolar recovery in magic-angle spinning nuclear magnetic resonance, *J. Chem. Phys.*, 101, 1805, 1994.

19. Verel, R., Baldus, M., Ernst, M. and Meier, B.H., A homonuclear spin-pair filter for solid-state NMR based on adiabatic-passage techniques, *Chem. Phys. Lett.*, 287, 421, 1998.

20. Verel, R., Ernst, M. and Meier, B.H., Adiabatic dipolar recoupling in solid-state NMR: the DREAM scheme, *J. Magn. Reson.*, 150, 81, 2001.

21. Tycko, R. and Dabbagh, G., Measurement of nuclear magnetic dipole-dipole couplings in magic-angle spinning NMR, *Chem. Phys. Lett.*, 173, 461, 1990.

22. Tycko, R. and Smith, S.O., Symmetry principles in the design of pulse sequences for structural measurements in magic-angle spinning nuclear-magnetic-resonance, *J. Chem. Phys.*, 98, 932, 1993.

23. Gregory, D.M., Mitchell, D.J., Stringer, J.A., Kiihne, S., Shiels, J.C., Callahan, J., Mehta, M.A. and Drobny, G.P., Windowless dipolar recoupling — the detection of weak dipolar couplings between spin-1/2 nucleic with large chemical-shift anisotropies, *Chem. Phys. Lett.*, 246, 654, 1995.

24. Sun, B.Q., Costa, P.R., Kocisko, D., Lansbury, P.T. and Griffin, R.G., Internuclear distance measurements in solid-state nuclear-magnetic-resonance — dipolar recoupling via rotor synchronized spin locking, *J. Chem. Phys.*, 102, 702, 1995.

25. Lee, Y.K., Kurur, N.D., Helmle, M., Johannessen, O.G., Nielsen, N.C. and Levitt, M.H., Efficient dipolar recoupling in the NMR of rotating solids. A sevenfold symmetric radiofrequency pulse sequence, *Chem. Phys. Lett.*, 242, 304, 1995.

26. Hohwy, M., Jakobsen, H.J., Edén, M., Levitt, M.H. and Nielsen, N.C., Broadband dipolar recoupling in the nuclear magnetic resonance of rotating solids: a compensated C7 pulse sequence, *J. Chem. Phys.*, 108, 2686, 1998.

27. Rienstra, C.M., Hatcher, M.E., Mueller, L.J., Sun, B.-Q., Fesik, S.W., Herzfeld, J. and Griffin, R.G., Efficient multispin homonuclear double-quantum recoupling for magic-angle spinning NMR: ^{13}C-^{13}C correlation spectroscopy of U-^{13}C-erythromycin A, *J. Am. Chem. Soc.*, 120, 10602, 1998.

28. Hohwy, M., Rienstra, C.M., Jaroniec, C.P. and Griffin, R.G., Fivefold symmetric homonuclear dipolar recoupling in rotating solids: application to double quantum spectroscopy, *J. Chem. Phys.*, 110, 7983, 1999.

29. Carravetta, M., Eden, M., Zhao, X., Brinkmann, A. and Levitt, M.H., Symmetry principles for the design of radiofrequency pulse sequences in the nuclear magnetic resonance of rotating solids, *Chem. Phys. Lett.*, 321, 205, 2000.

30. Kiihne, S., Mehta, M.A., Stringer, J.A., Gregory, D.M., Shiels, J.C. and Drobny, G.P., Distance measurements by dipolar recoupling two-dimensional solid-state NMR, *J. Phys. Chem. A*, 102, 2274, 1998.

31. Kiihne, S.R., Geahigan, K.B., Oyler, N.A., Zebroski, H., Mehta, M.A. and Drobny, G.P., Distance measurements in multiply labeled crystalline cytidines by dipolar recoupling solid state NMR, *J. Phys. Chem. A*, 103, 3890, 1999.

32. Hodgkinson, P. and Emsley, L., The accuracy of distance measurements in solid-state NMR, *J. Magn. Reson.*, 139, 46, 1999.

33. Jaroniec, C.P., Filip, C. and Griffin, R.G., 3D TEDOR NMR experiments for the simultaneous measurement of multiple carbon-nitrogen distances in uniformly C-13, N-15-labeled solids, *J. Am. Chem. Soc.*, 124, 10728, 2002.

34. Bjerring, M., Rasmussen, J.T., Krogshave, R.S. and Nielsen, N.C., Heteronuclear coherence transfer in solid-state nuclear magnetic resonance using a gamma-encoded transferred echo experiment, *J. Chem. Phys.*, 119, 8916, 2003.

35. Merrifield, R.B., Solid phase peptide synthesis. I. The synthesis of a tetrapeptide, *J. Am. Chem. Soc.*, 85, 2149, 1963.

36. Schnolzer, M. and Kent, S.B. H., Constructing proteins by dovetailing unprotected synthetic peptides — backbone-engineered HIV protease, *Science*, 256, 221, 1992.

37. Kochendoerfer, G.G. and Kent, S.B. H., Chemical protein synthesis, *Curr. Opin. Chem. Biol.*, 3, 665, 1999.

38. LeMaster, D.M., Isotope labeling in solution protein assignment and structural analysis, *Prog. Nucl. Magn. Reson. Spec.*, 26, 371, 1994.

39. Waugh, D.S., Genetic tools for selective labeling of proteins with alpha-N-15-amino acids, *J. Biomol. NMR*, 8, 184, 1996.

40. Hing, A.W., Vega, S. and Schaefer, J., Measurement of heteronuclear dipolar coupling by transferred-echo double-resonance NMR, *J. Magn. Reson. A*, 103, 151, 1993.

41. Schaefer, J. and Stejskal, E.O., Double cross polarization NMR of solids, *J. Magn. Reson.*, 34, 443, 1979.

42. Hediger, S., Meier, B.H. and Ernst, R.R., Adiabatic passage Hartmann-Hahn cross-polarization in NMR under magic-angle sample-spinning, *Chem. Phys. Lett.*, 240, 449, 1995.

43. Baldus, M., Geurts, D.G., Hediger, S. and Meier, B.H., Efficient N-15-C-13 polarization transfer by adiabatic-passage Hartmann-Hahn cross polarization, *J. Magn. Reson. A*, 118, 140, 1996.

44. Mueller, K.T., Analytic solutions for the time evolution of dipolar-dephasing NMR signals, *J. Magn. Reson. A* 113, 81, 1995.

45. Bak, M., Rasmussen, J.T. and Nielsen, N.C., SIMPSON: a general simulation program for solid-state NMR spectroscopy, *J. Magn. Reson.*, 147, 296, 2000.

46. McKay, R.A., Double-tuned single coil probe for NMR spectroscopy. US Patent 4,446,431, May 1, 1984.

47. McKay, R. A., Probes for special purposes, in *Encyclopedia of Magnetic Resonance*, John Wiley and Sons, NY, 1996, pp. 3768.

48. Nishimura, K., Naito, A., Tuzi, S., Saito, H., Hashimoto, C. and Aida, M., Determination of the three-dimensional structure of crystalline Leu-enkephalin dihydrate based on six sets of accurately determined interatomic distances from C-13-REDOR NMR and the conformation-dependent C-13 chemical shifts, *J. Phys. Chem. B*, 1998, 102, 7476.

49. Macholl, S., Sack, I., Limbach, H.H., Pauli, J., Kelly, M. and Buntkowsky, C., Solid-state NMR study of the SH3 domain of alpha-spectrin: application of C-13-N-15 TEDOR and REDOR, *Magn. Reson. Chem.*, 38, 596, 2000.

50. Petkova, A.T., Hatanaka, M., Jaroniec, C.P., Hu, J.G.G., Belenky, M., Verhoeven, M., Lugtenburg, J., Griffin, R.G. and Herzfeld, J., Tryptophan interactions in bacteriorhodopsin: a heteronuclear solid-state NMR study, *Biochemistry*, 41, 2429, 2002.

51. Holl, S.M., Marshall, G.R., Beusen, D.D., Kociolek, K., Redlinski, A.S., Leplawy, M.T., McKay, R.A., Vega, S. and Schaefer, J., Determination of an 8-Angstrom interatomic distance in a helical peptide by solid-state NMR-spectroscopy, *J. Am. Chem. Soc.*, 114, 4830, 1992.

52. McDowell, L.M., Lee, M.S., McKay, R.A., Anderson, K.S. and Schaefer, J., Inter-subunit communication in tryptophan synthase by carbon-13 and fluorine-19 REDOR NMR, *Biochemistry*, 35, 3328, 1996.

53. Kim, S.J., Cegelski, L., Studelska, D.R., O'Connor, R.D., Mehta, A.K. and Schaefer, J., Rotational-echo double resonance characterization of vancomycin binding sites in *Staphylococcus aureus*, *Biochemistry*, 41, 6967, 2002.

54. Merritt, M.E., Sigurdsson, S.T. and Drobny, G.P., Long-range distance measurements to the phosphodiester backbone of solid nucleic acids using P-31-F-19 REDOR NMR, *J. Am. Chem. Soc.*, 121, 6070, 1999.

55. Murphy, O.J., Kovacs, F.A., Sicard, E.L. and Thompson, L.K., Site-directed solid-state NMR measurement of a ligand-induced conformational change in the serine bacterial chemoreceptor, *Biochemistry*, 40, 1358, 2001.

56. Stringer, J.A. and Drobny, G.P., Methods for the analysis and design of a solid state nuclear magnetic resonance probe, *Rev. Sci. Instrum.*, 69, 3384, 1998.

57. Reif, B., Hohwy, M., Jaroniec, C.P., Rienstra, C.M. and Griffin, R.G., NH-NH vector correlation in peptides by solid-state NMR, *J. Magn. Reson.*, 145, 132, 2000.

58. Reif, B., Jaroniec, C.P., Rienstra, C.M., Hohwy, M. and Griffin, R.G., H-1-H-1 MAS correlation spectroscopy and distance measurements in a deuterated peptide, *J. Magn. Reson.*, 151, 320, 2001.

59. Andrew, E.R., Clough, S., Farnell, L.F., Gledhill, T.A. and Roberts, I., Resonant rotational broadening of NMR spectra, *Phys. Lett.*, 21, 505, 1966.

60. McDermott, A.E., Creuzet, F., Griffin, R.G., Zawakzke, L.E., Ye, Q.Z. and Walsh, C.T., Rotational resonance determination of the structure of an enzyme-inhibitor complex: phosphorylation of an (amino alkyl) phosphonate inhibitor of D-alanyl-D-alanine ligase by ATP, *Biochemistry*, 29, 5767, 1990.

61. Creuzet, F., McDermott, A.E., Gebhard, R., van der Hoef, K., Spijker-Assink, M.B., Herzfeld, J., Lugtenburg, J., Levitt, M.H. and Griffin, R.G., Determination of membrane protein structure by rotational resonance NMR: bacteriorhodopsin, *Science*, 251, 783, 1991.

62. Thompson, L.K., McDermott, A.E., Raap, J., van der Wielen, C.M., Lugtenberg, J., Herzfeld, J. and Griffin, R.G., Rotational resonance NMR study of the active site structure in bacteriorhodopsin: conformation of the schiff base linkage, *Biochemistry*, 31, 7931, 1992.

63. Griffiths, J.M., Ashburn, T.T., Auger, M., Costa, P.R., Griffin, R.G. and Lansbury, P.T., Jr., Rotational resonance solid-state NMR elucidates a structural model of pancreatic amyloid, *J. Am. Chem. Soc.*, 117, 3539, 1995.

64. Lansbury, P.T., Costa, P.R., Griffiths, J.M., Simon, E.J., Auger, M., Halverson, K.J., Kocisko, D.A., Hendsch, Z.S., Ashburn, T.T., Spencer, R.G. S., Tidor, B. and Griffin, R.G., Structural model for the beta-amyloid fibril based on interstrand alignment of an antiparallel-sheet comprising a C-terminal peptide, *Nat. Struct. Biol.*, 2, 990, 1995.

65. Costa, P.R., Sun, B.Q. and Griffin, R.G., Rotational resonance tickling: accurate internuclear distance measurement in solids, *J. Am. Chem. Soc.*, 119, 10821, 1997.

66. Karlsson, T. and Levitt, M.H., Longitudinal rotational resonance echoes in solid state nuclear magnetic resonance: investigation of zero quantum spin dynamics, *J. Chem. Phys.*, 109, 5493, 1998.

67. Helmle, M., Lee, Y.K., Verdegem, P.J. E., Feng, X., Karlsson, T., Lugtenburg, J., de Groot, H.J.M. and Levitt, M.H., Anomalous rotational resonance spectra in magic-angle spinning NMR, *J. Magn. Reson.*, 140, 379, 1999.

68. Balazs, Y.S. and Thompson, L.K., Practical methods for solid-state NMR distance measurements on large biomolecules: constant-time rotational resonance, *J. Magn. Reson.*, 139, 371, 1999.

69. Feng, X., Verdegem, P.J.E., Lee, Y.K., Helmle, M., Shekar, S.C., de Groot, H.J.M., Lugtenburg, J. and Levitt, M.H., Rotational resonance NMR of C-13(2)-labelled retinal: quantitative internuclear distance determination, *Solid State Nucl. Magn. Reson.*, 14, 81, 1999.

70. Costa, P.R., Sun, B.Q. and Griffin, R.G., Rotational resonance NMR: separation of dipolar coupling and zero quantum relaxation, *J. Magn. Reson.*, 164, 92, 2003.

71. Takegoshi, K., Nomura, K. and Terao, T., Rotational resonance in the tilted rotating-frame, *Chem. Phys. Lett.*, 232, 424, 1995.

72. Takegoshi, K., Nomura, K. and Terao, T., Selective homonuclear polarization transfer in the tilted rotating frame under magic-angle spinning in solids, *J. Magn. Reson.*, 127, 206, 1997.

73. Pauli, J., van Rossum, B., Forster, H., de Groot, H.J.M. and Oschkinat, H., Sample optimization and identification of signal patterns of amino acid side chains in 2D RFDR spectra of the alpha-spectrin SH3 domain, *J. Magn. Reson.*, 143, 411, 2000.

74. McDermott, A., Polenova, T., Bockmann, A., Zilm, K.W., Paulsen, E.K., Martin, R.W. and Montelione, G.T., Partial NMR assignments for uniformly (C-13, N-15)-enriched BPTI in the solid state, *J. Biomol. NMR*, 2000, 16, 209.

75. Igumenova, T.I., Wand, A.J. and McDermott, A.E., Assignment of the backbone resonances for microcrystalline ubiquitin, *J. Am. Chem. Soc.*, 126, 5323, 2004.

76. Bennett, A.E., Rienstra, C.M., Griffiths, J.M., Zhen, W.G., Lansbury, P.T. and Griffin, R.G., Homonuclear radio frequency-driven recoupling in rotating solids, *J. Chem. Phys.*, 108, 9463, 1998.

77. Bennett, A.E., Weliky, D.P. and Tycko, R., Quantitative conformational measurements in solid state NMR by constant-time homonuclear dipolar recoupling, *J. Am. Chem. Soc.*, 120, 4897, 1998.

78. Weliky, D.P., Bennett, A.E., Zvi, A., Anglister, J., Steinbach, P.J. and Tycko, R., Solid-state NMR evidence for an antibody-dependent conformation of the V3 loop of HIV-1 gp120, *Nat. Struct. Biol.*, 6, 141, 1999.

79. Goobes, G., Boender, G.J. and Vega, S., Spinning-frequency-dependent narrowband RF-driven dipolar recoupling, *J. Magn. Reson.*, 146, 204, 2000.

80. Goobes, G. and Vega, S., Improved narrowband dipolar recoupling for homonuclear distance measurements in rotating solids, *J. Magn. Reson.*, 154, 236, 2002.

81. Fujiwara, T., Khandelwal, P. and Akutsu, H., Compound radiofrequency-driven recoupling pulse sequences for efficient magnetization transfer by homonuclear dipolar interaction under magic-angle spinning conditions, *J. Magn. Reson.*, 145, 73, 2000.

82. Ishii, Y., Balbach, J.J. and Tycko, R., Measurement of dipole-coupled lineshapes in a many-spin system by constant-time two-dimensional solid state NMR with high-speed magic-angle spinning, *Chem. Phys.*, 266, 231, 2001.

83. Ishii, Y., C-13-C-13 dipolar recoupling under very fast magic-angle spinning in solid-state nuclear magnetic resonance: applications to distance measurements, spectral assignments, and high-throughput secondary-structure determination, *J. Chem. Phys.*, 114, 8473, 2001.

84. Leppert, J., Ohlenschlager, O., Gorlach, M. and Ramachandran, R., RFDR with adiabatic inversion pulses: application to internuclear distance measurements, *J. Biomol. NMR*, 2004, 28, 229.

85. Gilchrist, M.L., Monde, K., Tomita, Y., Iwashita, T., Nakanishi, K. and McDermott, A.E., Measurement of interfluorine distances in solids, *J. Magn. Reson.*, 152, 12001.

86. Suter, D. and Ernst, R.R., Spectral spin diffusion in the presence of an extraneous dipolar reservoir, *Phys. Rev. B*, 1982, 25, 6038.

87. Suter, D. and Ernst, R.R., Spin diffusion in resolved solid-state NMR spectra, *Phys. Rev. B*, 1985, 32, 5608.

88. Kubo, A. and McDowell, C.A., Spectral spin diffusion in polycrystalline solids under magic-angle spinning, *J. Chem. Soc., Faraday Trans.*, 84, 3713, 1988.

89. Tycko, R. and Dabbagh, G., A simple theory of 13-C nuclear spin diffusion in organic solids, *Israel J. Chem.*, 32, 179, 1992.

90. Griffiths, J.M., Bennett, A.E., Engelhard, M., Siebert, F., Raap, J., Lugtenburg, J., Herzfeld, J. and Griffin, R.G., Structural investigation of the active site in bacterior-hodopsin: geometric constraints on the roles of Asp-85 and Asp-212 in the proton-pumping mechanism from solid state NMR, *Biochemistry*, 39, 362, 2000.

91. Takegoshi, K., Nakamura, S. and Terao, T., C-13-H-1 dipolar-assisted rotational resonance in magic-angle spinning NMR, *Chem. Phys. Lett.*, 344, 631, 2001.

92. Takegoshi, K., Nakamura, S. and Terao, T., C-13-H-1 dipolar-driven C-13-C-13 recoupling without C-13 rf irradiation in nuclear magnetic resonance of rotating solids, *J. Chem. Phys.*, 118, 2325, 2003.

93. Crocker, E., Patel, A.B., Eilers, M., Jayaraman, S., Getmanova, E., Reeves, P.J., Ziliox, M., Khorana, H.G., Sheves, M. and Smith, S.O., Dipolar assisted rotational resonance NMR of tryptophan and tyrosine in rhodopsin, *J. Biomol. NMR*, 2004, 29, 11.

94. Stringer, J.A., Bronnimann, C.E., Mullen, C.G., Zhou, D.H., Stellfox, S.A., Li, Y., Williams, E.H. and Rienstra, C.M., Reduction of RF-induced sample heating with a scroll coil resonator structure for solid-state NMR probes, *J. Magn. Reson.*, 173, 40, 2005.

95. Jarrell, H.C., Lu, D.L. and Siminovitch, D.J., C-13-C-13 correlations and internuclear distance measurements with 2D-MELODRAMA, *J. Am. Chem. Soc.*, 120, 10453, 1998.

96. Gregory, D.M., Mitchell, D.J., Stringer, J.A., Kiihne, S., Shiels, J.C., Callahan, J., Mehta, M.A. and Drobny, G.P., Windowless dipolar recoupling — the detection of weak dipolar couplings between spin-1/2 nuclei with large chemical-shift anisotro-pies, *Chem. Phys. Lett.*, 246, 654, 1995.

97. Mehta, M.A., Gregory, D.M., Kiihne, S., Mitchell, D.J., Hatcher, M.E., Shiels, J.C. and Drobny, G.P., Distance measurements in nucleic acids using windowless dipolar recoupling solid state NMR, *Solid State Nucl. Magn. Reson.*, 7, 211, 1996.

98. Benzinger, T.L. S., Gregory, D.M., Burkoth, T.S., Miller-Auer, H., Lynn, D.G., Botto, R.E. and Meredith, S.C., Propagating structure of Alzheimer's β,-Amyloid(10) is parallel β-sheet with residues in exact register, *Proc. Natl. Acad. Sci. USA*, 95, 13407, 1998.

99. Gregory, D.M., Benzinger, T.L.S., Burkoth, T.S., Miller-Auer, H., Lynn, D.G., Meredith, S.C. and Botto, R.E., Dipolar recoupling NMR of biomolecular self-assemblies: determining inter- and intrastrand distances in fibrilized Alzheimer's β-amyloid peptide, *Solid State Nucl. Magn. Reson.*, 13, 149, 1998.

100. Benzinger, T.L.S., Gregory, D.M., Burkoth, T.S., Miller-Auer, H., Lynn, D.G., Botto, R.E. and Meredith, S.C., Two-dimensional structure of β,-amyloid(10) fibrils, *Biochemistry*, 39, 3491, 2000.

101. Long, J.R., Dindot, J.L., Zebroski, H., Kiihne, S., Clark, R.H., Campbell, A.A., Stayton, P.S. and Drobny, G.P., A peptide that inhibits hydroxyapatite growth is in an extended conformation on the crystal surface, *Proc. Natl. Acad. Sci. USA*, 95, 12083, 1998.

102. Long, J.R., Shaw, W.J., Slayton, P.S. and Drobny, G.P., Structure and dynamics of hydrated statherin on hydroxyapatite as determined by solid-state NMR, *Biochemistry*, 40, 15451, 2001.

103. Shaw, W.J., Long, J.R., Dindot, J.L., Campbell, A.A., Stayton, P.S. and Drobny, G.P., Determination of statherin N-terminal peptide conformation on hydroxyapatite crystals, *J. Am. Chem. Soc.*, 122, 1709, 2000.

104. Carravetta, M., Eden, M., Johannessen, O.G., Luthman, H., Verdegem, P.J.E., Lugtenburg, J., Sebald, A. and Levitt, M.H., Estimation of carbon-carbon bond lengths and medium-range internuclear: distances by solid-state nuclear magnetic resonance, *J. Am. Chem. Soc.*, 123, 10628, 2001.

105. Kristiansen, P.E., Carravetta, M., Lai, W.C. and Levitt, M.H., A robust pulse sequence for the determination of small homonuclear dipolar couplings in magic-angle spinning NMR, *Chem. Phys. Lett.*, 390, 12004.

106. Baldus, M., Geurts, D.G. and Meier, B.H., Broadband dipolar recoupling in rotating solids: a numerical comparison of some pulse schemes, *Solid State Nucl. Magn. Reson.*, 11, 157, 1998.

107. Karlsson, T., Popham, J.M., Long, J.R., Oyler, N. and Drobny, G.P., A study of homonuclear dipolar recoupling pulse sequences in solid-state nuclear magnetic resonance, *J. Am. Chem. Soc.*, 125, 7394, 2003.

108. Michal, C.A. and Jelinski, L.W., REDOR 3D: heteronuclear distance measurements in uniformly labeled and natural abundance solids, *J. Am. Chem. Soc.*, 119, 9059, 1997.

109. Schaefer, J., REDOR-determined distances from heterospins to clusters of C-13 labels, *J. Magn. Reson.*, 137, 272, 1999.

110. Mueller, K.T., Jarvie, T.P., Aurentz, D.J. and Roberts, B.W., The REDOR transform — direct calculation of internuclear couplings from dipolar-dephasing NMR data, *Chem. Phys. Lett.*, 242, 535, 1995.

111. Fyfe, C.A. and Lewis, A.R., Investigation of the viability of solid-state NMR distance determinations in multiple spin systems of unknown structure, *J. Phys. Chem. B*, 104, 48, 2000.

112. Gullion, T. and Pennington, C.H., Theta-REDOR: an MAS NMR methods to simplify multiple coupled heteronuclear spin systems, *Chem. Phys. Lett.*, 290, 88, 1998.

113. Liivak, O. and Zax, D.B., Multiple simultaneous distance determinations: application of rotational echo double resonance nuclear magnetic resonance to IS2 spin networks, *J. Chem. Phys.*, 113, 1088, 2000.

114. Vogt, F.G., Gibson, J.M., Mattingly, S.M. and Mueller, K.T., Determination of molecular geometry in solid-state NMR: rotational-echo double resonance of three-spin systems, *J. Phys. Chem. B*, 2003, 107, 1272.

115. Mueller, L.J. and Elliott, D.W., Correlated tensor interactions and rotational-echo double resonance of spin clusters, *J. Chem. Phys.*, 118, 8873, 2003.

116. Bennett, A.E., Rienstra, C.M., Lansbury, P.T. and Griffin, R.G., Frequency-selective heteronuclear dephasing by dipole couplings in spinning and static solids, *J. Chem. Phys.*, 105, 10289, 1996.

117. Jaroniec, C.P., Tounge, B.A., Rienstra, C.M., Herzfeld, J. and Griffin, R.G., Measurement of C-13-N-15 distances in uniformly C-13 labeled biomolecules: J-decoupled REDOR, *J. Am. Chem. Soc.*, 121, 10237, 1999.

118. Jaroniec, C.P., Tounge, B.A., Herzfeld, J. and Griffin, R.G., Frequency selective heteronuclear dipolar recoupling in rotating solids: accurate C-13-N-15 distance measurements in uniformly C-13,N-15-labeled peptides, *J. Am. Chem. Soc.*, 123, 3507, 2001.

119. Gavuzzo, E., Mazza, F., Pochetti, G. and Scatturin, A., Crystal-structure, conformation, and potential-energy calculations of the chemotactic peptide N-Formyl-L-Met-L-Leu-L- Phe-Ome, *Int. J. Pept. Protein Res.*, 34, 409, 1989.

120. Rienstra, C.M., Hohwy, M., Mueller, L.J., Jaroniec, C.P., Reif, B. and Griffin, R.G., Determination of multiple torsion-angle constraints in U-^{13}C,^{15}N-labeled peptides: 3D ^{1}H-^{15}N-^{13}C-^{1}H dipolar chemical shift spectroscopy in rotating solids, *J. Am. Chem. Soc.*, 124, 11908, 2002.

121. Rienstra, C.M., Tucker-Kellogg, L., Jaroniec, C.P., Hohwy, M., Reif, B., McMahon, M.T., Tidor, B., Lozano-Perez, T. and Griffin, R.G., *De novo* determination of peptide structure with solid-state magic-angle spinning NMR spectroscopy, *Proc. Natl. Acad. Sci. USA*, 99, 10260, 2002.

122. Gross, J.D., Costa, P.R. and Griffin, R.G., Tilted n-fold symmetric radio frequency pulse sequences: applications to CSA and heteronuclear dipolar recoupling in homonuclear dipolar coupled spin networks, *J. Chem. Phys.*, 108, 7286, 1998.

123. Hohwy, M., Jaroniec, C.P., Reif, B., Rienstra, C.M. and Griffin, R.G., Local structure and relaxation in solid-state NMR: accurate measurement of amide N-H bond lengths and H-N-H bond angles, *J. Am. Chem. Soc.*, 122, 3218, 2000.

124. Zhao, X., Eden, M. and Levitt, M.H., Recoupling of heteronuclear dipolar interactions in solid-state NMR using symmetry-based pulse sequences, *Chem. Phys. Lett.*, 342, 353, 2001.

125. Chan, J.C.C. and Eckert, H., C-rotational echo double resonance: heteronuclear dipolar recoupling with homonuclear dipolar decoupling, *J. Chem. Phys.*, 115, 6095, 2001.

126. Chan, J.C.C., C-REDOR: rotational echo double resonance under very fast magic-angle spinning, *Chem. Phys. Lett.*, 335, 289, 2001.

127. Bjerring, M. and Nielsen, N.C., Solid-state NMR heteronuclear dipolar recoupling using off-resonance symmetry-based pulse sequences, *Chem. Phys. Lett.*, 370, 496, 2003.

128. Schmidt-Rohr, K. and Hong, M., Measurements of carbon to amide-proton distances by C-H dipolar recoupling with N-15 NMR detection, *J. Am. Chem. Soc.*, 125, 5648, 2003.

129. Sinha, N. and Hong, M., X-H-1 rotational-echo double-resonance NMR for torsion angle determination of peptides, *Chem. Phys. Lett.*, 380, 742, 2003.

130. Yao, X.L. and Hong, M., Structure distribution in an elastin-mimetic peptide (VPGVG)(3) investigated by solid-state NMR, *J. Am. Chem. Soc.*, 126, 4199, 2004.

131. Williamson, P.T.F., Verhoeven, A., Ernst, M. and Meier, B.H., Determination of internuclear distances in uniformly labeled molecules by rotational-resonance solid-state NMR, *J. Am. Chem. Soc.*, 125, 2718, 2003.

132. Petkova, A.T. and Tycko, R., Rotational resonance in uniformly C-13-labeled solids: effects on high-resolution magic-angle spinning NMR spectra and applications in structural studies of biomolecular systems, *J. Magn. Reson.*, 168, 137, 2004.

133. Ramachandran, R., Ladizhansky, V., Bajaj, V.S. and Griffin, R.G., C-13-C-13 rotational resonance width distance measurements in uniformly C-13-labeled peptides, *J. Am. Chem. Soc.*, 125, 15623, 2003.

134. Nomura, K., Takegoshi, K., Terao, T., Uchida, K. and Kainosho, M., Determination of the complete structure of a uniformly labeled molecule by rotational resonance solid-state NMR in the tilted rotating frame, *J. Am. Chem. Soc.*, 121, 4064, 1999.

135. Nomura, K., Takegoshi, K., Terao, T., Uchida, K. and Kainosho, M., Three-dimensional structure determination of a uniformly labeled molecule by frequency-selective dipolar recoupling under magic-angle spinning, *J. Biomol. NMR*, 17, 111, 2000.

136. Hohwy, M., Rienstra, C.M. and Griffin, R.G., Band-selective homonuclear dipolar recoupling in rotating solids, *J. Chem. Phys.*, 117, 4973, 2002.

137. Matsuki, Y., Akutsu, H. and Fujiwara, T., Band-selective recoupling of homonuclear double-quantum dipolar interaction with a generalized composite 0 degrees pulse: application to C-13 aliphatic region-selective magnetization transfer in solids, *J. Magn. Reson.*, 162, 54, 2003.

138. Pauli, J., Baldus, M., van Rossum, B., de Groot, H. and Oschkinat, H., Backbone and side-chain ^{13}C and ^{15}N resonance assignments of the alpha-spectrin SH3 domain by magic-angle spinning solid state NMR at 17.6 Tesla, *ChemBioChem*, 2, 101, 2001.

139. Bockmann, A., Lange, A., Galinier, A., Luca, S., Giraud, N., Juy, M., Heise, H., Montserret, R., Penin, F. and Baldus, M., Solid state NMR sequential resonance assignments and conformational analysis of the 2 x 10.4 kDa dimeric form of the *Bacillus subtilis* protein Crh, *J. Biomol. NMR*, 27, 323, 2003.

140. Igumenova, T.I., McDermott, A.E., Zilm, K.W., Martin, R.W., Paulson, E.K. and Wand, A.J., Assignments of carbon NMR resonances for microcrystalline ubiquitin, *J. Am. Chem. Soc.*, 126, 6720, 2004.

141. Cornilescu, G., Delaglio, F. and Bax, A., Protein backbone angle restraints from searching a database for chemical shift and sequence homology, *J. Biomol. NMR*, 13, 289, 1999.

142. Gehman, J.D., Paulson, E.K. and Zilm, K.W., The influence of internuclear spatial distribution and instrument noise on the precision of distances determined by solid state NMR of isotopically enriched proteins, *J. Biomol. NMR*, 27, 235, 2003.

143. Lange, A., Luca, S. and Baldus, M., Structural constraints from proton-mediated rare-spin correlation spectroscopy in rotating solids, *J. Am. Chem. Soc.*, 124, 9704, 2002.

144. Tycko, R. and Ishii, Y., Constraints on supramolecular structure in amyloid fibrils from two-dimensional solid-state NMR spectroscopy with uniform isotopic labeling, *J. Am. Chem. Soc.*, 125, 6606, 2003.

145. Jaroniec, C.P., MacPhee, C.E., Bajaj, V.S., McMahon, M.T., Dobson, C.M., and Griffin, R.G. High-resolution molecular structure of a peptide in an amyloid fibril determined by magic-angle spinning NMR spectroscopy, *Proc. Natl. Acad. Sci. USA*, 101, 711, 2004.

146. Ladizhansky, V. and Griffin, R.G., Band-selective carbonyl to aliphatic side chain C-13-C-13 distance measurements in U-C-13,N-15-labeled solid peptides by magic-angle spinning NMR, *J. Am. Chem. Soc.*, 126, 948, 2004.

147. Sonnenberg, L., Luca, S. and Baldus, M., Multiple-spin analysis of chemical-shift-selective (C-13, C-13) transfer in uniformly labeled biomolecules, *J. Magn. Reson.*, 166, 100, 2004.

148. LeMaster, D.M., Dynamical mapping of *E-coli* thioredoxin via C-13 NMR relaxation analysis, *J. Am. Chem. Soc.*, 118, 9255, 1996.

149. Hong, M. and Jakes, K., Selective and extensive C-13 labeling of a membrane protein for solid-state NMR investigations, *J. Biomol. NMR*, 14, 71, 1999.

150. Castellani, F., van Rossum, B., Diehl, A., Schubert, M., Rehbein, K. and Oschkinat, H., Structure of a protein determined by solid-state magic-angle-spinning NMR spectroscopy, *Nature*, 420, 98, 2002.

151. Castellani, F., van Rossum, B.J., Diehl, A., Rehbein, K. and Oschkinat, H., Determination of solid-state NMR structures of proteins by means of three-dimensional N-15-C-13-C-13 dipolar correlation spectroscopy and chemical shift analysis, *Biochemistry*, 42, 11476, 2003.

152. Paulson, E.K., Morcombe, C.R., Gaponenko, V., Dancheck, B., Byrd, R.A. and Zilm, K.W., Sensitive high resolution inverse detection NMR spectroscopy of proteins in the solid state, *J. Am. Chem. Soc.*, 125, 15831, 2003.

153. Paulson, E.K., Morcombe, C.R., Gaponenko, V., Dancheck, B., Byrd, R.A. and Zilm, K.W., High-sensitivity observation of dipolar exchange and NOEs between exchangeable protons in proteins by 3D solid-state NMR spectroscopy, *J. Am. Chem. Soc.*, 125, 14222, 2003.

154. Reif, B., van Rossum, B.J., Castellani, F., Rehbein, K., Diehl, A. and Oschkinat, H., Characterization of H-1-H-1 distances in a uniformly H-2,N-15-labeled SH3 domain by MAS solid-state NMR spectroscopy, *J. Am. Chem. Soc.*, 125, 1488, 2003.

155. Lange, A., Seidel, K., Verdier, L., Luca, S. and Baldus, M., Analysis of proton-proton transfer dynamics in rotating solids and their use for 3D structure determination. *J. Am. Chem. Soc.*, 125, 12640, 2003.

156. Gardner, K.H. and Kay, L.E. The use of H-2, C-13, N-15 multidimensional NMR to study the structure and dynamics of proteins, *Annu. Rev. Biophys. Biomolec. Struct.*, 27, 357, 1998.

157. Goto, N.K., Gardner, K.H., Mueller, G.A., Willis, R.C. and Kay, L.E., A robust and cost-effective method for the production of Val, Leu, Ile (delta 1) methyl-protonated N-15-, C-13, H-2-labeled proteins, *J. Biomol. NMR*, 13, 369, 1999.

158. Nilges, M., Macias, M.J., Odonoghue, S.I. and Oschkinat, H., Automated NOESY interpretation with ambiguous distance restraints: the refined NMR solution structure of the pleckstrin homology domain from beta-spectrin, *J. Mol. Biol.*, 269, 408, 1997,

159. Linge, J.P., Habeck, M., Rieping, W. and Nilges, M., ARIA: automated NOE assignment and NMR structure calculation, *Bioinformatics*, 19, 315, 2003.

160. Lange, A., Becker, S., Seidel, K., Giller, K., Pongs, O. and Baldus, M., A concept for rapid protein-structure determination by solid-state NMR spectroscopy, *Angewandte Chemie–Int. Ed.*, 44, 2089, 2005.

161. Zech, S.G., Wand, A.J. and McDermott, A.E., Protein structure determined by high-resolution solid-state NMR spectroscopy: Application to microcrystalline ubiquitin, *J. Am. Chem. Soc.* (in press).

2 Spectral Assignment of (Membrane) Proteins under Magic-Angle Spinning

Marc Baldus

CONTENTS

ABSTRACT This chapter reports on recent progress to obtain sequential resonance assignments under magic-angle spinning conditions. These techniques provide the spectroscopic basis to study molecular structure and dynamics of multiply isotope-labeled (membrane) proteins.

KEY WORDS: *assignment, MAS, membrane protein, resonance, solid-state NMR*

2.1 INTRODUCTION

For a long time, magic-angle spinning (MAS)[1] has helped biomolecular nuclear magnetic resonance (NMR) applications in cases where slow molecular tumbling or susceptibility effects prohibit high-resolution spectroscopy under static conditions. For example, the beneficial effect of MAS has been observed for more than two decades on the study of membranes,[2-4] membrane proteins, protein complexes, or fibers.[5,6] Moreover, MAS has been applied to probe protein dynamics[7,8] and has become a standard technique in modern combinatorial chemistry.[9,10]

With the advent of efficient isotope-labeling schemes and solid-state NMR (SSNMR) instruments operating at 500 MHz or higher, significant progress is currently being made in the study of multiply or uniformly labeled proteins by MAS-based SSNMR methods. In a first stage, such a structural analysis usually requires resonance assignments of the isotope-labeled biomolecule of interest. Obviously, similar criteria apply to the study of soluble molecules where resonance assignment methods for (^{13}C,^{15}N) labeled proteins were pioneered by Bax and coworkers[11,12] and today represent an integral part of the toolbox of modern solution-state NMR.[13,14] Although these methods make extensive use of scalar couplings to direct polarization along the polypeptide chain, SSNMR mixing schemes can, as detailed below, employ through-bond or through-space transfer mechanisms to achieve sequential resonance assignments under MAS conditions.

In the following sections, the spectroscopic requirements for MAS-based assignment methods and recent applications to multiply labeled (membrane) proteins are discussed.

2.2 METHODS

2.2.1 SPECTRAL RESOLUTION

The problem of spectral assignments in a polypeptide is intimately related to the backbone structure depicted in Figure 2.1, leaving ^{13}C, ^{15}N, and ^{1}H resonances as the only NMR detectable nuclei. Spectral resolution is, in general, critical and may

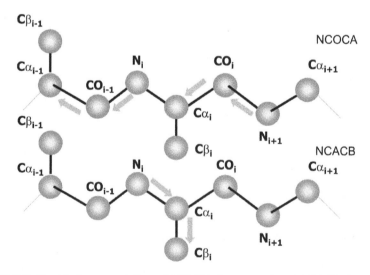

FIGURE 2.1 Graphical representation of NCOCA (upper row) and NCACB (lower row) correlation experiments that lead to spectral assignments under high-resolution solid-state NMR conditions.

be estimated from the ratio of the empirically observed chemical shift range and the NMR resonance line width. For ^{13}C, the spectral dispersion[14,15] in the side chain region spans up to 70 ppm between Thr Cβ and methyl carbons. For ^{1}H side chain resonances and ^{15}NH backbone shifts, typical values are 5 and 30 ppm, respectively. In well-ordered proteins, ^{13}C and ^{15}N resonance line widths are often found in the range of 0.5–1 ppm. A simple calculation using the available chemical-shift dispersion in protons hence would require ^{1}H line widths between 0.03 and 0.166 ppm to obtain a spectral separation of the NMR resonances comparable with ($^{13}C,^{13}C$) or ($^{13}C,^{15}N$) correlation spectroscopy. Although the combination of ultra-high-magnetic-fields NMR, proton dilution, fast MAS, and radiofrequency (rf) decoupling schemes have been shown to significantly improve the spectral resolution of ^{1}H spectroscopy, all protein resonance assignments reported to date are based on ($^{13}C,^{15}N$) correlation spectroscopy.

In addition to MAS, proton rf decoupling[16–18] is usually employed to enhance the resolution of ($^{13}C,^{15}N$) evolution and detection periods. Furthermore rf schemes that suppress broadening effects because of scalar ($^{13}C,^{13}C$) couplings have been shown to improve spectral resolution in uniformly labeled proteins.[19,20,21,22] Similar to the case of macroscopically oriented systems,[23,24] extending the MAS-NMR experiment to three spectral dimensions[25–30] enhances the possibilities for resonance assignments in larger proteins. A major drawback of three-dimensional (3D) NMR spectroscopy is the increased time requirement. Reduced dimensionality experiments[31] have been applied in solid-state MAS-NMR spectroscopy[32–34] and nonlinear sampling.[35] The maximization of the average signal-to-noise ratio (S/N) per scan without sacrificing resolution can also be achieved by a triangular sampling protocol[36] in which data points are only acquired in the lower triangle of the (t_1,t_2) plane (defined by the condition $t_1/t_{1max} + t_2/t_{2max} \leq 1$). Because relaxation leads to signal decay in both indirect evolution times, the experimental time can be reduced by up to a factor of two, without loss in spectral resolution, if t_{1max} and t_{2max} are chosen appropriately.[37]

Last but not least, the details of sample preparation will affect the line width of the system of interest. For example, techniques such as hydration,[33,38,39] flash freezing,[40,41] or the addition of precipitants such as 2-methyl-2,4-pentanediol (MPD)[8] or polyethyleneglycol (PEG)[20,41–43] have been shown to narrow SSNMR lines.

2.2.2 POLARIZATION TRANSFER SCHEMES

Once spectral resolution is optimized, ($^{13}C,^{15}N$) evolution and detection dimensions can be connected by through-space, through-bond, or relaxation-mediated transfer pathways along the polypeptide chain (Figure 2.1). Amino acid types and intraresidue interactions are, perhaps, most easily obtained from ($^{13}C,^{13}C$) broadband correlation spectra. Such experiments can involve through-bond[44–46] or through-space interactions. For the latter, ($^{13}C,^{13}C$) interactions may be actively recoupled (see Refs. 6, 47 for recent reviews), or they may rely on ($^{13}C,^{13}C$) transfer facilitated by multiple-(^{1}H) spin effects.[48,49] Discrimination of intraresidue or interresidue transfer can be easily established by rendering the dipolar (N,C) transfer chemical-shift selective. For this purpose, the conventional Hartmann-Hahn cross-polarization condition[50,51] can be modified by adjusting rf field strength and carrier offset frequency.[52] Because spectral resolution among CO resonances is usually limited, an additional homonuclear transfer step [i.e., $CO(i-1) \rightarrow C\alpha(i-1)$] is often mandatory. At ultra-high magnetic fields, mixing is required over a chemical shift difference $\Delta\delta = \delta(CO_{i1}) - \delta(C\alpha_{i1})$, ranging from about 18 kHz (600 MHz) to 27 kHz (900 MHz). This transfer step hence necessitates a very efficient suppression of chemical shift terms during a broadband polarization transfer or can rely on polarization transfer schemes such as RR,[53] RRTR,[54] or RFDR[55] that operate most efficiently[56] if $\Delta\delta$ and the MAS rate ω_R fulfill the following condition:

$$\Delta\delta(CO_{i-1}, C\alpha_{i-1}) \cong n\omega_R, \; n = 1,2,... \qquad (2.1)$$

Variations in MAS rate[57] or rf field[58] may enhance the efficiency of the polarization transfer and allow for the measurement of ($^{13}C,^{13}C$) distances in uniformly labeled peptides.[59,60,61] For rotating-frame polarization transfer schemes such as MELODRAMA,[62] RIL,[15] or members of the C7-type family,[63] higher spinning rates (and hence stronger rf recoupling fields) improve the offset compensation but may be limited by insufficient proton decoupling. In this case, polarization transfer methods that are selected on the basis of their heteronuclear decoupling performance in the absence of ^1H decoupling may be used.[64]

Intraresidue ($^{15}N,^{13}C$) assignments are most easily established if side-chain information is included. Even at ultrahigh fields, offset compensation can be usually established by a variety of rf schemes. The combination of these N-CO-Cα (NCOCA) and N-Cα-Cβ (NCACB) transfer schemes (Figure 2.1) provides the basis for sequential assignments. Examples are given later.

2.3 APPLICATIONS

For reasons of sensitivity and spectral resolution, an initial analysis of multiply labeled biomolecules under MAS conditions often calls for application of homonuclear ($^{13}C,^{13}C$) correlation experiments. Such studies have been reported on proteins such as Ubiquitin,[26,65,66] BPTI,[67] and the α spectrin SH3 domain.[68,69] Other fully labeled biomolecules include chlorophyll-a water aggregates,[70] erythromycin A,[71] TEE,[72] antamanide,[28,73] kaliotoxin,[74] and a variety of tri-peptides.[75–77] In the following, I concentrate on spectral assignments of multiply labeled (membrane) proteins or their bound peptide ligands.

2.3.1 GLOBULAR PROTEINS

NCOCA and NCACB-type experiments such as discussed in Section 2.2.2 were used to obtain an almost complete set of *de novo* resonance assignments of the SH3 domain of α-spectrin (62 residues, 7.2 kDa) under MAS conditions.[69] Correlation of these ($^{15}N,^{13}C$) chemical shift assignments with random coil values established in solution-state NMR subsequently showed[76] that secondary chemical shifts provide a straightforward and powerful instrument to monitor secondary structure in solid-phase proteins under MAS conditions (*vide infra*). Moreover, these resonance assignments represented the starting point for an investigation of the proton resonance frequencies[78] and for the structural analysis of the 3D structure of a set of ^{13}C block-labeled[79,80] protein variants.[81]

Similar to methodological studies using solution-state NMR, ubiquitin (76 amino acids, 8.6 kDa) is becoming an attractive model protein for high-resolution SSNMR investigations. The 3D structure of ubiquitin has been determined using x-ray crystallography[82] and solution-state NMR[83] and exhibits a rich nature in secondary structure (one α-helix, 3.5 turns, and a twisted antiparallel β-sheet). The first SSNMR studies were reported by Ernst and coworkers,[65] using two-dimensional (2D) ^{13}C/^{15}N chemical shift correlation spectroscopy. Subsequently, lyophilized ubiquitin was investigated by Hong et al.[26,80] Both studies yielded partial (~25%) site-specific assignments of ubiquitin at moderate applied magnetic field strengths. Recently, McDermott and coworkers reported backbone[30] and side-chain carbon[84] assignments obtained at 800 MHz. The uniformly [^{13}C,^{15}N]-enriched sample was crystallized by batch methods in 60% MPD, 20 mmol citrate buffer at pH 4.0-4.2.

High-resolution spectra of ubiquitin can also be obtained after rehydration[33,85] or after precipitation from PEG.[43] Using the latter method, a three-dimensional 2Q-1Q-1Q (^{13}C,^{13}C) correlation experiment was recorded under sensitivity-optimized conditions[37] and is depicted in Figure 2.2. The general transfer pathway hence involves

$$2QC \xrightarrow{t_1} 1QC \xrightarrow{t_2} CC_{mix} \xrightarrow{t_3} \qquad (2.2)$$

where double-quantum (2Q) excitation and reconversion was established using the SPC5[86] scheme and (^{13}C,^{13}C) mixing after t_2 relates to conventional ^{1}H driven spin diffusion. In Figure 2.2a, the (F1,F2) (2Q,1Q) plane is shown, while Figure 2.2b contains the (F2,F3) (1Q,1Q) section. In Figure 2.2c, the side-chain region of the resulting cube is shown. Notably, this spectrum was recorded on a narrow-bore 800 MHz instrument using a triple-resonance (^{1}H,^{13}C,^{15}N) 4-mm probe (Bruker Biospin, Germany) and SPINAL64[17] decoupling in evolution and detection periods.

Combined with a set of 2D and 3D (N,C) correlation experiments, resonance assignments for ubiquitin precipitated from PEG can be derived.[87] These spectral assignments differ from results reported for the MPD-derived form and may help to further investigate the influence of sample preparation on protein structure. Structural insight may also come from proteins such as the immunoglobin-binding domain B1 of streptococcal protein G[88] and thioredoxin,[89] for which sequential (^{13}C,^{15}N) assignments have been reported very recently.

2.3.2 PROTEIN FOLDING

Whereas solution-state NMR techniques have provided unprecedented insight into the structural details of unfolded or partially structured proteins,[90,91] recent progress in SSNMR spectroscopy offers new means to investigate protein folding and, in particular, aggregation and fibril formation. For example, a nearly complete assignment of ^{13}C NMR signals was obtained for amyloid fibrils formed from a uniformly

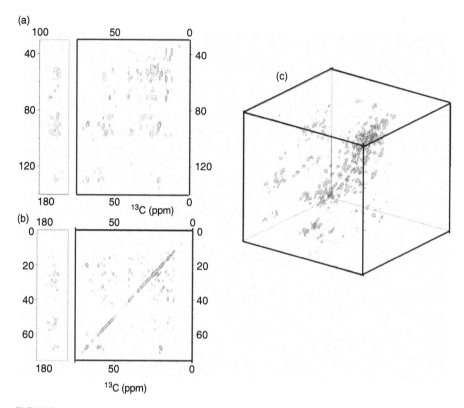

FIGURE 2.2 Three-dimensional 2Q-1Q-1Q (^{13}C,^{13}C) correlation experiment on a U-[^{13}C,^{15}N] labeled sample of ubiquitin at 5°C and 12.5 kHz MAS rate. In (a), the (F1,F2) (2Q,1Q) plane is shown, whereas (b) contains the (F2,F3) (1Q,1Q) section. In (c), the side-chain region of the resulting cube is depicted. The spectrum was recorded on a narrow-bore 800 MHz under sensitivity-optimized conditions,[37] using SPC5 for double-quantum (2Q) excitation and reconversion. (^{13}C,^{13}C) mixing after t_2 relates to conventional ^{1}H-driven spin diffusion. A triple-resonance (^{1}H,^{13}C,^{15}N) 4-mm MAS probe (Bruker Biospin, Germany) and SPINAL64[17] decoupling in evolution and detection periods was used.

labeled seven-residue fragment of the 40-residue Alzheimer Aβ peptide.[92] Recently, a structural model for the full-length Aβ amyloid fibrils could be proposed on the basis of SSNMR studies on samples containing selected five to seven uniformly labeled amino acids.[93] Full sequential assignment of ^{13}C and ^{15}N resonances was also reported for a 10-residue peptide fragment of transthyretin in an amyloid fibril based on a spectral analysis of samples containing different stretches of four consecutive uniformly labeled amino acids. Conformation-dependent chemical shifts provided the basis to predict the backbone conformation of fibrillized transthyretin.[94]

Together with A. Böckmann et al., we have begun investigating the structure and folding of the 85-residue catabolite repression histidine–containing protein (Crh) that regulates gene expression and can exist in a monomeric and a domain-swapped dimeric form. Domain swapping is defined as a process by which two or more protein molecules exchange parts of their structure to form intertwined oligomers. Similar folding

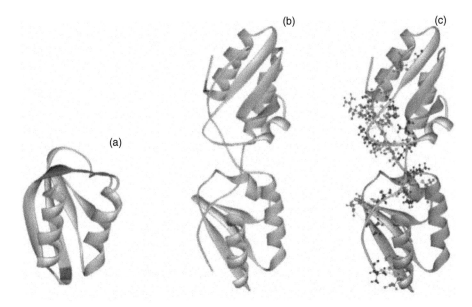

FIGURE 2.3 Comparison of backbone angles obtained for a microcrystalline sample of Crh using MAS-based NMR methods to (a) the monomer solution-state and (b) single-crystal dimer structure. Residues displaying large deviations in the dihedral angles are indicated in dark gray. (c) Monomer–monomer contacts (established by proton–proton transfer side-chain) detected by a 2D NHHC correlation experiment conducted on a (^{13}C:^{15}N) labeled sample of Crh.

events have been observed for amyloid-forming proteins, indicating that domain swapping and amyloid formation may share similar folding intermediates. In a first step, we have reported[20] SSNMR resonance assignments for the microcrystalline form of Crh in the solid state. In Figure 2.3, deviations between the SSNMR-based backbone structure of the microcrystalline state of Crh and the solution-state monomer and single-crystal dimer are indicated in dark gray. These results indicate that Crh also forms a domain-swapped dimer in the microcrystalline form.

To directly probe intermolecular monomer–monomer contacts, we have recently[95] suggested an approach that is based on the spectroscopic analysis of mixtures composed of different molecules, uniformly labeled with spin species X or Y (denoted X:Y). For example, to reliably detect intermolecular contacts in a (^{13}C:^{15}N)-labeled sample, ^{15}N and ^{13}C evolution and detection periods must be combined with a polarization transfer technique sensitive to the interface region only. Such mechanisms could involve a direct NC dipolar (cross polarization) transfer[96] or adiabatic[97] versions thereof, a NHC cross ("NHC") polarization transfer,[98] or a (^{1}H,^{1}H) transfer encoding NH-HC spin pairs (NHHC).[66,99] As shown in ref. 95, all considered NMR schemes are capable of probing intermolecular interactions in high spectral resolution. Application of the NHHC concept maximizes spectral resolution and the detectable distance between the two proteins of interest. Unlike the NC approach, the NHHC scheme does not necessitate careful optimization of CP transfer steps involving weak (intermolecular) dipolar interactions. Because NC and NHC transfers are based on a nonvanishing dipolar coupling element, they may be favorable in the case of applications under ultrafast MAS, where relaxation mediated transfer is attenuated,[99] or for the investigation of non-protonated target spins.

In the context of Crh, a (^{13}C:^{15}N)-labeled sample was prepared by mixing equal amounts of ^{15}N- and ^{13}C-labeled protein under denaturing conditions, followed by renaturation and microcrystallization. Application of the NHHC technique reveals a variety of intermolecular contacts (Figure 2.3c), strongly indicating that domain swapping also occurs in a microcrystalline state of Crh.

We have also begun to structurally characterize fibrillar states of α-synuclein (140 amino acids) and of the K19 domain of protein tau. Both systems are involved in possibly related[100] neurodegenerative diseases. Preliminary results of 2D and 3D correlation spectroscopy as outlined above reveals that many of the NMR detectable resonances exhibit narrow ^{13}C or ^{15}N line widths, consistent with the occurrence of well-ordered domains in both proteins. However, other segments of the considered proteins cannot readily be detected, indicating that they may display static or dynamic disorder under the experimental conditions considered.

2.3.3 MEMBRANE PEPTIDES AND PROTEINS

MAS has long been employed to obtain site-specific structural information in membrane proteins such as bacteriorhodopsin or rhodopsin (for recent reviews, see, e.g., refs. 101 and 102). In addition, MAS-based SSNMR studies were also conducted using selectively labeled membrane peptides[103,104] or peptides reconstituted into deuterated model lipids.[105–107] Such experimental conditions can also easily be modified for the case of macroscopically oriented systems.[108,109]

Uniformly [^{13}C,^{15}N]-labeled peptides studied recently by high-resolution correlation techniques include neurotensin,[110] mastoparan-X,[111] and a variety of fusion peptides.[112] Irrespective of whether they are measured in the gel or liquid-crystalline phase, secondary chemical shifts can report on local backbone conformation.[112–114]

Membrane protein (i.e., receptor) function often involves interactions with ligand molecules and represents an area of great pharmacological relevance. Until recently, no structural information of a high-affinity ligand bound to a G protein–coupled receptor (GPCR) was available. Unlike rhodopsin, the recombinant expression of GPCR in large quantities is usually difficult and must involve carefully optimized biochemical procedures. Restrictions regarding the availability of functional receptors also affect the ligand quantities that can be studied. Moreover, the chemical environment including lipids and receptor protein can hamper the unambiguous spectral identification of a bound ligand in a SSNMR experiment.[115]

We have recently shown[110] how 2D (double-quantum) SSNMR correlation experiments can be used to detect microgram quantities of bound neurotensin, a 13-residue neuropeptide that interacts in high affinity with its NTS-1 (101-kDa) receptor.[110] To elucidate the backbone conformation of neurotensin in complex with its receptor, we conducted a series of 2QF 2D correlation experiments to identify the peptide-bound signal, assign the side-chain resonances, and finally, derive the backbone

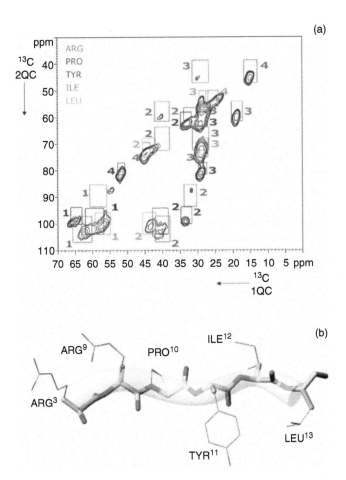

FIGURE 2.4 (a) Statistics-based chemical shift analysis of the 2D (2Q,1Q) correlation spectrum of NT(9–13). The experimental spectrum[108] is superimposed with rectangles that represent the expected chemical shift range for the individual resonances. The boxes are centered at the average chemical shifts, and their size is determined by two times the standard deviation. Values were taken from the most recent entries in "Restricted set of amino acid chemical shifts," found in BioMagResBank (http://www.bmrb.wisc.edu). Coupled spins appear at the same frequency in ω_1, further restricting the assignment. Grayscales of the rectangles denotes the individual residues and the digits relate to: 1, Cα; 2, Cβ; 3, Cγ; and 4, Cδ. (b) Representative backbone conformation of neurotensin(8–13) bound to its G-protein coupled receptor NTS-1.

structure using TALOS.[116] In Figure 2.4a, the general approach to identifying intraresidue correlations in the considered peptide fragment is outlined: Knowledge of amino acid–specific chemical shift ranges is used to establish a spectral grid that serves to identify the different residue types. As detailed in ref. 108, this analysis does not allow for discrimination of Arg8 and Arg9 in neurotensin but suffices to identify backbone and side-chain resonances of NT(9-13) (Figure 2.4a). A subsequent analysis of the resonance assignments obtained on receptor-bound NT(8-13) indicates that neurotensin adopts a β-strand conformation when bound to NTS-1 (Figure 2.4b). Further evidence that the SSNMR-consistent structure represents the bioactive conformation of neurotensin comes from a recent report by Gmeiner et al.,[117] who rigidized NT by chemical synthesis (using 4,4 spiro lactam) around the Pro10 backbone angle to (A) $\psi = 120°$ and (B) $\psi = -120°$. Although compound A, exhibiting a dihedral angle close to the value predicted from the 2D SSNMR data, reveals a nanomolar binding affinity to the NT-receptor, compound B is characterized by a binding affinity that is three orders of magnitude lower.

Not only can SSNMR report on the complete structure of a membrane protein, but it can also probe structural disorder in the polypeptide of interest. In the case of neurotensin, we have developed a general strategy[118] to relate conformational heterogeneity to 2D cross-peak patterns that provide a spectroscopic snapshot of all backbone conformations present in the polypeptide of interest. These dependencies reveal that the degree of backbone disorder detected for individual residues of neurotensin is in qualitative agreement with the residue-specific free energies ΔG calculated for a transfer from an aqueous to a lipid environment.

2D (^{13}C,^{15}N) correlation spectroscopy was also applied to a uniformly [^{13}C,^{15}N]-labeled version of the LH2 light-harvesting complex[119] from the purple nonsulfur photosynthetic bacterium *Rhodopseudomonas acidophila* 10050 strain,[120] one of the largest systems (~150 kDa) studied by SSNMR to date. Here, the application of ultrahigh magnetic fields (up to 750 MHz) was crucial to establish high-resolution conditions in the solid state. 2D (^{15}N,^{13}C) correlation experiments were conducted to assign the backbone and side-chain resonances at different temperatures. In Figure 2.5a, results of an NCO correlation experiment are shown. As expected, the spectral resolution is insufficient to resolve all interresidue backbone correlations. Nevertheless, a significant number of individual (^{15}N,^{13}C) correlations can be identified that are characterized by line widths below 1 ppm in both spectral dimensions. In Figure 2.5b, results of an NCACB-type experiment are shown, revealing a variety of intraresidue correlations. For (^{13}C,^{13}C) mixing, a proton-driven spin diffusion unit was applied that, as visible from Figure 2.5b, enhances polarization transfer for chemical shift differences comparable to the MAS rate (50 ppm [at 750 MHz] = 10.313 kHz) as a result of rotational-resonance effects. Combined with a NCOCA experiment, a variety of spectral assignments within the transmembrane sections of the protein were possible and are indicated in dark gray in Figure 2.5c. Additional correlations occur at temperatures below 10°C, consistent with the immobilization of flexible loop regions of the protein. A more detailed structural study in membrane proteins could include (^{13}C, ^{13}C) or indirect (^{1}H,^{1}H) correlation experiments, possibly in conjunction with advanced isotope labeling approaches, to determine the complete 3D structure of a membrane protein in the solid state.

2.4 CONCLUSIONS AND OUTLOOK

MAS-based SSNMR has recently made considerable progress in the assignment of multiply labeled protein samples. Such spectral assignments provide the basis for further investigations of 3D structure or dynamics. For example, structural information about backbone structure is readily available from an analysis of conformation-dependent

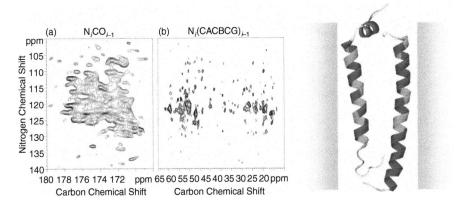

FIGURE 2.5 Combined NC-CC 2D transfer experiment on U-[^{13}C,^{15}N] labeled LH2 revealing a variety of N_iCO_{i-1} and (a) $N_i(CACBCG)_{i-1}$ (b) correlations. A schematic view of the backbone structure of the LH2 protomer complex embedded in a model membrane is shown. Assigned residues are indicated in dark gray.

chemical shifts. Additional approaches to obtain information about the 3D structure are discussed within this chapter or have been reviewed elsewhere.[121]

Advances regarding sample preparation (e.g., including modular labeling, *in vitro* expression, and intein technology[122]) and improvements in NMR hardware instrumentation could open up new areas of SSNMR research such as the investigation of large protein–protein complexes or the complete 3D characterization of larger membrane proteins. For such applications, further improvements of the applied spectral assignment methods will be beneficial and may result from numerically optimized decoupling and polarization transfer schemes. Likewise, conducting SSNMR experiments at higher magnetic fields and a further optimization of the macroscopic sample conditions will likely expand the utility of biological SSNMR.

Notably, such studies may not only target structural information but could also involve the investigation of protein dynamics. Here, spectral assignment methods that rely on one-bond interactions may offer a very efficient means to study protein folding and flexibility under biologically relevant conditions.

ACKNOWLEDGMENTS

I thank our collaborators and group members who contributed to work described here. Financial support by the MPG, the DFG, FCI (Fonds der Chemischen Industrie), the Sartorius AG, the Humboldt, and the Volkswagen foundation is gratefully acknowledged.

REFERENCES

1. Andrew, E.R., Bradbury, A. and Eades, R.G., Nuclear magnetic resonance spectra from a crystal rotated at high speed, *Nature*, 182, 1659, 1958.

2. Herzfeld, J., Roufosse, A., Haberkorn, R.A., Griffin, R.G. and Glimcher, M.J., Magic-angle sample spinning in inhomogeneously broadened biological-systems, *Philos. Trans. R. Soc. Lond. Ser. B–Biol. Sci.*, 289(1037), 459, 1980.

3. Oldfield, E., Bowers, J.L. and Forbes, J., High-resolution proton and C-13 NMR of membranes — why sonicate, *Biochemistry*, 26(22), 6919, 1987.

4. Yeagle, P.L. and Frye, J., Effects of unsaturation on H-2-NMR quadrupole splittings and C-13-NMR relaxation in phospholipid-bilayers, *Biochim. Biophys. Acta*, 899(2), 137, 1987.

5. Mcdowell, L.M. and Schaefer, J., High-resolution NMR of biological solids, *Curr. Opin. Struct. Biol.*, 6(5), 624, 1996.

6. Griffin, R.G., Dipolar recoupling in MAS spectra of biological solids, *Nat. Struct. Biol.*, 5, 508, 1998.

7. Torchia, D.A., Solid-state NMR-studies of protein internal dynamics, *Ann. Rev. Biophys. Bioeng.*, 13, 125, 1984.

8. Cole, H.B.R. and Torchia, D.A., An NMR-study of the backbone dynamics of *Staphylococcal* nuclease in the crystalline state, *Chem. Phys.*, 158(2–3), 271, 1991.

9. Keifer, P.A., High-resolution NMR techniques for solid-phase synthesis and combinatorial chemistry, *Drug Discov. Today*, 2(11), 468, 1997.

10. Shapiro, M.J. and Wareing, J.R., NMR methods in combinatorial chemistry, *Curr. Opin. Chem. Biol.*, 2(3), 372, 1998.

11. Ikura, M., Kay, L.E. and Bax, A., A novel-approach for sequential assignment of H-1, C-13, and N-15 spectra of larger proteins — heteronuclear triple-resonance 3-dimensional NMR-spectroscopy — application to calmodulin, *Biochemistry*, 29(19), 4659, 1990.

12. Ikura, M., Kay, L.E., Tschudin, R. and Bax, A., 3-Dimensional NOESY-HMQC spectroscopy of a C-13-labeled protein, *J. Magn. Reson.*, 86(1), 204, 1990.

13. Van De Ven, F.J. M., *Multidimensional NMR in Liquids: Basic Principles and Experimental Methods*; VCH Publishers, New York, 1995.

14. Cavanagh, J., Fairbrother, W.J., Palmer, A.G. and Skelton, N.J., *Protein NMR Spectroscopy, Principles and Practice*, Academic Press, San Diego, 1996.

15. Baldus, M., Tomaselli, M., Meier, B.H. and Ernst, R.R., Broad-band polarization-transfer experiments for rotating solids, *Chem. Phys. Lett.*, 230(4–5), 329, 1994.

16. Bennett, A.E., Rienstra, C.M., Auger, M., Lakshmi, K.V. and Griffin, R.G., Heteronuclear decoupling in rotating solids, *J. Chem. Phys.*, 103(16), 6951, 1995.

17. Fung, B.M., Khitrin, A.K. and Ermolaev, K., An improved broadband decoupling sequence for liquid crystals and solids, *J. Magn. Reson.*, 142(1), 97, 2000.

18. Detken, A., Hardy, E.H., Ernst, M. and Meier, B.H., Simple and efficient decoupling in magic-angle spinning solid-state NMR: the XiX scheme, *Chem. Phys. Lett.*, 356(3–4), 298, 2002.

19. Straus, S.K., Bremi, T. and Ernst, R.R. Resolution enhancement by homonuclear J decoupling in solid-state MAS NMR, *Chem. Phys. Lett.*, 262(6), 709, 1996.

20. Böckmann, A., Lange, A., Galinier, A., Luca, S., Giraud, N., Heise, H., Juy, M., Montserret, R., Penin, F. and Baldus, M., Solid-state NMR sequential resonance assignments and conformational analysis of the 2*10.4 kDa dimeric form of the *Bacillus subtilis* protein Crh, *J. Biomol. NMR*, 27(4), 323, 2003.

21. Carravetta, M., Zhao, X., Bockmann, A. and Levitt, M.H., Coherence transfer selectivity in two-dimensional solid-state NMR, *Chem. Phys. Lett.*, 376(3–4), 515, 2003.

22. Duma, L., Hediger, S., Brutscher, B., Bockmann, A. and Emsley, L., Resolution enhancement in multidimensional solid-state NMR spectroscopy of proteins using spin-state selection, *J. Am. Chem. Soc.*, 125(39), 11816, 2003.

23. Ramamoorthy, A., Gierasch, L.M. and Opella, S.J., 4-Dimensional solid-state NMR experiment that correlates the chemical-shift and dipolar-coupling frequencies of 2 heteronuclei with the exchange of dilute-spin magnetization, *J. Magn. Reson. Ser. B*, 109(1), 112, 1995.

24. Ramamoorthy, A., Marassi, F.M., Zasloff, M. and Opella, S.J., 3-Dimensional solid-state NMR-spectroscopy of a peptide oriented in membrane bilayers, *J. Biomol. NMR*, 6(3), 329, 1995.

25. Sun, B.Q., Rienstra, C.M., Costa, P.R., Williamson, J.R. and Griffin, R.G., 3D N-15-C-13-C-13 chemical shift correlation spectroscopy in rotating solids, *J. Am. Chem. Soc.*, 119(36), 8540, 1997.

26. Hong, M., Resonance assignment of C-13/N-15 labeled solid proteins by two- and three-dimensional magic-angle-spinning NMR, *J. Biomol. NMR*, 15(1), 1, 1999.

27. Rienstra, C.M., Hohwy, M., Hong, M. and Griffin, R.G., 2D and 3D N-15-C-13-C-13 NMR chemical shift correlation spectroscopy of solids: assignment of MAS spectra of peptides, *J. Am. Chem. Soc.*, 122(44), 10979, 2000.

28. Detken, A., Hardy, E.H., Ernst, M., Kainosho, M., Kawakami, T., Aimoto, S. and Meier, B.H., Methods for sequential resonance assignment in solid, uniformly C-13, N-15 labeled peptides: quantification and application to antamanide, *J. Biomol. NMR*, 20(3), 203, 2001.

29. Castellani, F., Van Rossum, B.J., Diehl, A., Rehbein, K. and Oschkinat, H., Determination of solid-state NMR structures of proteins by means of three-dimensional N-15-C-13-C-13 dipolar correlation spectroscopy and chemical shift analysis, *Biochemistry*, 42(39), 11476, 2003.

30. Igumenova, T.I., Wand, A.J. and McDermott, A.E., Assignment of the backbone resonances for microcrystalline ubiquitin, *J. Am. Chem. Soc.*, 126(16), 5323, 2004.

31. Szyperski, T., Wider, G., Bushweller, J.H. and Wuthrich, K., Reduced dimensionality in triple-resonance NMR experiments, *J. Am. Chem. Soc.*, 115(20), 9307, 1993.

32. Astrof, N.S., Lyon, C.E. and Griffin, R.G., Triple resonance solid state NMR experiments with reduced dimensionality evolution periods, *J. Magn. Reson.*, 152(2), 303, 2001.

33. Luca, S. and Baldus, M., Enhanced spectral resolution in immobilized peptides and proteins by combining chemical shift sum and difference spectroscopy, *J. Magn. Reson.*, 159(2), 243, 2002.

34. Leppert, J., Heise, B., Ohlenschlager, O., Gorlach, M. and Ramachandran, R., Triple resonance MAS NMR with (C-13, N-15) labeled molecules: reduced dimensionality data acquisition via C-13-N-15 heteronuclear two-spin coherence transfer pathways, *J. Biomol. NMR*, 28(2), 185, 2004.

35. Rovnyak, D., Filip, C., Itin, B., Stern, A.S., Wagner, G., Griffin, R.G. and Hoch, J.C., Multiple-quantum magic-angle spinning spectroscopy using nonlinear sampling, *J. Magn. Reson.*, 161(1), 43, 2003.

36. Aggarwal, K. and Delsuc, M.A., Triangular sampling of multidimensional NMR data sets, *Magn. Reson. Chem.*, 35(9), 593, 1997.

37. Heise, H., Seidel, K., Etzkorn, M., Becker, S. and Baldus, M., 3D NMR spectroscopy for resonance assignment and structure elucidation of proteins under MAS: novel pulse schemes and sensitivity considerations, *J. Magn. Reson.*, 173(1), 64, 2005.

38. Gregory, R.B., Gangoda, M., Gilpin, R.K. and Su, W., The influence of hydration on the conformation of bovine serum-albumin studied by solid-state C-13-NMR spectroscopy, *Biopolymers*, 33(12), 1871, 1993.

39. Gregory, R.B., Gangoda, M., Gilpin, R.K. and Su, W., The influence of hydration on the conformation of lysozyme studied by solid-state C-13-NMR spectroscopy, *Biopolymers*, 33(4), 513, 1993.

40. Evans, J.N.S., Appleyard, R.J. and Shuttleworth, W.A., Detection of an enzyme intermediate complex by time-resolved solid-state NMR-spectroscopy, *J. Am. Chem. Soc.*, 115(4), 1588, 1993.

41. Jakeman, D.L., Mitchell, D.J., Shuttleworth, W.A. and Evans, J.N.S., Effects of sample preparation conditions on biomolecular solid-state NMR lineshapes, *J. Biomol. NMR*, 12(3), 417, 1998.

42. Tomita, Y., Oconnor, E.J. and Mcdermott, A., A method for dihedral angle measurement in solids — rotational resonance NMR of a transition-state inhibitor of triose phosphate isomerase, *J. Am. Chem. Soc.*, 116(19), 8766, 1994.

43. Martin, R.W. and Zilm, K.W., Preparation of protein nanocrystals and their characterization by solid state NMR, *J. Magn. Reson.*, 165(1), 162, 2003.

44. Ramamoorthy, A., Fujiwara, T. and Nagayama, K., An rf pulse sequence optimized for homonuclear-j cross-polarization under magic-angle-spinning conditions in solids, *J. Magn. Reson. Ser. A,* 104(3), 366, 1993.

45. Baldus, M. and Meier, B.H., Total correlation spectroscopy in the solid state. The use of scalar couplings to determine the through-bond connectivity, *J. Magn. Reson. Ser. A,* 121(1), 65, 1996.

46. Baldus, M., Iuliucci, R.J. and Meier, B.H., Probing through-bond connectivities and through-space distances in solids by magic-angle-spinning nuclear magnetic resonance, *J. Am. Chem. Soc.*, 119(5), 1121, 1997.

47. Baldus, M., Correlation experiments for assignment and structure elucidation of immobilized polypeptides under magic-angle spinning, *Prog. Nucl. Magn. Reson. Spectrosc.*, 41(1–2), 1, 2002.

48. Bloembergen, N., On the interaction of nuclear spins in a crystalline lattice, *Physica*, 15, 386, 1949.

49. Takegoshi, K., Nakamura, S. and Terao, T., C-13-H-1 dipolar-assisted rotational resonance in magic-angle spinning NMR, *Chem. Phys. Lett.*, 344(5–6), 631, 2001.

50. Hartmann, S.R. and Hahn, E.L., Nuclear double resonance in rotating frame, *Phys. Rev.*, 128(5), 2042, 1962.

51. Pines, A., Gibby, M.G. and Waugh, J.S., Proton-enhanced NMR of dilute spins in solids, *J. Chem. Phys.*, 59(2), 569, 1973.

52. Baldus, M., Petkova, A.T., Herzfeld, J. and Griffin, R.G., Cross polarization in the tilted frame: assignment and spectral simplification in heteronuclear spin systems, *Mol. Phys.*, 95(6), 1197, 1998.

53. Raleigh, D.P., Levitt, M.H. and Griffin, R.G., Rotational resonance in solid-state NMR, *Chem. Phys. Lett.*, 146(1–2), 71, 1988.

54. Takegoshi, K., Nomura, K. and Terao, T., Rotational resonance in the tilted rotating-frame, *Chem. Phys. Lett.*, 232(5–6), 424, 1995.

55. Bennett, A.E., Ok, J.H., Griffin, R.G. and Vega, S., Chemical-shift correlation spectroscopy in rotating solids — radio frequency-driven dipolar recoupling and longitudinal exchange, *J. Chem. Phys.*, 96(11), 8624, 1992.

56. Baldus, M., Geurts, D.G. and Meier, B.H. Broadband dipolar recoupling in rotating solids: a numerical comparison of some pulse schemes, *Solid State Nucl. Magn. Reson.*, 11(3–4), 157, 1998.

57. Verel, R., Baldus, M., Nijman, M., Van Os, J.W.M. and Meier, B.H., Adiabatic homonuclear polarization transfer in magic-angle-spinning solid-state NMR, *Chem. Phys. Lett.*, 280(1–2), 31, 1997.

58. Verel, R., Baldus, M., Ernst, M. and Meier, B.H., A homonuclear spin-pair filter for solid-state NMR based on adiabatic-passage techniques, *Chem. Phys. Lett.*, 287(3–4), 421, 1998.
59. Ramachandran, R., Ladizhansky, V., Bajaj, V.S. and Griffin, R.G., C-13-C-13 rotational resonance width distance measurements in uniformly C-13-labeled peptides, *J. Am. Chem. Soc.*, 125(50), 15623, 2003.
60. Ladizhansky, V. and Griffin, R.G., Band-selective carbonyl to aliphatic side chain C-13-C-13 distance measurements in U-C-13,N-15-labeled solid peptides by magic-angle spinning NMR, *J. Am. Chem. Soc.*, 126(3), 948, 2004.
61. Sonnenberg, L., Luca, S. and Baldus, M., Multiple-spin analysis of (13C,13C) chemical-shift selective transfer in uniformly labeled biomolecules, *J. Magn. Reson.*, 166 100, 2004.
62. Sun, B.Q., Costa, P.R., Kocisko, D., Lansbury, P.T. and Griffin, R.G., Internuclear distance measurements in solid-state nuclear-magnetic-resonance — dipolar recoupling via rotor synchronized spin locking, *J. Chem. Phys.*, 102(2), 702, 1995.
63. Lee, Y.K., Kurur, N.D., Helmle, M., Johannessen, O.G., Nielsen, N.C. and Levitt, M.H., Efficient dipolar recoupling in the NMR of rotating solids — A sevenfold symmetrical radiofrequency pulse sequence, *Chem. Phys. Lett.*, 242(3), 304, 1995.
64. Hughes, C.E., Luca, S. and Baldus, M., Radio-frequency-driven polarization transfer without heteronuclear decoupling in rotating solids, *Chem. Phys. Lett.*, 385 435, 2004.
65. Straus, S.K., Bremi, T. and Ernst, R.R., Experiments and strategies for the assignment of fully C-13/N-15-labeled polypeptides by solid state NMR, *J. Biomol. NMR*, 12(1), 39, 1998.
66. Lange, A., Luca, S. and Baldus, M., Structural constraints from proton-mediated rare-spin correlation spectroscopy in rotating solids, *J. Am. Chem. Soc.*, 124(33), 9704, 2002.
67. Mcdermott, A., Polenova, T., Bockmann, A., Zilm, K.W., Paulsen, E.K., Martin, R.W. and Montelione, G.T., Partial NMR assignments for uniformly (C-13, N-15)-enriched BPTI in the solid state, *J. Biomol. NMR*, 16(3), 209, 2000.
68. Pauli, J., Van Rossum, B., Forster, H., De Groot, H.J.M. and Oschkinat, H., Sample optimization and identification of signal patterns of amino acid side chains in 2D RFDR spectra of the alpha-spectrin SH3 domain, *J. Magn. Reson.*, 143(2), 411, 2000.
69. Pauli, J., Baldus, M., Van Rossum, B., De Groot, H. and Oschkinat, H., Backbone and side-chain C-13 and N-15 signal assignments of the alpha-spectrin SH3 domain by magic-angle spinning solid-state NMR at 17.6 Tesla, *Chembiochem*, 2(4), 272, 2001.
70. Boender, G.J., Raap, J., Prytulla, S., Oschkinat, H. and Degroot, H.J.M., MAS NMR structure refinement of uniformly c-13 enriched chlorophyll-a water aggregates with 2d dipolar correlation spectroscopy, *Chem. Phys. Lett.*, 237(5–6), 502, 1995.
71. Rienstra, C.M., Hatcher, M.E., Mueller, L.J., Sun, B.Q., Fesik, S.W. and Griffin, R.G., Efficient multispin homonuclear double-quantum recoupling for magic-angle spinning NMR: C-13-C-13 correlation spectroscopy of U-C-13-erythromycin A, *J. Am. Chem. Soc.*, 120(41), 10602, 1998.
72. Helluy, X. and Sebald, A., Structure and dynamic properties of solid l-tyrosine-ethylester as seen by C-13 MAS NMR, *J. Phys. Chem. B*, 107(14), 3290, 2003.
73. Straus, S.K., Bremi, T. and Ernst, R.R., Side-chain conformation and dynamics in a solid peptide: CP-MAS NMR study of valine rotamers and methyl-group relaxation in fully C-13-labeled antamanide, *J. Biomol. NMR*, 10(2), 119, 1997.

74. Lange, A., Becker, S., Giller, K., Seidel, K., Pongs, O. and Baldus, M., A concept for rapid protein-structure determination by solid-state NMR spectroscopy, *Angew. Chem.-Int. Ed.*, 44(14), 2089, 2005.

75. Hong, M. and Griffin, R.G., Resonance assignments for solid peptides by dipolar-mediated C-13/N-15 correlation solid-state NMR, *J. Am. Chem. Soc.*, 120(28), 7113, 1998.

76. Nomura, K., Takegoshi, K., Terao, T., Uchida, K. and Kainosho, M., Determination of the complete structure of a uniformly labeled molecule by rotational resonance solid-state NMR in the tilted rotating frame, *J. Am. Chem. Soc.*, 121(16), 4064, 1999.

77. Luca, S., Filippov, D.V., Van Boom, J.H., Oschkinat, H., De Groot, H.J.M. and Baldus, M., Secondary chemical shifts in immobilized peptides and proteins: a qualitative basis for structure refinement under magic-angle spinning, *J. Biomol. NMR*, 20(4), 325, 2001.

78. Van Rossum, B.J., Castellani, F., Rehbein, K., Pauli, J. and Oschkinat, H., Assignment of the nonexchanging protons of the alpha-spectrin SH3 domain by two- and three-dimensional H-1-C-13 solid-state magic-angle spinning NMR and comparison of solution and solid-state proton chemical shifts, *Chembiochem*, 2(12), 906, 2001.

79. Lemaster, D.M. and Kushlan, D.M., Dynamical mapping of *E-coli* thioredoxin via C-13 NMR relaxation analysis, *J. Am. Chem. Soc.*, 118(39), 9255, 1996.

80. Hong, M. and Jakes, K., Selective and extensive C-13 labeling of a membrane protein for solid-state NMR investigations, *J. Biomol. NMR*, 14(1), 71, 1999.

81. Castellani, F., Van Rossum, B., Diehl, A., Schubert, M., Rehbein, K. and Oschkinat, H., Structure of a protein determined by solid-state magic-angle-spinning NMR spectroscopy, *Nature*, 420(6911), 98, 2002.

82. Vijaykumar, S., Bugg, C.E. and Cook, W.J., Structure of ubiquitin refined at 1.8 Å resolution, *J. Mol. Biol.*, 194(3), 531, 1987.

83. Cornilescu, G., Marquardt, J.L., Ottiger, M. and Bax, A., Validation of protein structure from anisotropic carbonyl chemical shifts in a dilute liquid crystalline phase, *J. Am. Chem. Soc.*, 120(27), 6836, 1998.

84. Igumenova, T.I., Mcdermott, A.E., Zilm, K.W., Martin, R.W., Paulson, E.K. and Wand, A.J., Assignments of carbon NMR resonances for microcrystalline ubiquitin, *J. Am. Chem. Soc.*, 126(21), 6720, 2004.

85. Luca, S., Solid-state NMR studies of globular and membrane proteins, Ph.D. Thesis, Goettingen, 2003.

86. Hohwy, M., Rienstra, C.M., Jaroniec, C.P. and Griffin, R.G., Fivefold symmetric homonuclear dipolar recoupling in rotating solids: application to double quantum spectroscopy, *J. Chem. Phys.*, 110(16), 7983, 1999.

87. Seidel, K., Etzkorn, M., Heise, H., Becker, S. and Baldus, M., High-resolution solid-state NMR studies on uniformly [^{13}C, ^{15}N] labeled Ubiquitin, *Chembiochem.*, in press.

88. Rienstra, C.M., Franks, T., Donghua Zhou, D., Wylie, B., Money, B., Graesser, D. and Sahota, G., 3D Magic-Angle Spinning NMR Studies of the Immunoglobin-Binding Domain B1 of Streptococcal Protein G (GB1): Chemical Shift Assignments and Structural Constraints, paper presented at the 45th Exper. Nucl. Magnet. Reson. Conf., Asilomar, 2004.

89. Polenova, T., Tasayco, M.L., Mcdermott, A.E., Marulanda, D., Cataldi, M., Arriaran, V. and Bai, S., Homonuclear 13C Solid-State NMR Spectroscopy of the Uniformly (13C, 15N)-Enriched *E. coli* Thioredoxin at 17.6 Tesla. Spin System Identification and Sidechain Resonance Assignments, paper presented at the 45th Exper. Nucl. Magnet. Reson. Conf., Asilomar, 2004.

90. Dyson, H.J. and Wright, P.E., Equilibrium NMR studies of unfolded and partially folded proteins, *Nat. Struct. Biol.*, 5 499, 1998.

91. Dobson, C.M., The structural basis of protein folding and its links with human disease. *Philos. Trans. R. Soc. Lond. Ser. B–Biol. Sci.*, 356(1406), 133, 2001.

92. Balbach, J.J., Ishii, Y., Antzutkin, O.N., Leapman, R.D., Rizzo, N.W., Dyda, F., Reed, J. and Tycko, R., Amyloid fibril formation by A beta(16-22), a seven-residue fragment of the Alzheimer's beta-amyloid peptide, and structural characterization by solid state NMR, *Biochemistry*, 39(45), 13748, 2000.

93. Petkova, A.T., Ishii, Y., Balbach, J.J., Antzutkin, O.N., Leapman, R.D., Delaglio, F. and Tycko, R., A structural model for Alzheimer's beta-amyloid fibrils based on experimental constraints from solid state NMR, *Proc. Natl. Acad. Sci. USA*, 99(26), 16742, 2002.

94. Jaroniec, C.P., Macphee, C.E., Astrof, N.S., Dobson, C.M. and Griffin, R.G., Molecular conformation of a peptide fragment of transthryretin in an amyloid fibril, *Proc. Natl. Acad. Sci. USA*, 99, 16748, 2002.

95. Etzkorn, M., Böckmann, A., Lange, A. and Baldus, M., Probing molecular interfaces using 2D magic-angle-spinning NMR on protein mixtures with different uniform labeling, *J. Am. Chem. Soc.*, 126, 14746, 2004.

96. Schaefer, J., McKay, R.A. and Stejskal, E.O., Double-cross-polarization NMR of solids, *J. Magn. Reson.*, 34(2), 443, 1979.

97. Baldus, M., Geurts, D.G., Hediger, S. and Meier, B.H., Efficient N-15-C-13 polarization transfer by adiabatic-passage Hartmann-Hahn cross polarization, *J. Magn. Reson. Ser. A*, 118(1), 140, 1996.

98. Lange, A., ^{1}H-NMR-spectroscopy in the solid-state: Methods and biomolecular applications, Diploma thesis, University Goettingen, 2002.

99. Lange, A., Seidel, K., Verdier, L., Luca, S. and Baldus, M., Analysis of proton-proton transfer dynamics in rotating solids and their use for 3D structure determination, *J. Am. Chem. Soc.*, 125(41), 12640, 2003.

100. Giasson, B.I., Forman, M.S., Higuchi, M., Golbe, L.I., Graves, C.L., Kotzbauer, P.T., Trojanowski, J.Q. and Lee, V.M.Y., Initiation and synergistic fibrillization of tau and alpha-synuclein, *Science*, 300(5619), 636, 2003.

101. Smith, S.O., Aschheim, K. and Groesbeek, M., Magic-angle spinning NMR spectroscopy of membrane proteins, *Q. Rev. Biophys.*, 29(4), 395, 1996.

102. Davis, J.H. and Auger, M., Static and magic-angle spinning NMR of membrane peptides and proteins, *Prog. Nucl. Magn. Reson. Spectrosc.*, 35(1), 1, 1999.

103. Peersen, O.B., Yoshimura, S., Hojo, H., Aimoto, S. and Smith, S.O., Rotational resonance NMR measurements of internuclear distances in an alpha-helical peptide, *J. Am. Chem. Soc.*, 114(11), 4332, 1992.

104. Hirsh, D.J., Hammer, J., Maloy, W.L., Blazyk, J. and Schaefer, J., Secondary structure and location of a magainin analogue in synthetic phospholipid bilayers, *Biochemistry*, 35(39), 12733, 1996.

105. Bouchard, M., Davis, J.H. and Auger, M., High-speed magic-angle spinning solid-state H-1 nuclear magnetic resonance study of the conformation of gramicidin a in lipid bilayers, *Biophys. J.*, 69(5), 1933, 1995.

106. Davis, J.H., Auger, M. and Hodges, R.S., High resolution H-1 nuclear magnetic resonance of a transmembrane peptide, *Biophys. J.*, 69(5), 1917, 1995.

107. Zhang, W.Y., Crocker, E., Mclaughlin, S. and Smith, S.O., Binding of peptides with basic and aromatic residues to bilayer membranes — phenylalanine in the myristoylated alanine-rich C kinase substrate effector domain penetrates into the hydrophobic core of the bilayer, *J. Biol. Chem.*, 278(24), 21459, 2003.

108. Glaubitz, C., An introduction to MAS NMR spectroscopy on oriented membrane proteins, *Concepts Magn. Reson.*, 12(3), 137, 2000.

109. Sizun, C. and Bechinger, B., Bilayer sample for fast or slow magic-angle oriented sample spinning solid-state NMR spectroscopy, *J. Am. Chem. Soc.*, 124(7), 1146, 2002.

110. Luca, S., White, J.F., Sohal, A.K., Filippov, D.V., Van Boom, J.H., Grisshammer, R. and Baldus, M., The conformation of neurotensin bound to its G protein-coupled receptor, *Proc. Natl. Acad. Sci. USA*, 100(19), 10706, 2003.

111. Fujiwara, T., Todokoro, Y., Yanagishita, H., Tawarayama, M., Kohno, T., Wakamatsu, K. and Akutsu, H., Signal assignments and chemical-shift structural analysis of uniformly C-13, N-15-labeled peptide, mastoparan-X, by multidimensional solid-state NMR under magic-angle spinning, *J. Biomol. NMR*, 28(4), 311, 2004.

112. Bodner, M.L., Gabrys, C.M., Parkanzky, P.D., Yang, J., Duskin, C.A. and Weliky, D.P., Temperature dependence and resonance assignment of C-13 NMR spectra of selectively and uniformly labeled fusion peptides associated with membranes, *Magn. Reson. Chem.*, 42(2), 187, 2004.

113. Barre, P., Zschornig, O., Arnold, K. and Huster, D., Structural and dynamical changes of the bindin B18 peptide upon binding to lipid membranes. A solid-state NMR study, *Biochemistry*, 42(27), 8377, 2003.

114. Andronesi, O.C., Pfeifer, J.R., Al-Momani, L., Özdirekcan, S., Rijkers, D.T.S., Angerstein, B., Luca, S., Koert, U., Killian, J.A. and Baldus, M., Probing membrane protein structure and orientation under fast magic-angle-spinning, *J. Biomol. NMR*, 30(3), 253, 2004.

115. Lange, A. and Baldus, M., Novel solid-state NMR methods for structural studies on GPCRs, *G-Protein Coupled Receptors in Drug Discovery*, Lundstrom, K. and Chiu, M., Eds., Marcel Dekker, New York (in press).

116. Cornilescu, G., Delaglio, F. and Bax, A., Protein backbone angle restraints from searching a database for chemical shift and sequence homology, *J. Biomol. NMR*, 13(3), 289, 1999.

117. Bittermann, H., Einsiedel, J., Hubner, J. and Gmeiner, P., Evaluation of lactam-bridged neurotensin analogous adjusting psi(Pro10) close to the experimentally derived bioactive conformation of NT(8-13), *J. Med. Chem.*, 47(22), 5587, 2004.

118. Heise, H., Luca, S., deGroot, B., Grubmüller, H. and Baldus, M., Probing conformational disorder in neurotensin by 2D solid-state NMR and comparison to MD simulations, *Biophys. J.*, in press.

119. Egorova-Zachernyuk, T.A., Hollander, J., Fraser, N., Gast, P., Hoff, A.J., Cogdell, R., De Groot, H.J.M. and Baldus, M., Heteronuclear 2D-correlations in a uniformly [C-13, N-15] labeled membrane-protein complex at ultra-high magnetic fields, *J. Biomol. NMR*, 19(3), 243, 2001.

120. McDermott, G., Prince, S.M., Freer, A.A., Hawthornthwaitelawless, A.M., Papiz, M.Z., Cogdell, R.J. and Isaacs, N.W., Crystal-structure of an integral membrane light-harvesting complex from photosynthetic bacteria, *Nature,* 374(6522), 517, 1995.

121. Luca, S., Heise, H. and Baldus, M., High-resolution solid-state NMR applied to polypeptides and membrane proteins, *Accounts Chem. Res.*, 36(11), 858, 2003.

122. Staunton, D., Owen, J. and Campbell, I.D., NMR and structural genomics, *Accounts Chem. Res.*, 36(3), 207, 2003.

3 Resonance Assignments and Secondary Structure Determination in Uniformly and Differentially Enriched Proteins and Protein Reassemblies by Magic-Angle Spinning Nuclear Magnetic Resonance Spectroscopy

Tatyana Polenova

CONTENTS

ABSTRACT General approaches for resonance assignments and secondary structure determination in uniformly and differentially enriched proteins and protein assemblies by magic-angle spinning (MAS) nuclear magnetic resonance (NMR) spectroscopy are discussed. Using a 108-amino acid residue *Escherichia coli* thioredoxin as an example, it is demonstrated that a combination of multidimensional homo- and heteronuclear one- and two-bond correlation experiments yields nearly complete resonance assignments for the uniformly enriched full-length protein as well as for the differentially enriched non-covalent complex formed by its complementary U-^{15}N(1–73) and U-^{13}C,^{15}N(74–108) fragments. The practical considerations for high-resolution solid-state NMR studies of uniformly enriched polypeptides are outlined, including sample preparation protocol, instrumentation and pulse sequences. The potential of employing differential labeling strategies for high-resolution structural studies of protein interfaces and protein assemblies by solid-state NMR spectroscopy is addressed.

KEY WORDS: *solid-state NMR, magic-angle spinning, proteins, peptides, resonance assignments, recoupling, microcrystalline, precipitate, thioredoxin, heteronuclear, homonuclear*

3.1 INTRODUCTION

The inherent capability of solid-state nuclear magnetic resonance (SSNMR) spectroscopy to probe noncrystalline systems intractable by other high-resolution techniques determines the increasing interest toward its structural applications. A number of recent reports demonstrate the potential of SSNMR for analysis of proteins and complex biological assemblies, such as membrane proteins,[1,2] amyloid fibrils,[3,4] biomaterials,[5] and intact cells.[6] For the intrinsically soluble macromolecules, very often additional information inaccessible from solution experiments can be inferred, such as internal protein and ligand dynamics on slow timescales,[7] electronic structure and geometry of paramagnetic centers[8] or quadrupolar metal sites,[9,10] ionization states and hydrogen bonding geometry.[11,12]

In recent years, because of advances in hardware and pulse sequences, structural and dynamics investigations of uniformly and extensively isotopically enriched proteins by SSNMR have become active and growing areas of research. In comparison with the traditional selective enrichment approaches employed in solids NMR, introduction of uniform or extensive isotopic labels allows for determination of multiple intramolecular structural constraints with a limited number of sample preparations. However, because of the large number of signals in the spectra of uniformly enriched proteins, resonance assignments are a major challenge and a prerequisite for subsequent site-specific identification of structural constraints.

For analysis of interfaces formed by interacting proteins, differential isotopic enrichment presents an alternative and complementary strategy,[13] expected to be especially advantageous for large complexes. Incorporating different isotopic labels for each of the individual interacting partners enables isotope-edited magnetization transfers through the interface, yielding intermolecular contacts only. The specific isotopic labeling protocols are tailored to highlight the desired intermolecular interactions (e.g., hydrogen bonds or long-range tertiary contacts). An additional advantage of this approach is a significant reduction in spectral complexity, allowing for interfaces formed by large protein assemblies to be addressed at an atomic level of detail.

Significant progress has been recently made by several laboratories in establishing the experimental protocols for resonance assignments and the determination of structural constraints from multidimensional solid-state magic-angle spinning (MAS) spectra. The concrete methodology employed by different investigators depends on the molecular weight and isotopic labeling schemes feasible for a particular protein, as well as on the hardware available for the studies. In contrast to solution NMR, in which resonance assignments are performed according to the standard procedures, even for fairly large proteins (>15 kDa), the best experimental routes are still a subject of debate within the SSNMR community. Partial or nearly complete resonance assignments by MAS NMR have been reported to date for six uniformly enriched proteins, encompassing ubiquitin,[14-18] bovine pancreatic trypsin inhibitor (BPTI),[19] the α-spectrin Src homology region3 (SH$_3$) domain,[20,21] the catabolite repression histidine-containing phosphocarrier protein corticotropin-releasing hormone (Crh),[22] reassembled thioredoxin,[13] and the immunoglobulin-binding domain B1 of Streptococcal Protein G (GB1).[23] Despite the fact that several complementary approaches worked well for the above systems, their applicability to larger proteins remains to be demonstrated, and additional efforts will be required to standardize the technologies. Whatever the concrete choice of pulse sequences, however, high magnetic fields are expected to be beneficial for structural analysis of uniformly enriched large proteins, because of the explicit dependence of resolution and sensitivity on the magnetic field strength.

In this chapter, we outline the general strategies for resonance assignments and secondary structure determination by multidimensional MAS techniques, using *Escherichia coli* thioredoxin as an example. Parent thioredoxin and its reassemblies is an excellent system for establishing methodologies for high-resolution structural and dynamic studies by MAS NMR. Reassembled thioredoxin presents an exciting system for design of differential enrichment schemes and development of SSNMR protocols for interface analysis.[13] Full-length 108–amino acid residue thioredoxin, being the largest uniformly enriched protein analyzed in detail by SSNMR to date, offers an appropriate increase in the level of complexity compared with our earlier studies of BPTI,[19] reassembled thioredoxin,[13] and the work on SH3 domain.[20]

We demonstrate that rapid precipitation of intact and reassembled thioredoxin with polyethylene glycol under controlled conditions yields excellent-quality solid-state MAS spectra, from which an overwhelming majority of ^{13}C and ^{15}N resonances are readily assigned. We discuss the advantages and limitations of employing high magnetic fields and two-dimensional spectroscopy. We address the potential of employing differential labeling schemes for high-resolution structural studies of protein assemblies and protein interfaces by SSNMR spectroscopy. The 108–amino acid residue thioredoxin is the largest protein assigned by SSNMR spectroscopy to date, and as is detailed later, the quality of the spectra indicates that substantially larger uniformly and extensively enriched systems will be amenable to detailed characterization by SSNMR.

3.2 SAMPLE PREPARATION STRATEGIES FOR HIGH-RESOLUTION STRUCTURAL STUDIES OF PROTEINS BY MAS NMR SPECTROSCOPY

3.2.1 *E. COLI* THIOREDOXIN: GENERAL BACKGROUND

E. coli thioredoxin is a disulfide reductase and a member of a large superfamily of multifunctional protein modules. Thioredoxins are universally present in organisms from Archea to humans and are the key elements in the redox regulation of protein function and signaling via thiol redox center.[24] Mammalian thioredoxins play multifaceted roles encompassing defense against oxidative stress, regulation of growth and apoptosis, immunomodulation, embryonic implantation, and developmental biology.[24] Thioredoxin is therefore considered the ultimate "moonlighting protein,"[24,25] with new functions being discovered. Because of its key roles in a number of different metabolic pathways, the thioredoxin/thioredoxin reductase system has been explored as a potential target for a number of chemotherapeutic applications ranging from infectious diseases to cancer therapy.[26,27]

A remarkable property of *E. coli* thioredoxin in solution is its ability to reassemble *in vitro* from the complementary fragments (generated by either proteolytic or chemical cleavage of the intact protein at a specific position) to yield a molecule with preserved tertiary structure.[28–30] This behavior has been found for a number of different cleavage sites in thioredoxin. Notably, the thioredoxin fold is conserved in a wide variety of organisms, and the protein is very stable thermally and chemically.[31,32] Moreover, some of the thioredoxin reassemblies by fragment complementation also display high thermodynamic stability.[33-35]

It is interesting in this context to investigate how the local structure or dynamic behavior of various thioredoxin reassemblies are altered with respect to the native protein and whether these changes can be correlated to the thermodynamic stability of these complexes. In a related system, a human–*E. coli* chimeric thioredoxin, the origin of the decreased thermodynamic stability was recently probed via structure and dynamics measurements by solution NMR.[36] Curiously, the overall structure and fast timescale dynamics were found to be very similar to the parent thioredoxin, and subtle multiple structural changes were thus postulated to be responsible for the reduced stability. A broader range of timescales will be accessible via SSNMR measurements, including slower motional modes that may be perturbed in different thioredoxin reassemblies.

We address the full-length U-^{13}C,^{15}N thioredoxin and the C-terminal portion of the reassembled (U-^{15}N-1–73/U-^{13}C,^{15}N-74–108) thioredoxin in the solid state by a combination of homo- and heteronuclear two-dimensional correlation spectroscopy.

(a)

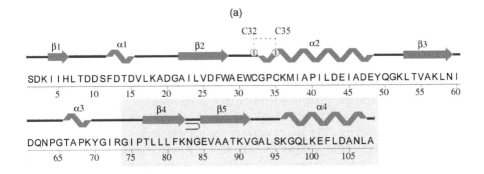

(b)

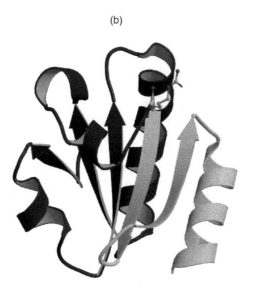

FIGURE 3.1 (a) Amino acid sequence and secondary structure of *E. coli* thioredoxin generated by PDBsum[37] using the PDB file 2trx.ent.[50] The C-terminal fragment encompassing amino acid residues 74–108 is highlighted in gray. (b) Tertiary structure generated using Molscript[51] and rendered using Raster3D.[52,53] The N-terminal fragment (residues 1–73) is shown in dark grey; the C-terminus (residues 74–108) is in light gray. Arg-73 at the cleavage site is depicted using the stick representation. The interface consists of two beta strands β2 and β4 formed by residues 23–29 and 76–82, belonging to the complementary N- and C-terminal fragments, respectively.

In Figure 3.1, the primary amino acid sequence, the secondary structure generated using PDBsum,[37] and the tertiary fold of thioredoxin are shown. The interface of the 1–73/74–108 thioredoxin complex consists of two beta strands formed by residues 23–29 and 76–82, belonging to the complementary N- and C-terminal fragments resulting from the cleavage of the intact protein at Arg-73.

3.2.2 Sample Morphology and Precipitation Conditions

Adequate spectral resolution is the main prerequisite for successful structural studies of uniformly enriched proteins by MAS SSNMR. The pioneering work from the McDermott and Oschkinat groups as well as subsequent investigations reveals that microcrystalline preparations for several globular proteins yield extremely well resolved MAS spectra suitable for resonance assignments.[16,19,22,38] Smaller-sized crystals termed nanocrystals also displayed similarly highly resolved MAS spectra, and it was proposed by Zilm and coworkers that a lower degree of long-range order can be tolerated.[39] These approaches, although apparently generally applicable to globular soluble proteins, pose several challenges. A large matrix of crystallization conditions is screened first, followed by a scale-up to generate sufficient amounts of protein microcrystals for SSNMR analysis. The first step is laborious and requires large amounts of protein for initial screens. Moreover, as we have observed for reassembled thioredoxin, microscale crystallization conditions are not necessarily directly transferable to the batch mode and often vary with protein preparation. In addition, batch-mode crystallization is a very slow process, usually ranging between 10 and 30 days. Finally, the best conditions for a particular protein (e.g., nature and concentration of precipitating agent, pH, ionic strength) typically correspond to those in which diffraction-quality crystals are formed. It is thus unclear whether samples of difficult-to-crystallize proteins could be generally prepared for high-resolution SSNMR studies using the above approach.

Rapid precipitation of proteins and protein complexes under controlled conditions is an alternative approach, which is potentially applicable to a broader range of proteins and protein complexes. Hydrated precipitates are generated via very slow addition of 40–50% polyethylene glycol (PEG) solution to the concentrated protein solution. Precipitates can be obtained in a fairly broad range of pH values. The highest yields appear to be attained when pH is substantially different from the isoelectric point of the protein, pI, in contrast to micro- and nanocrystalline preparations, in which pH is kept close to pI. Compared with micro- and nanocrystalline preparations, this protocol requires only minimal screening of precipitation conditions, and 95–99% yields are obtained in less than 24 h. Moreover, prior studies indicate that PEG-precipitated proteins are functionally competent.[11,40]

In this study, hydrated precipitates of U-^{13}C,^{15}N full-length intact and (U-^{15}N-1–73/U-^{13}C,^{15}N-74–108) reassembled thioredoxin were generated using PEG-4000 as a precipitating agent.[13] In Figure 3.2, photographs of reassembled thioredoxin precipitates under light microscope are shown and contrasted with the microcrystals obtained by hanging drop and batch methods. For controlled PEG precipitation, 50% w/v solution of PEG-4000 in 10 mmol NaCH$_3$COO buffer (pH 3.5) was slowly added to the solution containing 1 ml of 70 mg/ml of reassembled thioredoxin, 10 mmol NaCH$_3$COO, and 1 mmol NaN$_3$, pH 3.5, until no further protein precipitation was observed (~2 h). It appears that the precipitate is in neither a micro- nor a nanocrystalline state, the latter judged by comparison with the images of nanocrystalline proteins recorded with similar magnification by Zilm and coworkers.[39] Whether yet-smaller crystalline domains are formed remains to be established by x-ray powder diffraction, this work is under way.

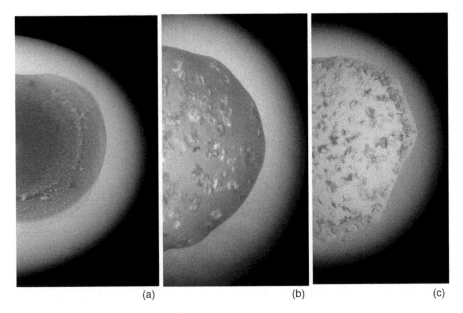

(a)   (b)   (c)

FIGURE 3.2 (a) PEG-4000 precipitate of *E. coli* thioredoxin. 50% w/v solution of PEG-4000 in 10 mmol NaCH$_3$COO buffer (pH 3.5) was slowly added to the solution containing 1 ml of 70 mg/ml of reassembled thioredoxin, 10 mmol NaCH$_3$COO, and 1 mmol NaN$_3$, pH 3.5, until no further protein precipitation was observed (~2 h). The PEG precipitate was centrifuged and transferred after the supernatant removal into the 4-mm Bruker HRMAS rotor assembly. The samples were sealed using the upper spacer and the top spinner, according to the standard procedures. (b) Microcrystals of *E. coli* thioredoxin grown in a hanging drop at 4°C. The reservoir solutions contained 1 ml of 10 mmol NaCH$_3$COO, 1 mmol NaN$_3$, 25% of PEG-4000, pH 5. On a siliconized coverslip, to form the hanging drop, 0.5 μl of 70 mg/ml thioredoxin solution was mixed with 0.5 μl of the corresponding reservoir solution. (c) Microcrystals of *E. coli* thioredoxin grown in a batch, by slow addition of PEG-4000 over a period of 6 h. The rest of the conditions are the same as in (a). Microcrystals formed within 1 week.

In Figure 3.3, the ^{15}N and ^{13}C CPMAS spectra for the PEG-precipitated thioredoxin are shown. The line widths for the outlying resonances are 15–38 Hz (0.2–0.5 ppm) and 35–94 Hz (0.5–0.8 ppm) for the ^{15}N and ^{13}C data sets, respectively, indicating a high degree of conformational homogeneity. The somewhat depressed signal intensities in the carbonyl region of the ^{13}C CPMAS spectrum as compared with the aliphatic region are the result of cross-polarization dynamics (contact times of 1.5 ms were optimized to yield the maximum transfer in the aliphatic region of the spectrum for this particular spectrum) and moderate spinning speeds. The spinning speed of 10 kHz was selected to avoid the overlap between the carbonyl side bands and the aliphatic cross peaks in the homonuclear ^{13}C-^{13}C correlation spectra. Alternatively, very high speeds exceeding 25 kHz would be required at 17.6 T. These speeds were neither feasible with the current hardware configuration nor anticipated to be advantageous because of the smaller rotor sizes and, hence, significantly

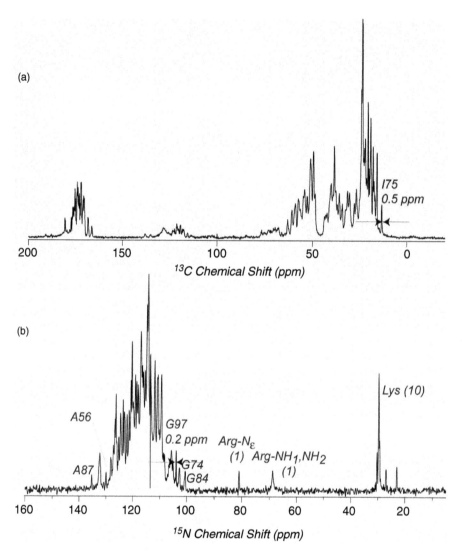

FIGURE 3.3 ^{13}C (a) and ^{15}N (b) CPMAS spectra of PEG-precipitated *E. coli* thioredoxin reassembled from complementary fragments U-^{15}N(1-73)/U-^{13}C,^{15}N(74-108). The line widths are 0.2–0.5 ppm (^{15}N) and 0.5–0.8 ppm (^{13}C). The ^{13}C spectrum was acquired with 512 transients; the recycle delay was 2 s. For ^{15}N spectrum, 128 transients were added. For cross polarization, a 1.5-ms contact time was used. The ^{13}C (or ^{15}N) amplitude was 46 kHz, and the ^{1}H amplitude was linearly ramped from 80 to 100%, with the center of the ramp corresponding to the first Hartmann-Hahn spinning sideband (56 kHz). During detection, 70 kHz X inverse-X (XiX) decoupling was applied. The spinning frequency was 10 kHz.

reduced sample amounts required for the ultra-high MAS applications (leading to the lower sensitivity of the experiments).

As is discussed in the following sections, the resolution of the corresponding two-dimensional spectra for full-length intact and reassembled thioredoxin is such that the overwhelming majority of resonance assignments could be readily accomplished. Our ongoing efforts on a number of other systems, including azurin, vanadium chloroperoxidase, and Max, indicate that this sample preparation protocol may generally work for a number of globular proteins. For example, in precipitated azurin (128 residues, 13.9 kDa), both the ^{15}N line widths of the resolved sites (~0.3–0.8 ppm) and the overall sensitivity were found to be quite similar to those for thioredoxin.

3.2.3 DIFFERENTIALLY ENRICHED THIOREDOXIN REASSEMBLIES FOR HIGH-RESOLUTION MAS STUDIES OF PROTEIN INTERFACES

The ability of *E. coli* thioredoxin to reassemble into a molecule with a preserved structure can be conveniently exploited for design of differential isotopic-labeling strategies for the complementary fragments forming a noncovalent complex. This approach is expected to be beneficial for high-resolution structural studies of protein interfaces and protein assemblies by SSNMR, especially in larger systems. One advantage of differential labeling is significantly simplified spectral assignments, allowing one binding partner to be addressed at a time. For example, as will be discussed in the following sections, assignments of the C-terminal portion of reassembled differentially labeled (U-^{15}N-1–73/U-^{13}C,^{15}N-74–108) thioredoxin were straightforward because of the smaller number of carbon resonances compared with the full-length uniformly enriched protein. Another advantage is isotope editing of polarization transfers, highlighting only the interactions of interest. Specifically, the mixed labeling scheme discussed in this work was developed for analysis of hydrogen-bonding patterns at the reassembled thioredoxin interface via the ^{13}C-(^{1}H)-^{15}N experiments and is applicable for assessing intermolecular interactions in protein interfaces formed by any arbitrary pair of proteins. Introduction of ^{13}C,^{15}N/^{15}N labels according to this protocol would enable both the resonance assignments of one binding partner and the evaluation of interfacial contacts between the individual molecules. This is envisioned to be useful in a wide variety of systems; for example, protein assemblies in which only the interface structure is of interest and the complete assignment of the complementary partner may not be required or not be feasible. Preparation of a ^{13}C/^{15}N mixed-label complex would allow for directing the ^{15}N-^{13}C magnetization transfers via the interface and highlighting the corresponding interface residues only. These residues then can be readily assigned via a combination of ^{15}N-^{13}C-^{13}C experiments establishing intraresidue and sequential ^{13}C-^{13}C correlations. A complementary labeling scheme would yield ^{13}C assignments for the interface residues of the binding partner. The assignment of ^{15}N resonances could be accomplished via a combination of ^{13}C-^{15}N and ^{15}N-^{15}N correlation experiments. Additional protocols employing a combination of uniform ^{13}C,^{15}N and sparse ^{13}C labels have been also devised for analysis of long-range tertiary contacts at the reassembled thioredoxin interface.

3.3 EXPERIMENTAL PROTOCOLS FOR RESONANCE ASSIGNMENTS

3.3.1 GENERAL CONSIDERATIONS

Several successful *de novo* resonance assignment protocols have been presented in the recent literature.[13,16,19,20,22] On the basis of these studies, salient experimental considerations can be outlined for high-resolution MAS studies of uniformly enriched polypeptides. As will be discussed below, using thioredoxin as an example, two-dimensional spectroscopy at high magnetic fields (17.6 T) yields nearly complete resonance assignments. For moderate-size proteins, backbone assignments for a fraction of residues can be performed at modest field strengths (7.0 and 9.4 T);[14,15] however, three-dimensional experiments and a field strength of at least 9.4 T are required for nearly complete assignments.[16] Higher magnetic fields appear to be necessary for adequate resolution in homonuclear ^{13}C-^{13}C experiments employed for side-chain assignments.[19] Sensitivity is an additional important factor. At 17.6 T (750 MHz), 0.1–0.5 μmol of protein are sufficient for the measurements, and the experimental time ranges between 8 and 30 h. Larger amounts of sample and significantly longer experiment times are necessary at lower fields. For example, for thioredoxin we observed about a 40% reduction in the sensitivity when experiments were conducted at 14.1 T (600 MHz) using a similar experimental setup.

Conceptually, the resonance assignment strategies by solids NMR are very similar to the solution NMR protocols, except that ^1H dimension is not commonly employed. Figure 3.4 summarizes the polarization transfer pathways used in the two- and three-dimensional SSNMR experiments. The individual experiments for backbone and side-chain assignments are discussed below.

3.3.2 BACKBONE CORRELATIONS

For intraresidue and sequential backbone assignments, NCA, NCO NCACB, NCACX, NCOCX, and NN experiments are generally used. The corresponding pulse sequences are summarized in Figure 3.5. In the NCA and NCO experiments, polarization is transferred from the backbone amide nitrogen atom to the alpha- or carbonyl-carbons, thus establishing intraresidue or sequential backbone one-bond correlations. A band-selective spectrally induced filtering in combination with CP (SPECIFIC-CP) mixing sequence[41] with weak radiofrequency fields centered at ^{15}N and ^{13}C (Cα or Co) frequencies and $\omega_1(^{15}N) = \omega_1(^{13}C) + /- \omega_r$, (which is an $n = 1$ DCP recoupling condition for on-resonance magnetization) yielded the desired transfer selectivity. During the initial 3-ms cross-polarization period, the ^1H amplitude was linearly ramped from 80 to 100%, with the center of the ramp corresponding to $\omega_1(^1H) = 35$ kHz; $\omega_1(^{15}N) = 28$ kHz. During the ^{15}N chemical shift evolution period in f_1, CW heteronuclear ^1H-^{13}N decoupling was used. The polarization was subsequently transferred from the amide nitrogen atoms to either the Cα (NCA experiment) or the Co (NCO experiment) carbons via selective double-cross polarization, with the ^{15}N amplitude being linearly ramped from 80 to 100%. The cross-polarization mixing time was 6 ms; $\omega_1(^{15}N) = 25$ kHz (center of the ramp), $\omega_1(^{13}C)$ = 18 kHz. Because of the probe restrictions on the maximum amount of power, the

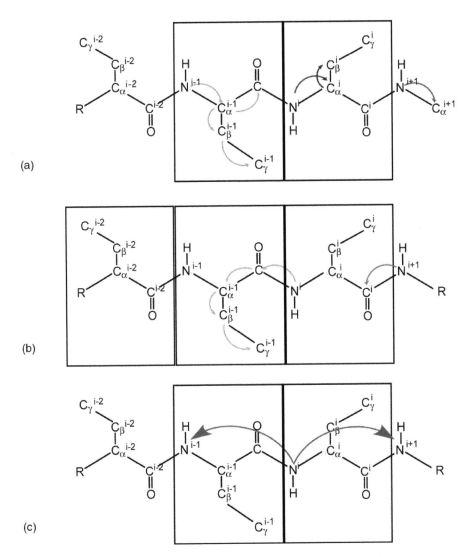

FIGURE 3.4 Polarization transfer pathways for resonance assignments employed in different experiments and demonstrating (a) intraresidue NCA, NCACB, and NCACX correlations; (b) sequential NCO and NCOCX correlations; and (c) sequential NN correlations.

heteronuclear decoupling field strength was kept at 50 kHz throughout the experiment, and the XiX decoupling scheme was employed. The ^{15}N frequency was centered at 119 ppm, and the ^{13}C frequency at 51 ppm (NCA experiment) or 180 ppm (NCO experiment). Either 208 (NCO) or 224 (NCA) scans were added for the final free-induction-decays (FIDs) in each t_1 transient, a total of 185 t_1 transients were acquired, and recycle delays of 3 s were employed. The total experiment times were 32 and 34 h for the NCO and NCA experiments, respectively.

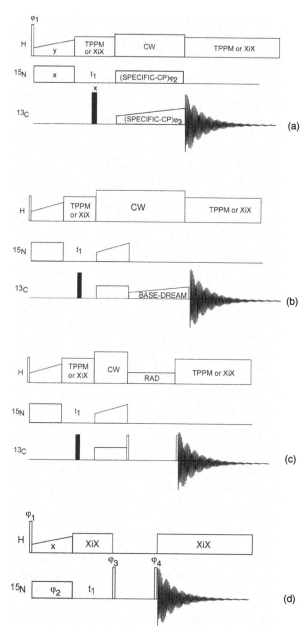

FIGURE 3.5 Pulse sequences for two-dimensional NCA/NCO (a), NCACB (b), NCACX/NCOCX (c), and NN (d) experiments used in the backbone assignments of reassembled and intact thioredoxin. Open rectangles represent π/2 pulses; filled rectangles are π pulses. SPECIFIC-CP with 8-ms mixing time was used for the N-Cα(Co) transfers. Band-selective double-quantum DREAM mixing sequence (BASE-DREAM) with a 2-ms mixing time was employed to selectively transfer magnetization from Cα to Cβ carbons in the NCACB experiment. DARR mixing (10 ms mixing time) was used to record the NCACX and NCOCX spectra. ^{15}N-^{15}N sequential correlations were achieved using the PDSD mixing sequence; $t_{mix} = 4$ s.

For small-size proteins, such as the 35-residue C-terminal portion of the reassembled (U-^{15}N-1–73/U-^{13}C,^{15}N-74–108) thioredoxin, the resolution in the NCA/NCO spectra was such that the majority of backbone assignments were accomplished with this pair of experiments. However, because of the signal overlap in the NCA/NCO spectra of the full-length protein, two additional correlation experiments were necessary. In the NCACB experiment, the backbone amide nitrogens were correlated with C_α and C_β resonances, incorporating an additional mixing period with a band-selective dipolar recoupling enhancement through amplitude modulation (DREAM) mixing sequence.[20,42] The double-quantum matching condition $n\omega_r = (\omega_{rf}^2 + \Omega_1^2)^{1/2} + (\omega_{rf}^2 + \Omega_2^2)^{1/2}$ in conjunction with the chemical shift difference for the C_α-C_β pair ($\Delta\Omega = 20$–40 ppm) dictates the experimental conditions, which for 17.6 T translate into an approximately 3–6-kHz radiofrequency (rf) field ramp for $n = 1$, and the spectrometer frequency centered around 40 ppm. For NCACX/NCOCX experiments, RAD[43] C-C mixing followed the first N-Cα/N-Co transfer step, correlating the backbone amide resonances with the side-chain carbons.

The homonuclear NN experiment employs the proton-driven spin diffusion (PDSD) mixing period to transfer polarization between backbone amide nitrogens. Because of the small magnitude of the homonuclear ^{15}N dipole–dipole interaction (25 or 60 kHz, corresponding to the average distances of 2.7 or 3.5 Å in alpha-helices or beta-sheets, respectively), only sequential N(i)–N(i+1) correlations are present in the spectra of thioredoxin. At the mixing time of 4 s, cross-peak intensities were 10 to 30% of the diagonal and were found to be largely independent of the secondary structure. Sequential nitrogen assignments based on these spectra alone are usually not feasible because of the large degree of spectral overlap. However, the NN experiment was useful in corroborating a number of nitrogen backbone assignments based on the NCO and NCA data.

3.3.3 SIDE-CHAIN CORRELATIONS

C-C correlations are the basis for side-chain resonance assignments. These are commonly established with radiofrequency-driven dipolar recoupling (RFDR),[44] rotary-assisted spin diffusion (DARR),[43] or proton spin diffusion (PDSD)[45] sequences, summarized in Figure 3.6. The three recoupling sequences in practice yield similar information. However, in contrast to RFDR, the recoupling efficiencies in DARR are typically largely insensitive to the chemical shift offset, and uniform polarization transfer is achieved for the aliphatic side-chain carbons in thioredoxin. The experiment is readily tuned for establishing dipolar interactions of the desired strength through the appropriate choice of the mixing period. At short mixing times of 1–2 ms, predominantly one-bond correlations are established that are used for intraresidue side-chain assignments. With increasing the mixing time to 10 ms, two-bond correlations emerge, and at mixing times of 100 ms or longer, a number of multibond correlations are present. These correspond to medium-range ^{13}C-^{13}C distances and are useful for confirming sequential assignments in thioredoxin. Interestingly, the intensities of the cross peaks corresponding to the two-bond correlations were found to be sensitive to the dihedral angles: The residues belonging to β-sheets exhibited stronger C_β-C_O cross peaks than those composing turns and α-helices.

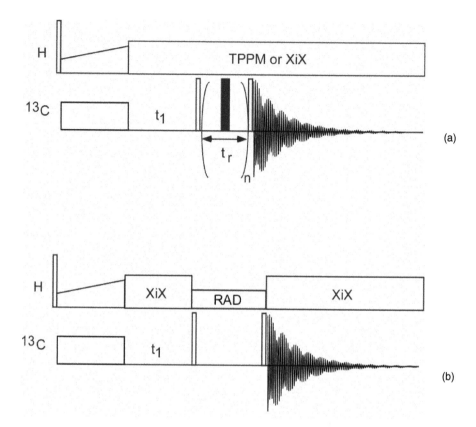

FIGURE 3.6 Pulse sequences for two-dimensional CC correlation experiments used in the side-chain assignments of reassembled and intact thioredoxin: (a) RFDR, (b) DARR. Open rectangles represent $\pi/2$ pulses; filled rectangle is π pulse. DARR experiments with 2- and 10-ms mixing time were used to establish the one- and some two-bond correlations. DARR mixing times of 100 ms were employed to yield medium-range correlations. In the PDSD experiment, no rf field is applied during the mixing period.

These intensity differences might potentially serve as an independent qualitative indicator of the secondary structure context of a particular residue; however, more experimental evidence is needed to test the generality of this approach.

3.3.4 Resonance Assignments in Reassembled Thioredoxin

Intraresidue one- and two-bond CC correlations observed in the DARR experiment with a mixing time of 10 ms yielded the majority of the amino acid types in the C-terminal fragment of reassembled thioredoxin. Five alanines, two valines, two threonines, three out of four glycines, two out of four lysines, two phenylalanines, one serine, and one isoleucine were identified readily on the basis of the characteristic fingerprint region and their expected intraresidue cross peaks, especially in the aliphatic region. In the carbonyl region, a number of C_β-C_o and C_γ-C_o intraresidue connectivities were observed as well. Several examples of site-specific assignments for individual residues are presented below.

Site-specific assignments of I75 and S95 were accomplished based on the above experiment. I75 is the unique isoleucine in the sequence and was easily identified by the multiple correlations among its aliphatic carbons, especially the methyl groups, which were the most shielded signals in the aliphatic region of the spectrum. S95 was identified by its characteristically deshielded C_β resonance (64 ppm). The unambiguous assignment of I75 in addition revealed the identity of the unique proline, P76, via the multiple cross peaks between its aliphatic carbons, and the carbonyl carbon of I75 found in the carbonyl region of the DARR experiments acquired with long mixing times of 100 and 300 ms. These three residues, I75, P76, and S95, served as the starting point in the assignments.

For sequential connectivities, a combination of four experiments was employed: NCO, NCA, and NN experiments, as well as DARR with long mixing times (100 and 300 ms) (Figures 3.7–3.9). The last experiment was found particularly useful in corroborating the sequential heteronuclear assignments (Figures 3.7 and 3.8), as well as in resolving the ambiguities for the poorly dispersed leucine and lysine residues.

Glycine residues were identified readily because of their low abundance in the sequence (four) and their characteristic upfield C_α and C_o carbon chemical shifts. Three of the four glycines were found on the basis of their cross peaks in the carbonyl regions of the DARR experiments at all mixing times. The typical N-C correlations corresponding to all four glycines were present in the NCA experiment.

Site-specific assignments for glycines were accomplished via sequential correlations between each of the four glycines and their neighboring residues. For example, G84 was identified from its multiple CC sequential correlations with E85 and V86 (Figure 3.8). In the glutamate residues, the C_δ carbon is deshielded (~180–186 ppm). Therefore, the characteristic C_γ-C_δ correlations of the two glutamate residues E85 and E101 appeared in the isolated region of the spectrum and were used to report on these two residues. C_δ carbon of E85 exhibited sequential correlations to C_α of G84, as well as to $C_{\gamma 1}$,$C_{\gamma 2}$ of V86 residues in the DARR spectra at long mixing times. Sequential cross peaks between the aliphatic and the carbonyl carbons of G84, E85, and N83 were also observed under these conditions: G84C_α-E85C_α, E85C_β-G84C_α, G84C_α-E85C_δ, G84C_α-N83C_α, G84C_α-N83C_α, and G84C_α-N83C_β.

Several sequential cross peaks between G74 and I75 in the aliphatic and carbonyl regions of the 100- and 300-ms DARR spectra allowed for site-specific assignment of G74. In Figure 3.8, the following sequential correlations are indicated: G74C_o-I75C_α, G74C_o-I75C_β, G74C_o- I75$C_{\gamma 1}$, G74C_o-I75$C_{\gamma 2}$, G74C_o-I75$C_{\delta 1}$, G74C_α-I75C_α, G74C_α-I75C_β, and G74C_α-I75$C_{\delta 1}$. Absence of a cross peak to the preceding residue in the NCO experiment confirmed this assignment (R73 belongs to the complementary thioredoxin fragment containing no ^{13}C labels).

Assignment of the two remaining glycines, G92 and G97, was straightforward based on the various sequential cross peaks found in the NCO, NN, and 100- and 300-ms CC DARR experiments. The observed G92C_α-A93C_o, G92C_α-A93C_β, G92C_α-V91C_o, G92C_α-V91C_α, G92C_α-V91$C_{\gamma 1}$, G92C_α-V91$C_{\gamma 2}$, G92N-C91C_o, and G92N-V91N correlations in conjunction with the intraresidue V91N-C_α cross peaks

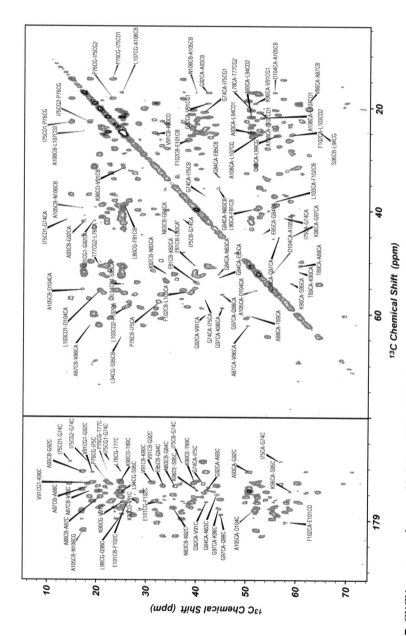

FIGURE 3.7 CXCY spectrum of reassembled *E. coli* thioredoxin demonstrating examples of sequential ¹³C-¹³C correlations. The experiment was recorded with DARR sequence and a mixing time of 100 ms. The spectrum was processed using a 60° shifted sine filter in both dimensions for resolution enhancement, zero filling to 4096 points in the f_2 dimension and 1024 points in the f_1 dimension, and automatic baseline correction was used in the f_2 dimension.

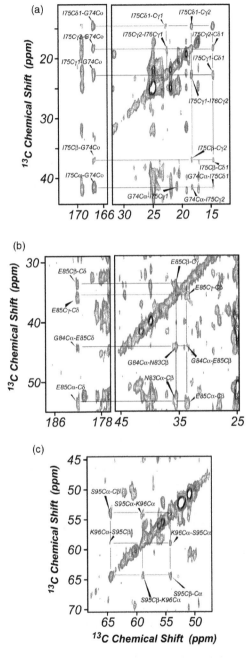

FIGURE 3.8 Expansions of the DARR spectrum acquired with a mixing time of 100 ms, demonstrating the sequential cross peaks between (a) G74 and I75, (b) G84-E85, and (c) K96 and S95.

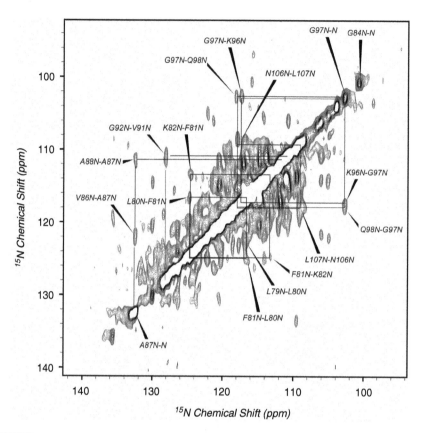

FIGURE 3.9 N-N correlation spectrum of reassembled *E. coli* thioredoxin demonstrating examples of sequential backbone walks. The ^{15}N-^{15}N correlations were established via a PDSD mixing sequence with a 12-ms mixing time. The spectrum was processed with linear prediction up to 384 points in f_1, cosine apodization in both dimensions; zero filling up to 2048 points in f_2, and 1024 points in f_1; and automatic baseline correction in f_2.

yielded the ^{13}C and ^{15}N backbone and side-chain chemical shifts for G92. For G97, G97C$_\alpha$-K96C$_\alpha$, G97C$_\alpha$-Q98C$_o$, G97C$_\alpha$-Q98C$_\alpha$, G97C$_\alpha$-K96C$_\alpha$, G97N-K96N, and G97N-K96N, correlations were employed to derive the ^{13}C and ^{15}N chemical shifts. Assignment procedures for the remaining residue types are reported elsewhere.[13]

CXCY (Figure 3.10a and b), NCO (Figure 3.10c), and NCA (Figure 3.10d) spectra of reassembled *E. coli* thioredoxin demonstrating examples of intraresidue and sequential backbone and side-chain walks are presented in Figure 3.10. In summary, we were able to unambiguously assign 93% of the ^{13}C and ^{15}N chemical shifts of the C-terminal of reassembled thioredoxin by using a combination of two-dimensional homo- and heteronuclear correlation experiments. The unassigned signals belong mostly to the side-chain ^{13}C resonances, which exhibit substantial degree of overlap in the spectra. Increasing dimensionality of the experiments is expected to resolve these ambiguities.

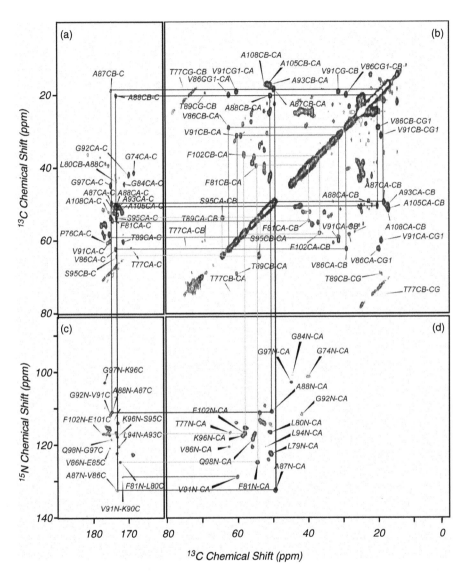

FIGURE 3.10 CXCY (a) aliphatic region; (b) carbonyl region), (c) NCO, and (d) NCA and spectra of reassembled *E. coli* thioredoxin demonstrating examples of intraresidue and sequential backbone and side-chain assignment walks. The CXCY spectrum was recorded with the DARR mixing element and a mixing time of 10 ms, resulting in predominantly one-bond correlations and a limited number of two-bond intraresidue correlations. The spectra were processed with cosine apodization in both dimensions, zero filling up to 4096 points in direct dimension and 512 points in the indirect dimension.

3.3.5 RESONANCE ASSIGNMENTS IN FULL-LENGTH INTACT THIOREDOXIN

The general assignment procedure for the full-length thioredoxin was similar to that for the reassembled protein, employing NCA, NCO, and CC correlations. Because of the large number of signals in the NCA and NCO experiments, additional NCACB and NCOCX experiments were conducted, which helped resolve the ambiguous cross peaks. In Figure 3.11, the NCACB and NCOCX spectra illustrate examples of backbone assignments.

The ^{13}C-^{13}C correlation spectra acquired with both DARR and RFDR mixing sequences display excellent resolution, allowing for assignment of the majority of the amino acid spin systems (Figure 3.12), which is remarkable considering the large size of thioredoxin.

In Figure 3.13, NCO (Figure 3.13A), NCA (Figure 3.13B), and CXCY (Figure 3.13C) spectra are depicted demonstrating examples of intraresidue and sequential backbone and side-chain assignment walks. In summary, we have currently completed assignments for 96% of the ^{15}N, C_α, and C_β; 65% of C_o; and 70% of all side-chain resonances. Similar to the reassembled thioredoxin, the unassigned signals are the result of spectral overlap, especially in the ^{13}C resonances of carbonyls and side chains.

3.4 SECONDARY STRUCTURE DETERMINATION

Secondary carbon and nitrogen chemical shifts are sensitive to the polypeptide backbone conformation.[46] Empirical predictions of the dihedral ψ and ϕ angles based on the statistical analysis of secondary chemical shifts as implemented in torsion angle likelihood obtained from shift and sequence similarity (TALOS)[47] have been widely applied in solution NMR for deriving the secondary structure, and more recently, this approach has been demonstrated to yield accurate results for solid-state chemical shifts of several peptides and proteins.[13,22,48,49]

In Figure 3.14, the dihedral ϕ and ψ angles for full-length and reassembled thioredoxin predicted from solid-state state shifts are compared against those predicted from solution data. The overwhelming majority of ψ and ϕ angles in both thioredoxin samples deviates by less than 20° and correctly reflects the secondary structure of thioredoxin. The discrepancies of 20° or more are displayed by one residue in the reassembled thioredoxin and by 10 residues of the full-length protein; 9 of these 10 residues, as expected, are found at the termini of the individual secondary structure elements (Figure 3.15). Surprisingly, the dihedral angle ψ of Ala-87 differs significantly from that expected based on its secondary structure (beta sheet); this deviation is observed in both the intact and the reassembled thioredoxin. The reason for such discrepancy is not clear at this point. An additional body of evidence for other proteins will be necessary to evaluate whether the observed discrepancy for Ala-87 is statistically meaningful. Overall, the secondary structure motifs are correctly inferred from SSNMR data, providing additional evidence to the general applicability of TALOS prediction algorithm to SSNMR chemical shifts.

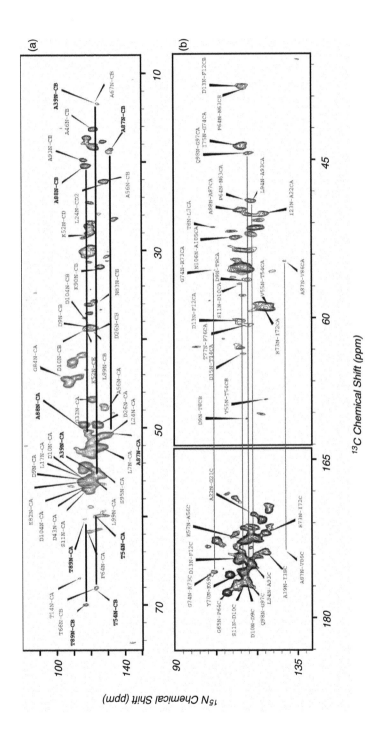

FIGURE 3.11 NCACB (a) and NCOCX (b) spectra of full-length *E. coli* thioredoxin demonstrating examples of backbone assignments. The NCACB spectrum was recorded with the BASE-DREAM mixing element and a mixing time of 2 ms. DARR mixing of 2 ms was used to acquire the NCOCX spectrum. The spectra were processed with cosine apodization in both dimensions, zero filling up to 4096 points in direct dimension and 512 points in the indirect dimension.

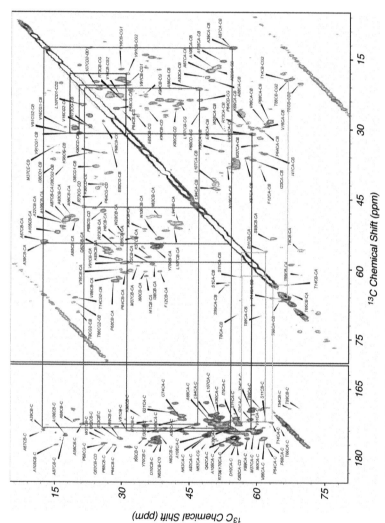

FIGURE 3.12 CXCY spectra of full-length *E. coli* thioredoxin demonstrating intraresidue side-chain walks for identification of amino acid residue types. The CXCY spectrum was recorded with the DARR mixing element and a mixing time of 2 ms, resulting in predominantly one-bond correlations. The spectra were processed with cosine apodization in both dimensions, zero filling up to 4096 points in direct dimension and 512 points in the indirect dimension. Examples of chemical shift topologies are shown for the following amino acids: Ala, Ser, Thr, and Val.

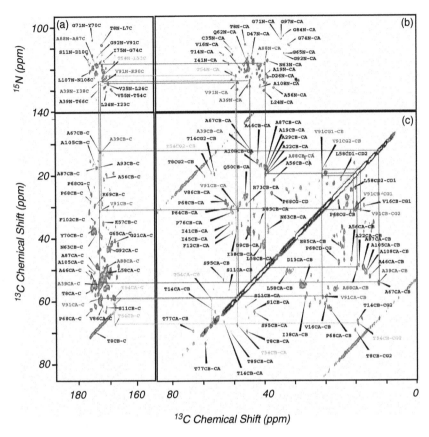

FIGURE 3.13 NCO (a), NCA (b), and CXCY (c) spectra of full-length *E. coli* thioredoxin demonstrating examples of intraresidue and sequential backbone and side-chain assignment walks. The CXCY spectrum was recorded with the DARR mixing element and a mixing time of 10 ms, resulting in predominantly one-bond correlations and a limited number of two-bond intraresidue correlations. The spectra were processed with cosine apodization in both dimensions, zero filling up to 4096 points in direct dimension and 512 points in the indirect dimension.

3.5 CONCLUSIONS AND FUTURE PROSPECTS

Resonance assignments for full-length 108–amino acid residue *E. coli* thioredoxin and its uniformly $^{13}C,^{15}N$-enriched C-terminal portion reassembled from complementary fragments have been accomplished by two-dimensional solid-state MAS NMR spectroscopy at 17.6 T. The sample preparation procedure developed for thioredoxin results in spectra of excellent resolution. Statistical analysis of ^{13}C and ^{15}N chemical shifts accurately predicts backbone dihedral angles and secondary structure for the overwhelming majority of amino acid residues in thioredoxin and indicates no significant perturbations in the secondary structure of the complex with respect to the full-length native protein. The idea of using differentially isotopically labeled fragments for the preparation of reassembled thioredoxin is anticipated to be generally applicable for

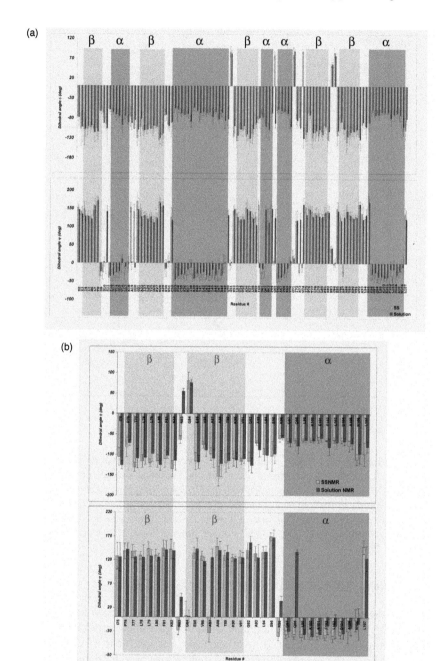

FIGURE 3.14 Comparison of the dihedral ψ angles predicted from the observed solid-state and solution chemical shifts of full-length intact (a) and reassembled (b) thioredoxin, using TALOS.[47] The lighter rectangles represent solid state, and the darker rectangles solution values. The α-helical and β-sheet regions in the secondary structure are highlighted in dark and light grey, respectively.

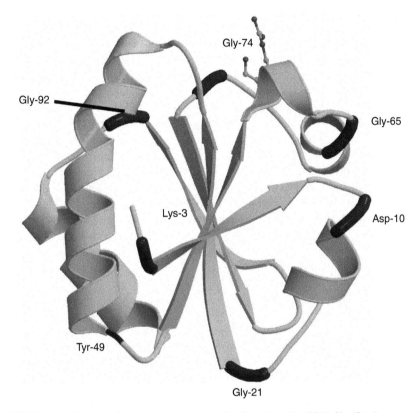

FIGURE 3.15 Thioredoxin structure representation based on the PDB file (2trx), generated in WebLabViewerPro (Accelrys, Inc.). Residues whose dihedral angles predicted from solid-state and solution NMR chemical shifts differ by more than 20° are highlighted in dark grey. These residues are located in the termini of the individual α-helices, β-sheets, or flexible loops.

structural studies of protein interfaces and protein assemblies by SSNMR. Resonance assignments of reassembled thioredoxin presented here are the first step in our forth-coming efforts in this area.

This work also demonstrates that at high magnetic fields, resonance assignments of moderate-size proteins and protein assemblies can be established in a straight-forward and time-efficient way via a combination of two-dimensional homo- and heteronuclear correlation experiments. This approach is anticipated to be especially advantageous for systems in which increasing dimensionality of the experiments is undesirable. This involves systems where limited sample quantities may be available, resulting in compromised sensitivity of multidimensional experiments, or proteins undergoing conformational exchange in the intermediate timescales. These experi-mental protocols are expected to be generally suited for structural studies of larger proteins and protein interfaces by high-resolution SSNMR spectroscopy.

ACKNOWLEDGMENTS

The work reviewed here would have been impossible without the seminal role of Maria Lusia Tasayco in initiating this collaborative study. The author gratefully acknowledges her collaborators, group members, and colleagues, who have contributed to this chapter via experimental work or valuable scientific discussions: Maria Luisa Tasayco, Dabeiba Marulanda, Marcela Cataldi, Vilma Arrairan, and Ann McDermott. The author, as adjunct faculty of the City University of New York, is a member of the New York Structural Biology Center (supported in part by the National Institutes of Health [P41 GM066354] and its Member Institutions), and acknowledges instrument time on its 750-MHz SSNMR spectrometer. The author thanks the National Institutes of Health (5S06GM060654-04 under SCORE program, P20-17716 under COBRE program, and 2 P20 016472-04 under INBRE program of NCRR), the National Science Foundation–CAREER development award (CHE 0350385), the American Chemical Society Petroleum Research Fund (PRF grant 39827-G5M), and the University of Delaware for support of this and other ongoing research in the Polenova group.

REFERENCES

1. Park, S.H., Mrse, A.A., Nevzorov, A.A., Mesleh, M.F., Oblatt-Montal, M., Montal, M. and Opella, S.J., Three-dimensional structure of the channel-forming trans-membrane domain of virus protein "u" (Vpu) from HIV-1, *J. Mol. Biol.,* 333, 409, 2003.
2. Nishimura, K., Kim, S.G., Zhang, L. and Cross, T.A., The closed state of a H+ channel helical bundle combining precise orientational and distance restraints from solid state NMR-1, *Biochemistry,* 41, 13170, 2002.
3. Tycko, R., Applications of solid state NMR to the structural characterization of amyloid fibrils: methods and results, *Prog. NMR Spec.,* 42, 53, 2003.
4. Petkova, A.T., Ishii, Y., Balbach, J.J., Antzutkin, O.N., Leapman, R.D., Delaglio, F. and Tycko, R. A structural model for Alzheimer's beta-amyloid fibrils based on experimental constraints from solid state NMR, *Proc. Natl. Acad. Sci. USA,* 99, 16742, 2002.
5. Drobny, G.P., Long, J.R., Karlsson, T., Shaw, W., Popham, J., Oyler, N., Bower, P., Stringer, J., Gregory, D., Mehta, M. and Stayton, P.S., Structural studies of biomaterials using double-quantum solid-state NMR spectroscopy, *Annu. Rev. Phys. Chem.,* 54, 531, 2003.
6. Cegelski, L., Kim, S.J., Hing, A.W., Studelska, D.R., O'Connor, R.D., Mehta, A.K. and Schaefer, J., Rotational-echo double resonance characterization of the effects of vancomycin on cell wall synthesis in *Staphylococcus aureus, Biochemistry,* 41, 13053, 2002.
7. Krushelnitsky, A.G., Hempel, G. and Reichert, D., Simultaneous processing of solid-state NMR relaxation and 1D-MAS exchange data: the backbone dynamics of free vs. binase-bound barstar. *Biochem. Biophys. Acta Proteins Proteom.,* 1650(1-2), 117–127, 2003.
8. Lee, H., de Montellano, P.R.O. and McDermott, A.E., Deuterium magic-angle spinning studies of substrates bound to cytochrome P450, *Biochemistry,* 38, 10808, 1999.

9. Lipton, A.S., Heck, R.W. and Ellis, P.D., Zinc solid-state NMR spectroscopy of human carbonic anhydrase: implications for the enzymatic mechanism, *J. Am. Chem. Soc.,* 126, 4735, 2004.

10. Lipton, A.S., Buchko, G.W., Sears, J.A., Kennedy, M.A. and Ellis, P.D., Zn-67 solid-state NMR spectroscopy of the minimal DNA binding domain of human nucleotide excision repair protein XPA, *J. Am. Chem. Soc.,* 123, 992, 2001.

11. Gu, Z.T., Drueckhammer, D.G., Kurz, L., Liu, K., Martin, D.P. and McDermott, A., Solid-state NMR studies of hydrogen bonding in citrate synthase inhibitor complex, *Biochemistry,* 38, 8022, 1999.

12. Emmler, T., Gieschler, S., Limbach, H.H. and Buntkowsky, G., A simple method for the characterization of OHO-hydrogen bonds by H-1-solid state NMR spectroscopy, *J. Mol. Struct.,* 700, 29, 2004.

13. Marulanda, D., Tasayco, M.L., McDermott, A., Cataldi, M., Arriaran, V. and Polenova, T., Magic-angle spinning solid-state NMR spectroscopy for structural studies of protein interfaces. Resonance assignments of differentially enriched *E. coli* thioredoxin reassembled by fragment complementation, *J. Am. Chem. Soc.,* 126, 16608, 2004.

14. Straus, S.K., Bremi, T. and Ernst, R.R., Experiments and strategies for the assignment of fully C-13/N-15-labelled polypeptides by solid state NMR, *J-Bio. NMR,* 12, 39, 1998.

15. Hong, M., Resonance assignment of C-13/N-15 labeled solid proteins by two- and three-dimensional magic-angle-spinning NMR, *J. Bio. NMR,* 15, 1, 1999.

16. Igumenova, T.I., Wand, A.J. and McDermott, A.E., Assignment of the backbone resonances for microcrystalline ubiquitin, *J. Am. Chem. Soc.,* 126, 5323, 2004.

17. Igumenova, T.I., Assignment of uniformly carbon-13-enriched proteins and optimization of their carbon lineshapes, Ph.D. thesis, Columbia University, New York, 2003.

18. Igumenova, T.I., McDermott, A.E., Zilm, K.W., Martin, R.W., Paulson, E.K. and Wand, A.J., Assignments of carbon NMR resonances for microcrystalline ubiquitin, *J. Am. Chem. Soc.,* 126, 6720, 2004.

19. McDermott, A.E., Polenova, T., Bockmann, A., Zilm, K., Martin, R., Paulson, E. and Montellione, G., Partial NMR assignments for uniformly (^{13}C, ^{15}N)-enriched BPTI in the solid state, *J. Bio. NMR,* 16, 209, 2000.

20. Pauli, J., Baldus, M., van Rossum, B., de Groot, H. and Oschkinat, H., Backbone and side-chain C-13 and N-15 signal assignments of the alpha-spectrin SH3 domain by magic-angle spinning solid-state NMR at 17.6 Tesla, *Chembiochem,* 2, 272, 2001.

21. Castellani, F., van Rossum, B., Diehl, A., Schubert, M., Rehbein, K. and Oschkinat, H., Structure of a protein determined by solid-state magic-angle-spinning NMR spectroscopy, *Nature,* 420, 98, 2002.

22. Bockmann, A., Lange, A., Galinier, A., Luca, S., Giraud, N., Juy, M., Heise, H., Montserret, R., Penin, F. and Baldus, M., Solid state NMR sequential resonance assignments and conformational analysis of the 2 x 10.4 kDa dimeric form of the *Bacillus subtilis* protein Crh, *J. Bio. NMR,* 27, 323, 2003.

23. Rienstra, C.M., 3D Magic-angle spinning NMR studies of the immunoglobulin-binding domain B1 of streptococcal protein G (GB1): chemical shift assignments and structural constraints, paper presented at the Experimental Nuclear Magnetic Resonance Conference (ENC), Pacific Grove, CA, 2004.

24. Arnér, E.S. J. and Holmgren, A., Physiological functions of thioredoxin and thioredoxin reductase. *Eur. J. Biochem.,* 267, 6102, 2000.

25. Jeffery, C., Moonlighting proteins, *Trends Biochem. Sci.,* 24, 8, 1999.

26. Gromer, S., Urig, S. and Becker, K., The thioredoxin system — from science to clinic, *Med. Res. Rev.*, 24, 40, 2004.

27. Hirt, R.P., Muller, S., Embley, T.M. and Coombs, G.H., The diversity and evolution of thioredoxin reductase: new perspectives, *Trends Parasitol.*, 18, 302, 2002.

28. Yu, W.-F., Tung, C.-C., Wang, H. and Tasayco, M.L., NMR analysis of cleaved *E. coli* thioredoxin (1, 74) and its P76A variant: cis/trans peptide isomerization, *Protein Sci.*, 9, 20, 2000.

29. Tasayco, M.L. and Chao, K., NMR study of the reconstitution of the beta-sheet of thioredoxin by fragment complementation, *Proteins*, 22, 41, 1995.

30. Yang, X.M., Yu, W.F., Li, J.H., Fuchs, J., Rizo, J. and Tasayco, M.L., NMR evidence for the reassembly of an alpha/beta domain after cleavage of an alpha-helix: implications for protein design, *J. Am. Chem. Soc.*, 120, 7985, 1998.

31. Eklund, H., Gleason, F.K. and Holmgren, A., Structural and functional relations among thioredoxins of different species. *Proteins: Struct. Funct. Genet.*, 11, 13, 1991.

32. Martin, J.L., Thioredoxin — a fold for all reasons, *Structure*, 3, 245, 1995.

33. Yang, X.M., Georgescu, R.E., Li, J.H., Yu, W.F., Haierhan, and Tasayco, M.L., Recognition between Disordered Polypeptide Chains from Cleavage of an Alpha/Beta Domain: Self- versus Non-Self-Association, *Proceedings of the Pacific Symposium in Biocomputing*, Hawaii, 1999, p. 590.

34. Tasayco, M.L., Fuchs, J., Yang, X.-M., Dyalram, D. and Georgescu, R.E., Interaction between two discontiguous chain segments from the β-sheet of *E. coli* thioredoxin suggests initiation site for folding, *Biochemistry*, 39, 10613, 2000.

35. Georgescu, R.E., García-Mira, M.D.M., Tasayco, M.L. and Sánchez-Ruiz, J.M., Heat capacity analysis of oxidized *E. coli* thioredoxin fragments (1, 74) and their non-covalent complex: evidence for the burial of apolar surface in protein unfolded states, *Eur. J. Biochem.*, 268, 1477, 2001.

36. Dangi, B., Dobrodumov, A.V., Louis, J.M. and Gronenborn, A.M., Solution structure and dynamics of the human — *Escherichia coli* thioredoxin chimera: insights into thermodynamic stability, *Biochemistry*, 41, 9376, 2002.

37. Laskowski, R.A., Hutchinson, E.G., Michie, A.D., Wallace, A.C., Jones, M.L. and Thornton, J.M., PDBsum: a web-based database of summaries and analyses of all PDB structures, *Trends Biochem. Sci.*, 22, 488, 1997.

38. Pauli, J., van Rossum, B., Forster, H., de Groot, H.J.M. and Oschkinat, H., Sample optimization and identification of signal patterns of amino acid side chains in 2D RFDR spectra of the alpha-spectrin SH3 domain, *J. Magn. Reson.*, 143, 411, 2000.

39. Martin, R.W. and Zilm, K.W., Preparation of protein nanocrystals and their characterization by solid state NMR, *J. Magn. Reson.*, 165, 163, 2003.

40. Williams, J.C. and McDermott, A.E., Dynamics of the flexible loop of triosephosphate isomerase — the loop motion is not ligand-gated, *Biochemistry*, 43, 8309, 1995.

41. Baldus, M., Petkova, A.T., Herzfeld, J. and Griffin, R.G., Cross polarization in the tilted frame: assignment and spectral simplification in heteronuclear spin systems, *Mol. Phys.*, 95, 1197, 1998.

42. Verel, R., Baldus, M., Ernst, M. and Meier, B.H., A homonuclear spin-pair filter for solid-state NMR based on adiabatic passage techniques, *Chem. Phys. Lett.*, 287, 421, 1998.

43. Takegoshi, K., Nakamura, S. and Terao, T., ^{13}C-^{1}H dipolar-assisted rotational resonance in magic-angle spinning NMR, *Chem. Phys. Lett.*, 344, 631, 2001.

44. Bennett, A.E., Ok, J.H., Griffin, R.G. and Vega, S., Chemical shift correlation spectroscopy in rotating solids: radio-frequency dipolar recoupling and longitudinal exchange, *J. Chem. Phys.*, 96, 8624, 1992.

45. Bloembergen, N., On the interaction of nuclear spins in crystalline lattice, *Physica*, 15, 386, 1949.

46. Wishart, D.S. and Sykes, B.D., The C-13 chemical-shift index — a simple method for the identification of protein secondary structure using C-13 chemical-shift data, *J-Bio. NMR*, 4, 171, 1994.

47. Cornilescu, G., Delaglio, F. and Bax, A., Protein backbone angle restraints from searching a database for chemical shift and sequence homology, *J-Bio. NMR*, 13, 289, 1999.

48. Ishii, Y., C-13-C-13 dipolar recoupling under very fast magic-angle spinning in solid-state nuclear magnetic resonance: applications to distance measurements, spectral assignments, and high-throughput secondary-structure determination, *J. Chem. Phys.*, 114, 8473, 2001.

49. Castellani, F., van Rossum, B.-J., Diehl, A., Rehbein, K. and Oschkinat, H., Determination of solid-state NMR structures of proteins by means of three-dimensional ^{15}N-^{13}C-^{13}C dipolar correlation spectroscopy and chemical shift analysis, *Biochemistry*, 42, 11476, 2003.

50. Katti, S.K., LeMaster, D.M. and Eklund, H., Crystal structure of thioredoxin from *Escherichia coli* at 1.68 Å resolution, *J. Mol. Biol.*, 212, 167, 1990.

51. Kraulis, P.J., Molscript — a program to produce both detailed and schematic plots of protein structures, *J. Appl. Crystallogr.*, 24, 946, 1991.

52 Merritt, E.A. and Bacon, D.J., Raster3D: photorealistic molecular graphics, *Method Enzymol.*, 277, 505, 1997.

53. Merritt, E.A. and Murphy, M.E.P., Raster3d Version-2.0 — a program for photorealistic molecular graphics, *Acta Crystallogr. D*, 50, 869, 1994.

4 Torsion Angle Determination in Biological Solids by Solid-State Nuclear Magnetic Resonance

Mei Hong and Sungsool Wi

CONTENTS

ABSTRACT Molecular torsion angles in peptides and proteins encode important secondary structure (backbone) and tertiary structure (side-chain) information. Solid-state nuclear magnetic resonance spectroscopy is a powerful tool for determining these torsion angles because of the inherent orientation dependence of nuclear spin interactions. This review summarizes the large number of techniques developed in the last decade for determining molecular torsion angles. Techniques are available for both non-spinning biopolymers, from which exquisite angular resolution can be obtained, and for magic-angle spinning samples, from which high site resolution is available to yield multiple torsion angles in complex proteins. Many of these techniques rely on correlating two spin interaction tensors across the torsion bond of interest and extracting the torsion angle from the sum and difference frequencies of the two interactions. The other techniques measure distances between two nuclei separated by three bonds or more.

KEY WORDS: *torsion angles, tensor correlation, dipolar recoupling*

4.1 STATIC TENSOR CORRELATION TECHNIQUES

Tensor correlation under nonspinning conditions produces broad but featureful line shapes that depend sensitively on the relative orientation of the two interactions of interest. This class of techniques is suitable for synthetic or biological polymers with relatively simple chemical structures, where site resolution can be achieved by specific labeling and sensitivity is high because of large sample amounts. Although earlier static nuclear magnetic resonance (NMR) experiments for correlating tensor orientations were available,[1] the DOQSY (double-quantum spectroscopy) technique[2] was one of the first static techniques used in peptides and proteins for determining torsion angles. It correlates the chemical shift anisotropy of two adjacent carbons and uses a simple Hahn echo sequence to excite the double-quantum (DQ) coherence, which evolves under the sum chemical shift tensor in the indirect dimension. After reconverting the DQ coherence back to single-quantum magnetization by an identical Hahn echo period, the direct dimension detects the chemical shifts and the ^{13}C-^{13}C dipolar coupling of the two carbons (Figure 4.1a). The two-dimensional (2D) spectra can be represented by $(\omega_1, \omega_2) = (\omega_a + \omega_b, \omega_{a,b} \pm \omega_D)$. The torsion angle around the C-C bond is contained in the sum chemical shift tensor $\omega_a + \omega_b$. For two identical functional groups such as the methylene groups in polyethylene, the trans conformation with a torsion angle of 180° gives a large sum chemical shift tensor, while the gauche conformer with a torsion angle of ±60° has a much reduced chemical shift span in the ω_1 dimension. Because of the sizable one-bond ^{13}C-^{13}C dipolar coupling, the chemical shift anisotropy pattern in the direct dimension is split by the dipolar coupling. This dipolar splitting can be removed by ^{13}C homonuclear decoupling during detection, using a modified magic-sandwich echo sequence.[3] The ^{13}C-decoupled DOQSY experiment was applied to poly(ethylene terephthalate) to determine the O-CH$_2$-CH$_2$-O torsion angle.[4] As a result of the ^{13}C dipolar decoupling, the trans conformation exhibits a distinct sharp ridge and the gauche conformer yields a compact line shape broadened in both dimensions. Trans-gauche statistics are thus readily obtained from the superposition of the two subspectra.

In principle, the DOQSY technique can be applied to the directly bonded Cα and C′ sites in peptides to determine the ψ torsion angle. In practice, however, the Cα CSA tensor orientation is itself sensitive to the backbone conformation, and thus the spectra would depend not only on the ψ angle of interest but also on the Cα CSA tensor orientation in the molecule.[5,6] A related technique that correlates two interaction

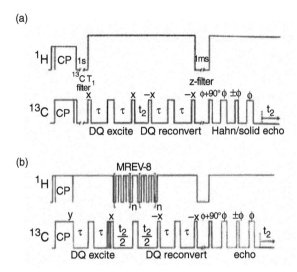

FIGURE 4.1 (a) DOQSY pulse sequence for correlating two ^{13}C CSA tensor orientations. A Hahn echo sequence is used to excite the DQ coherence, which evolves under the sum chemical shift. The DQ coherence is reconverted to single-quantum observable magnetization for detection during t_2 (Adapted from ref. 2). (b) SELFIDOQ pulse sequence for correlating the C-H dipolar coupling of a C-H group with the CSA tensor of an adjacent carbon. ^{13}C DQ coherence evolves under the C-H dipolar coupling, whereas ^{1}H homonuclear couplings are removed by MREV-8 (Adapted from Ref. 7.)

tensors at Cα and C′ without requiring information on the Cα chemical shift tensor is the SELFIDOQ (separated-local-field double-quantum) experiment.[7] It correlates the Cα-Hα dipolar coupling in the ω_1 dimension with the C′ chemical shift anisotropy in the ω_2 dimension, thus yielding the ψ torsion angle. The Cα-C′ DQ coherence evolves under C-H heteronuclear dipolar coupling during t_1, while ^{1}H homonuclear decoupling is achieved by multiple-pulse sequences (Figure 4.1b). The undesirable DOQSY effect — evolution under the sum chemical shift frequency — is removed by a 180° ^{13}C pulse in the middle of the evolution period. To prevent refocusing of the C-H dipolar coupling, a ^{1}H 180° pulse is applied at the same time. These two π pulses also refocus the unwanted ^{13}C′-^{14}N and ^{1}H-^{14}N dipolar couplings. The original SELFIDOQ experiment does not contain ^{13}C homonuclear decoupling in the direct dimension, although this can be readily added to simplify the spectra. The SELFIDOQ spectra are symmetric in the ω_1 dimension, as represented by the characteristic spectra shown in Figure 4.2. It can be seen that the 2D patterns for $\psi = 120°$ (β-sheet) and $\psi = -60°$(α-helix) differ significantly. In both the DOQSY and SELFIDOQ experiments, the DQ excitation time (τ) imposes an orientational dependence, $\sin^2(2\omega_D\tau)$, to the spectra, which needs to be taken into account in the simulations.

Although the DOQSY technique is not particularly useful for Cα-C′ tensor correlation, it is a powerful method for correlating the orientations of two consecutive C′ CSA tensors in proteins. The relative orientation of two consecutive C′ CSA tensors (Figure 4.3a) reflects all three torsion angles between them — ω, ϕ, and ψ — but the peptide-plane torsion angle ω is usually assumed to be 180°. Simulated

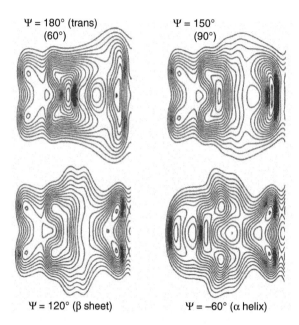

$\Psi = 180°$ (trans)
(60°)

$\Psi = 150°$
(90°)

$\Psi = 120°$ (β sheet) $\Psi = -60°$ (α helix)

FIGURE 4.2 Simulated SELFIDOQ spectra in the C′ region for various ψ angles. Note the different patterns for the β-sheet and α-helix conformations (Adapted from Ref. 7.)

C′-C′ 2D DOQSY spectra for all combinations of (φ, ψ) angles are shown in Figure 4.3b. To excite the DQ coherence between the two weakly coupled carbonyl carbons with large chemical shift anisotropies, the excitation sequence was modified from the simple Hahn echo sequence of the original DOQSY experiment, using a method developed by Antzutkin and Tycko.[8] The sequence consists of a two-quantum selective 90° pulse train that is interleaved with a train of π pulses spaced sufficiently close to refocus the CSA. The modified C′-C′ DOQSY technique was successfully applied to spider dragline silk, in which the Ala-rich domain is ^{13}C labeled at 15%.[9] The native silk fiber showed a predominantly (70%) β-sheet DOQSY pattern with (φ, ψ) = (135°, 150°). In contrast, silk film from the liquid extracted directly from the silk gland exhibited mostly (60%) α-helical DOQSY pattern with (φ , ψ) = (±60°, ±45°) (Figure 4.4).

The main limitation of the static tensor correlation techniques for torsion-angle determination is their low spectral sensitivity. High-resolution magic-angle spinning (MAS) experiments that recouple the anisotropic interactions bypass this problem and allow torsion angles to be determined in peptides and proteins with non-repeating amino acid sequences.

4.2 HIGH-RESOLUTION MAS TENSOR CORRELATION TECHNIQUES

A previous review of the application of solid-state NMR to biological solids has already included some of the MAS torsion-angle experiments.[10] Since then, new

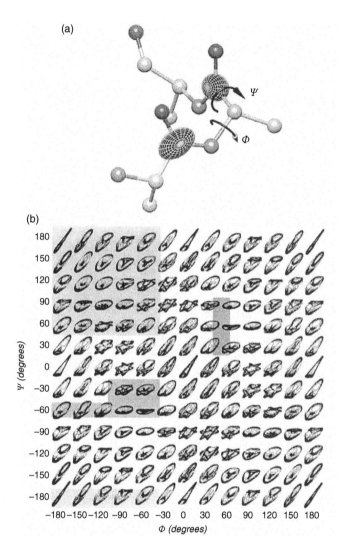

FIGURE 4.3 (a) Peptide fragment illustrating the two carbonyl CSA tensors that allow the determination of the (ϕ, ψ) angles. (b) Simulated DOQSY spectra of two consecutive C′ carbons in peptides as a function of (ϕ, ψ) angles. Energetically favorable areas are shaded. (Adapted from Ref. 9.)

developments in torsion angle determination have focused on incorporating tensor correlation elements into 2D and three-dimensional (3D) experiments to obtain multiple torsion angles simultaneously. New correlation schemes have been developed to determine the side-chain torsion angles, to provide higher angular resolution in specific regions of the Ramachandran diagram, and to enable faster spinning speeds for these experiments. Finally, new distance techniques have also been developed to measure torsion angles indirectly.

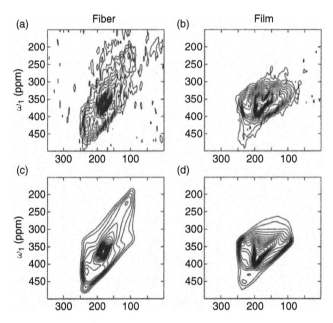

FIGURE 4.4 Experimental (a, b) and simulated (c, d) $^{13}C'$ DOQSY spectra of the Ala residues of silk from *Samia Cynthia ricini* silkworms. Spectra for silk fiber (a, c) and silk film (b, d) cast on a flat polystyrene plate differ significantly. The most probable conformation for the fiber sample is $(\phi, \psi) = (135°, 150°)$ based on the best-fit simulation (c). The film sample exhibits a bimodal conformation at $(\phi, 4) = (\pm 60°, \pm 45°)$, which corresponds to left- and right-handed α-helices. (Adapted from Ref. 9.)

4.2.1 HCCH Experiment

The first tensor correlation technique developed under MAS measures the torsion angle around a C-C bond by correlating the two C-H dipolar tensors along the C-H bonds.[11] The experiment excites the ^{13}C-^{13}C DQ coherence using a homonuclear recoupling sequence C7 [12] or its variants (Figure 4.5). The DQ coherence evolves under the C-H dipolar coupling for the t_1 period, whereas ^1H-^1H homonuclear coupling is averaged by a multiple-pulse sequence such as MREV-8. Similar to most techniques that use ^1H homonuclear decoupling to define a heteronuclear dipolar period,[13] the t_1 period in this experiment is confined to a constant time of one rotor period, which not only reduces the number of t_1 points to be sampled but also removes T_2 relaxation effects in the t_1 signal. The DQ coherence is reconverted to observable single-quantum magnetization by an identical period of the C7 sequence, except that the phase of the reconversion block is incremented relative to the excitation block by 90° in successive scans. The receiver phase is incremented concomitantly in the opposite sense to select the magnetization that has passed through the DQ filter. To refocus the sum isotropic chemical shifts, the C-H dipolar evolution period can be extended to two rotor periods, where the second rotor period is bracketed by two π pulses and is subject to continuous-wave ^1H decoupling. The

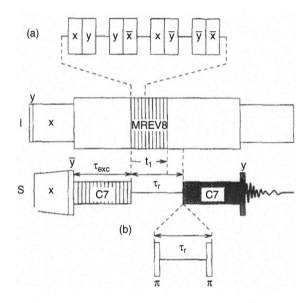

FIGURE 4.5 HCCH pulse sequence. C7 is used to excite the ^{13}C DQ coherence, which evolves under the C-H heteronuclear dipolar coupling for a period t_1. The reconverted DQ coherence is detected. (Adapted from Ref. 14.)

first π pulse serves to refocus the isotropic shift, while the second π pulse maintains the phase correspondence between the DQ excitation and reconversion periods.

The H1-C1-C2-H2 torsion angle is encoded in the sum and difference frequencies of the time-dependent C-H dipolar couplings, $\omega_{jk}^{IS}(t)$

$$\omega_{jk}^{IS}(t) = \sum_{m',m} b_{jk}^{IS} D_{0m'}^2 \left(\Omega_{PM}^{jk} \right) D_{m'm}^2 \left(\Omega_{MR} \right) \times \exp\left(im\omega_r t \right) d_{m0}^2 \left(\beta_{RL} \right) \qquad (4.1)$$

Here b_{jk}^{IS} is the distance-dependent dipolar coupling constant between I_j and S_k, β_{RL} is the magic-angle between the rotor axis and the laboratory frame, and Ω_{MR} are the powder angles describing the orientation of the molecules in the rotor frame. The crucial angular parameters that determine the torsion angle are Ω_{PM}^{jk}, which represent the orientation of the I_j-S_k vector relative to a molecule-fixed frame. The main frequencies that determine the angle-dependent spectral line shape are $\omega_{11}^{IS}(t)$ and $\omega_{22}^{IS}(t)$. Choosing the C1-C2 bond as the z-axis of the molecular frame, then $\beta_{PM}^{11} = \pi - \beta_{PM}^{22}$ corresponds to the H-C-C bond angle, while the difference of two γ angles, $\gamma_{PM}^{11} - \gamma_{PM}^{22}$, is the torsion angle of interest. Cross terms $\omega_{12}^{IS}(t)$ and $\omega_{21}^{IS}(t)$ resulting from two-bond C–H dipolar couplings are readily included in the simulations.

The HCCH experiment was initially demonstrated on organic compounds containing a C=C double bond to determine the cis and trans configuration.[11] It was later applied to the retinylidene chromophore in rhodopsin to determine the H-C10-C11-H torsion angle.[14] It was found that the H-C10-C11-H angle deviates from the planar 10-11-s-trans conformation (180°) and is, instead, 160 ± 10°.

The HCCH technique has several features common to other tensor correlation methods developed subsequently. The isotropic shift detection provides site resolution so that multiple torsion angles can be measured from the same experiment. The use of DQ coherence to measure two C-H dipolar couplings simultaneously suppresses the natural-abundance ^{13}C background, which is critical for applying these techniques to specifically labeled samples. The uniaxial nature of the dipolar interaction gives a double degeneracy in the resulting angle, so that only the magnitude of the angle can be measured. The most sensitive angular regime of the technique occurs when the two internuclear vectors are parallel or antiparallel, as under this condition, the difference in dipolar frequency nearly vanishes, giving rise to a slowly decaying component in the t_1 signal or a zero-frequency peak in the spectrum that change sensitively with the dihedral angle. In addition to double-bond or conjugated molecules, the HCCH technique can be used to determine side-chain torsion angle χ_1 in β-branched amino acids using the Hα-Cα-Cβ-Hβ spin topology.

4.2.2 HNCH TECHNIQUE: φ ANGLE DETERMINATION

The HNCH technique[15] correlates the H^N-N dipolar coupling with the Cα-Hα coupling to determine the ϕ_H = H^N-N-Cα-Hα angle. This is related to the backbone-defined φ angle (C′-N-Cα-C′) according to $\phi_H = \phi - 60°$ for L-amino acids, assuming standard covalent geometry. The experiment shares the same conceptual framework as the HCCH experiment, but instead of exciting ^{13}C-^{13}C homonuclear DQ coherence, it uses ^{13}C-^{15}N heteronuclear DQ and zero-quantum (ZQ) coherence C_yN_x. After CP from ^{1}H to ^{13}C, the heteronuclear coherence is generated by a ^{13}C-^{15}N REDOR pulse train and a 90° ^{15}N pulse (Figure 4.6a). The DQ and ZQ coherences then evolve under the C-H and N-H heteronuclear dipolar couplings for a period t_1, defined by the length of the ^{1}H-^{1}H homonuclear decoupling sequence. The X-^{1}H (X=^{13}C, ^{15}N) dipolar evolution has a maximum of one rotor period (τ_r) and is followed by a ^{13}C π pulse and another rotor period with ^{1}H heteronuclear decoupling to refocus the ^{13}C chemical shift evolution. Subsequently, the C_yN_x coherence is reconverted to observable ^{13}C magnetization by a mirror symmetric application of the excitation block containing the ^{15}N 90° pulse and the REDOR sequence.

The indirect dimension of the HNCH experiment generates the sum and difference diplar phases $\cos[(\Phi_{CH}(t_1) \pm \Phi_{NH}(t_1))/2]$, where $\Phi_{XH}(t_1) = \int_0^{t_1} dt \omega_{XH}(t)$. Because of the orientation dependence of the dipolar coupling as shown by Equation (4.1), the sum and difference spectra depend on the relative orientation of the two dipolar vectors. To simulate the angle-dependent time signal or frequency spectrum, the C-H and N-H vectors are transformed to a convenient common molecular frame, defined with the z-axis along the N-Cα bond. Using similar nomenclature to that defined in Equation (4.1), the β_{PM}^{CH} and β_{PM}^{NH} angles are related to the Hα-Cα-N and the H^N-N-Cα bond angles, respectively, and $\gamma_{PM}^{CH} - \gamma_{PM}^{NH} = \phi_H$.

Similar to the HCCH experiment, the HNCH technique has the highest angular resolution when the two dipolar vectors are antiparallel, as under this condition, the difference frequency is the smallest and the time signal is weakly decaying, and both

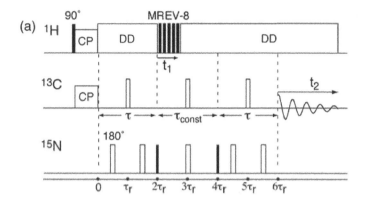

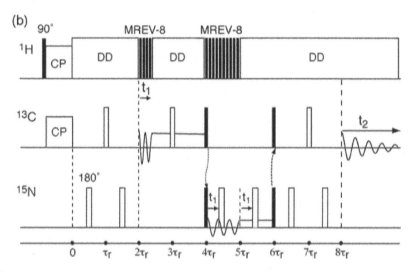

FIGURE 4.6 (a) HNCH pulse sequence. ^{15}N-^{13}C DQ and ZQ coherences are excited using a REDOR pulse train and evolve under the C-H and N-H dipolar couplings for a period t_1. The coherences are reconverted to single-quantum ^{13}C magnetization for detection during t_2. (Adapted from ref. 15.) (b) HNx2CH pulse sequence. The C-H heteronuclear dipolar evolution is separate but synchronous with the N-H dipolar evolution. The N-H evolution is incremented twice as fast as the C-H evolution period. The same REDOR sequence is used to generate ^{13}C and ^{15}N antiphase magnetization. (Adapted from Ref. 13.)

change sensitively with the torsion angle. Because a ϕ_H of 180° corresponds to ϕ = −120°, the technique has the highest angular resolution for the β-sheet conformation.

Because the one-bond N-H dipolar coupling is about half the one-bond C-H coupling, higher angular resolution for the HNCH technique can be obtained if the N-H dipolar dephasing can be doubled to become comparable to the C-H dephasing. This can be understood in the limit of drastically different coupling strengths for the two interactions: in that case, the sum and difference frequencies are the same as the larger coupling and are thus independent of the relative orientation of the two tensors. Conversely, when the two constituent dipolar couplings have the same strengths, maximal angular resolution can be achieved. The doubling of N-H dipolar dephasing is achieved by modifying the N-H dipolar evolution: Instead of incrementing the ^{1}H homonuclear decoupling period to define t_1, the multiple-pulse sequence is applied for the full rotor period, and a moving ^{15}N π pulse defines the active heteronuclear coupling period.[13] The N-H dipolar phase $\Phi_{NH}^{2x}(t_1)$ is thus

$$\Phi_{NH}^{2x}\left(t_1\right) = \int_0^{t_1} dt\omega_{NH}\left(t\right) - \int_{t_1}^{\tau_r} dt\omega_{NH}\left(t\right) = 2\int_0^{t_1} dt\omega_{NH}\left(t\right) = 2 \cdot \Phi_{NH}\left(t_1\right) \qquad (4.2)$$

which is twice the dipolar phase accumulated without the moving ^{15}N π pulse, $\Phi_{NH}(t_1)$.

To double the N-H dipolar phase while maintaining the same C-H dipolar phase, the simultaneous C-H and N-H dipolar encoding approach of the original HNCH experiment has to be changed. Specifically, the C_yN_z coherence created by the REDOR pulse train first evolves under the C-H dipolar coupling for a t_1 period defined in the normal way; then a pair of ^{13}C and ^{15}N 90° pulses creates C_zN_x coherence, which evolves under the doubled N-H dipolar coupling[13] (Figure 4.6b). Thus, the C-H and N-H dipolar evolutions are sequential but synchronous. The simulated ϕ-angle dependent HNx2CH spectra are shown in Figure 4.7. The spinning sideband patterns are similar to the HNCH experiment, except that they are more sensitive to angle changes. The enhanced angular resolution of the HNx2CH technique over the HNCH technique is shown by the 2D RMSD comparison in Figure 4.8.

Because dipolar interactions create an avoidable twofold degeneracy around $\phi_H = 180°$ or $\phi = -120°$, correlation of the ^{15}N chemical shift anisotropy with the C-H dipolar coupling has been used to determine the sign of the ϕ angle.[16] In general, as long as a non-uniaxial interaction tensor is used in the correlation and the tensor does not have any principal axis perpendicular to the torsion bond, then the correlation spectra will depend on both the sign and the magnitude of the torsion angle. If the molecular frame is defined with the z-axis as the N-Cα bond, then the orientation of the ^{15}N chemical shift tensor is described by a polar angle β_{PM}^N of 137° and an azimuthal angle $\alpha_{PM}^N = -20°$. The out-of-plane angle of $-20°$ is the critical parameter that breaks the degeneracy between positive and negative ϕ_H angles. If the x-axis of the molecular frame is defined as the Hα-Cα bond, then $\gamma_{PM}^N = 240°-\phi$.

4.2.3 Multiple ϕ, ψ, and χ_1 Angles by 3D HNCH Experiments

The multiplex advantage of ϕ angle determination can be further exploited by adding a second isotropic shift dimension in addition to the directly detected ^{13}C dimension. ^{15}N chemical shift is a natural choice, and its evolution can be readily incorporated into the pulse sequence. The torsion-angle-dependent dipolar coupling can be sampled either completely, thereby making the experiment three-dimensional, or partially with two points, making it two 2D experiments to reduce the signal-averaging

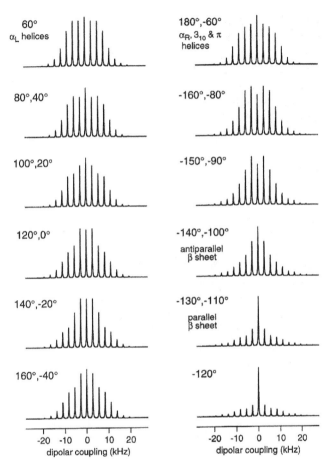

FIGURE 4.7 Simulated HNx2CH spectra as a function of the ϕ angle at a spinning speed of 2778 Hz. Note the increasing angular resolution of the spectral patterns near $\phi = -120°$. (Adapted from (Ref. 13.)

time. The latter was demonstrated as a β-sheet filter experiment.[17] It takes advantage of the fact that the angle-dependent HNCH dipolar oscillations for α-helical ($\phi_H = -120°$ or $\phi = -60°$) and β-sheet ($\phi_H = -180°$, $\phi = -120°$) conformations differ significantly. The β-sheet time signal decays less than the helix signal because of the near-antiparallel orientation of the N-H and Cα-Hα bonds. As a result, in the middle of the rotor period, the helix HNCH signal is nearly zero, whereas the β-sheet signal remains significant (Figure 4.9). This distinction is used to filter out the α-helical signals while retaining the β-sheet signals of proteins. The β-sheet filter has been demonstrated on selectively and extensively labeled ubiquitin (Figure 4.10), and indeed the filtered 2D spectrum shows peaks consistent with the crystal-structure predicted β-sheet residue positions.[17]

Griffin and coworkers extended the HNCH method to effectively three dimensions to measure the torsion angles of multiple residues and to determine not only

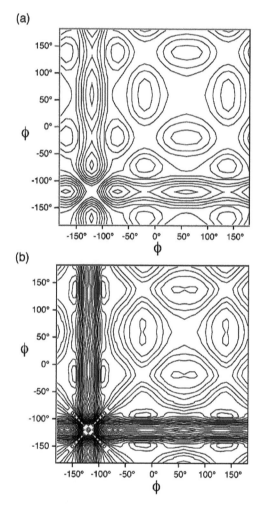

FIGURE 4.8 Comparison of the φ-angle resolution of the HNCH and HNx2CH techniques. Contour levels represent the RMSD between two simulated spectra $S(\phi_1)$ and $S(\phi_2)$, and are plotted from 0 and 100% at 5% increments. The largest RMSD occurs for $\phi = -120°$ for both experiments. The RMSD values of the N-H doubled experiment are more than twice those of the NHCH experiment. (Adapted from Ref. 13).

the φ angles but also ψ and χ_1 angles.[18] [13]C and ^{15}N isotropic shifts constitute two dimensions for resolving and assigning the resonances. The third dimension is the angle-dependent dipolar coupling. The ψ and χ_1 angles are obtained by correlating the N-H coupling with C-H couplings that are more than one bond removed from the amide [15]N. Specifically, for the ψ angle, the two dipolar vectors are N-H of residue $i + 1$ and Cα-Hα of residue i. For the χ_1 angle, the two tensors are the N-H and Cβ-Hβ couplings of the same residue (Figure 4.11b). These angle-dependent dipolar spectra are resolved in the 2D [13]C-^{15}N plane as the [15]N_{i+1}-[13]$C\alpha_i$ cross peaks and the [15]N_i-[13]$C\beta_i$ cross peaks. The pulse sequence, shown in Figure 4.11a, adopts

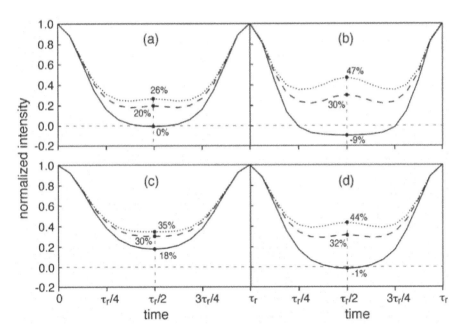

FIGURE 4.9 Principle of the β-sheet filter experiment. (a, c) HNCH curves and (b, d) HNx2CH curves for selected φ angles. (a, b) Spinning speed: 4252 Hz, with MREV-8 homonuclear decoupling. (c, d) Spinning speed: 6600 Hz, with FSLG homonuclear decoupling. Dotted lines: φ = –120° (parallel β-sheet). Dashed lines: φ = –140° (antiparallel β-sheet). Solid lines: φ = –60° (α-helix). (Adapted from Ref. 17.)

the scheme of separate and synchronous evolution for the N-H and C-H dipolar periods. The transfer of polarization between ^{15}N and ^{13}C is achieved by frequency-specific CP.[19] The main difference from the original HNCH experiment is the method of ^{1}H homonuclear decoupling. The new homonuclear decoupling sequence, T-MREV,[20] uses semi-windowless MREV-8 as the basic element but concatenates n elements with phase shifts of $2\pi/n$ to build a Cn-type pulse sequence.[12] In this way, X-^{1}H heteronuclear coupling is actively recoupled rather than passively detected, as in most separated-local-field (SLF) experiments. This active recoupling creates a dipolar line shape that is independent of the MAS frequency, so that higher spinning speeds can be used and dipolar evolution needs not be confined to one rotor period. The recoupled heteronuclear coupling has a transverse effective field, in contrast to normal MREV-8. The T-MREV sequence is γ-encoded, giving rise to two singularities in the spectra spaced at the scaled dipolar coupling.

The higher spinning speed allowed by T-MREV makes 3D HNCH experiment possible on uniformly ^{13}C, ^{15}N-labeled peptides, for which the minimization of rotational resonance conditions and residual ^{13}C-^{13}C dipolar broadening is important. The demonstration of the 3D HNCH technique on the tripeptide formyl-MLF[18,21] used spinning speeds of about 9 kHz, which avoided rotational resonance between ^{13}C spins, improved the chemical shift resolution by suppressing the residual ^{13}C dipolar coupling, and moved undesirable sidebands away from the torsion-angle-dependent centerband region.

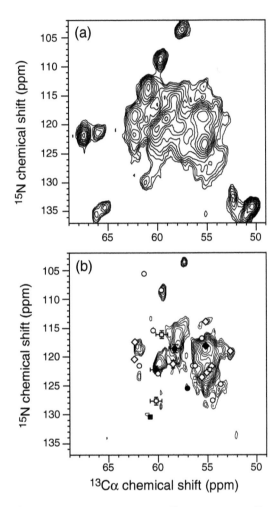

FIGURE 4.10 2D ^{13}C-^{15}N correlation spectra of [2-^{13}C] glycerol and ^{15}N uniformly labeled ubiquitin without (a) and with (b) with the HNCH β-sheet filter. The solution NMR chemical shifts of the β-sheet residues of ubiquitin are overlaid on top of the experimental spectrum. Diamonds: $\phi = -120° \pm 5°$. Squares: $\phi = 110° \pm 5°$. Circles: $\phi = 100° \pm 5°$. Filled and open symbols represent 100% and 50% Cα-labeled residues, respectively. (Adapted from Ref. 17.)

It should be noted, however, that the T-MREV line shape is sensitive to differential ^{1}H and ^{15}N relaxation, weaker couplings of the heteronuclear spin to remote protons, and the directional angle between the multiple X-^{1}H vectors. For example, the T-MREV N-H line shape of a residue depends both on the N-H^N and the N-Hα coupling and on the H^N-N-Hα angle. Although the H^N-N-Hα angle dependence can be exploited for determining the dihedral angle ϕ, the same information is already encoded in the HNCH dipolar correlation pattern. Thus, it is important to conduct control T-MREV experiments without the dipolar correlation to define these adjustable parameters before obtaining precise torsion angle values from the HNCH correlation pattern.

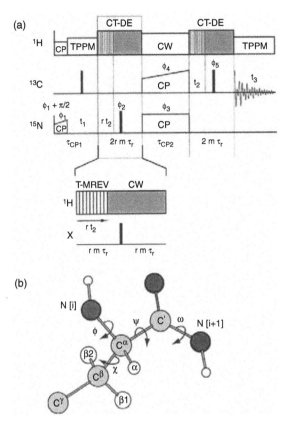

FIGURE 4.11 (a) Pulse sequence for the 3D HNCH experiment using T-MREV for ^{1}H homonuclear decoupling and X-^{1}H heteronuclear recoupling. The ratio of the N-H to C-H dipolar evolution (r) is a fixed integer, optimally 2. (b) Peptide fragment illustrating the torsion angles that can be measured by the 3D HNCH technique. These include ϕ, ψ, and χ_1. (Adapted from Ref. 18.)

The 3D T-MREV-HNCH experiment readily incorporates the coupling amplification scheme for the N-H dipolar interaction to enhance the torsion angle resolution. It was shown that N-H doubling indeed yields the best angular resolution compared to no amplification and threefold amplification.[18] Among the three torsion angles (ϕ, ψ, χ_1) that are detected from the 3D HNCH experiments, the angular precision is the lowest for the ψ angle because the N_{i+1} to $C\alpha_i$ polarization transfer is much less efficient compared to the intraresidue N_i to $C\alpha_i$ transfer. Fortunately, other more direct ψ angle determination techniques are available, as discussed below.

4.2.4 NCCN TECHNIQUE: ψ ANGLE DETERMINATION

The most direct dipolar correlation approach for measuring ψ_i angles is to correlate the dipolar couplings associated with the two bonds adjacent to the central Cα-C' bond (i.e., the N_i-Cα_i coupling and the C'_i-N_{i+1} coupling). This elegant NCCN method was simultaneously developed in two laboratories,[14,22] using different dipolar

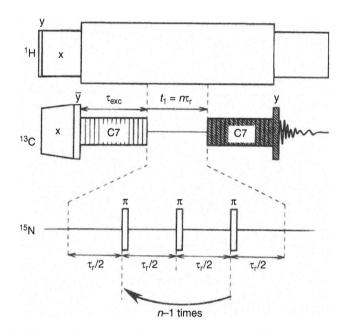

FIGURE 4.12 NCCN pulse sequence incorporating C7 as the ^{13}C DQ excitation sequence and REDOR for recoupling the ^{15}N-^{13}C dipolar interaction. (Adapted from Ref. 14.)

recoupling sequences. Both involved the excitation of the Cα-C′ DQ coherence and evolution of the DQ coherence under C-N dipolar coupling (Figure 4.12). The C-C dipolar recoupling is achieved using C7 in one case[12] and MELODRAMA[23] in the other. The C-N recoupling is achieved by REDOR[14] in one case and SPI-R[3] (synchronous phase inversion rotary resonance recoupling) in the other.[22] The exact recoupling sequences, however, do not affect the ψ-angle dependent signals. As usual, the highest ψ angle resolution is found when the N_i-$Cα_i$ and C'_i-N_{i+1} bonds are collinear. Because the N-Cα-C′-N topology coincides with the definition of the ψ angle, the highest angular resolution, at ψ = 180°, is closest to the antiparallel β-sheet conformation. Below a |ψ| of 120°, the time oscillations or spectral line shapes are poorly resolved (Figure 4.13). Fortunately, this angular insensitivity can be overcome by alternative techniques (see below).

Similar to the HNCH experiment, the NCCN experiment has been extended to incorporate two chemical shift dimensions to better separate the resonances of uniformly labeled proteins.[24] In the INADEQUATE-NCCN experiment (Figure 4.14a), the Cα-C′ DQ frequency is correlated with the directly detected ^{13}C single-quantum chemical shift. The DQ evolution is inserted after the ^{13}C-^{15}N REDOR period, which encodes the relative orientation of the two N-C bonds. Thus, the spin-pair intensities in the DQ spectra as a function of the C-N mixing time yield the ψ angle. In the NCOCA-NCCN experiment (Figure 4.14b), N_{i+1} – $Cα_i$ correlation peaks are obtained from a two-step polarization transfer through C'_i. The ψ-angle-dependent N_i-$Cα_i$ and C'_i-N_{i+1} correlation is measured by evolving the Cα and C′ coherences separately and synchronously during the polarization transfer pathway:

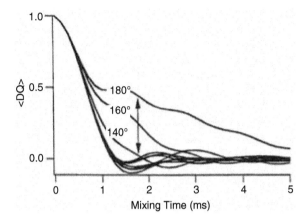

FIGURE 4.13 Simulated NCCN curves as a function of the ψ torsion angle. Note the limited angular resolution below $|\psi| = 120°$. (Adapted from Ref. 22.)

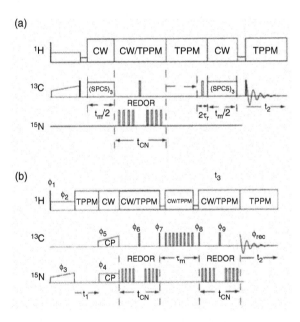

FIGURE 4.14 NCCN experiments incorporating (a) INADEQUATE and (b) NCOCA chemical shift correlation for determining multiple ψ torsion angles simultaneously. Solid and open rectangles represent 90° and 180° pulses, respectively. (Adapted from Ref. 24.)

$N_{i+1} (t_1) \rightarrow C'_i$ (NC) $\rightarrow C\alpha_i$ (NC) (t_2). The C'_i to $C\alpha_i$ transfer is accomplished using RFDR[25] or similar homonuclear recoupling sequences. The two techniques are complementary in two ways. The INADEQUATE-NCCN experiment has higher spectral sensitivity because of the one-step DQ excitation (30–35%) and is not affected by ^{13}C-^{13}C J-coupling, whereas the NCOCA-NCCN experiment suffers from lower efficiency (~20%) because of the two-step polarization transfer, and homonuclear J-coupling is active during the C-N REDOR dipolar evolution. However, the NCOCA-NCCN technique provides better chemical shift resolution and thus yields more ψ angles than the INADEQUATE-NCCN experiment.

The 3D NCCN experiment was demonstrated on the 62-residue model protein α-spectrin SH3 domain.[24] Thirteen ψ angles were extracted from the INADE-QUATE-NCCN experiment, and 22 ψ angles were obtained from the NCOCA-NCCN experiment. Among these angles, the β-sheet torsion angles were determined with higher precision as expected from the NCCN symmetry, but over 50% of the residues fall outside this favorable conformational region, and thus their ψ angle values cannot be determined exactly, but only a boundary condition (less than ~145°) can be given. This limitation is addressed by the HCCN experiment discussed below.

4.2.5 HCCN Technique: α-Helical ψ Angles

Because the NCCN technique is insensitive to ψ angles below ~120°, and α-helical residues have ψ angles around −60°, it is desirable to develop an alternative tensor correlation scheme with an angular sensitivity that is complementary to the NCCN experiment. This is shown to be possible[26] by taking advantage of the Hα-Cα-C′-N spin topology. The Hα-Cα-C′-N torsion angle, ζ, is related to the ψ angle by $\psi = \zeta + 120°$. As before, the maximal angular resolution occurs at the trans conformation for the four spins involved, $\zeta = 180°$. Accordingly, the highest HCCN angular resolution occurs at ψ = 60°, coinciding with the α-helical conformation.

The strategy of the HCCN experiment is to recouple the Cα-Hα coupling using LG-CP and to correlate it with the REDOR-recoupled C'_i-N_{i+1} interaction (Figure 4.15a). The C′ to Cα polarization transfer is achieved using RFDR or rotational-resonance in the tilted frame (R2TR).[27] One can also explicitly measure the ^{13}C isotropic shift by inserting a chemical shift evolution period. To maximize the ψ angle resolution, the dephasing resulting from the weak C'_i-N_{i+1} coupling is amplified relative to the C-H dephasing by incrementing the C-N REDOR period faster than the increment of the LG-CP period. This is analogous to the HNx2CH approach described above.

It should be noted that the ^{13}C site couples both to its directly bonded ^{15}N of the next residue and to the intraresidue ^{15}N two bonds away. The latter is undesirable for the ψ angle determination. Removing the intraresidue ^{15}N effect by isotopic labeling either biosynthetically or chemically is impractical. The two-bond intraresidue $^{13}C'_i$-$^{15}N_i$ dipolar coupling (225 Hz) is only about four times weaker than the one-bond C'_i-N_{i+1} coupling of interest and cannot be neglected. Thus, accurate extraction of the HCCN angle requires the inclusion of this second coupling, which is fortunately fixed by the covalent geometry and is independent of any torsion angle.

Simulations taking into account all the geometric parameters, including the two-bond C′-N coupling, the orientation-dependent C′-Cα polarization transfer efficiency, and the torsion angle of interest, confirm that the most sensitive angular region is $|\zeta| = 150° - 170°$. This corresponds to ψ angles of −30° to −90°, encompassing the α-helical region (Figure 4.15b). Experimental demonstration on formyl-MLF bears out the design of the experiment.

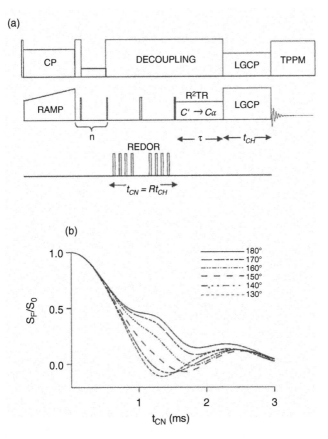

FIGURE 4.15 (a) HCCN pulse sequence for determining α-helical ψ angles with high angular resolution. The $^{13}C'$-^{15}N dipolar evolution under the REDOR sequence is correlated with Cα-Hα dipolar coupling under LG-CP evolution. (b) HCCN curves as a function of the Hα-Cα-C'-N angle, which is equal to ψ − 120°. (Adapted from Ref. 26.)

4.2.6 OCCH TECHNIQUES: ψ ANGLE DETERMINATION

Two MAS techniques that correlate the C' CSA tensor with the Cα-Hα dipolar interaction of the same residue have been developed to determine ψ torsion angles. The first is the relayed anisotropy correlation (RACO) technique,[28] suitable under intermediate spinning speeds of ~5 kHz. The technique is unique in that it recouples both the dipolar and CSA interactions in a quasistatic fashion and detects this quasistatic line shape directly. The experiment, whose pulse sequence is shown in Figure 4.16b, selects C' transverse magnetization over Cα by combining a hard and a soft 90° pulse, and then evolves it under the CSA interaction, which is recoupled by six π-pulses per rotor period.[29] This forms the indirect dimension of the 2D experiment. The C' magnetization is then transferred to Cα using the R2TR sequence according to the resonance condition $\Delta\omega_{cs} \dfrac{\omega_f}{\sqrt{\omega_1^2 + \omega_f^2}} = n\omega_r$, where ω_f is the carrier

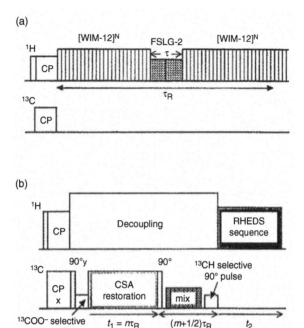

FIGURE 4.16 Pulse sequence for the RACO experiment. (a) RHEDS pulse sequence for observing ^{13}C-^{1}H dipolar powder pattern under MAS. (b) RACO pulse sequence for correlating ^{13}C CSA with the C-H dipolar coupling. (Adapted from Ref. 28.)

offset.[27] The use of frequency offset allows the C'-Cα polarization transfer to be turned on during the mixing time but turned off at other times. A soft 90° pulse then creates transverse Cα magnetization, which evolves under the C-H dipolar coupling during direct detection t_2. C-H recoupling is achieved by applying a ^{1}H WIM (windowless isotropic mixing) multiple-pulse sequence,[30] the middle of which is interrupted by a short FSLG sequence (Figure 4.16a). The combination provides ^{1}H homonuclear decoupling throughout the rotor period while retaining the C-H dipolar interaction only for a fraction of τ_r. The ^{13}Cα signal is sampled several times per rotor period to provide a large spectral width. This rotor-synchronous dipolar switching (RHEDS) sequence (Figure 4.16a)[28] yields static-like dipolar line shapes except for a small isotropic peak and a scaling factor that depends on the duration of the FSLG sequence in each rotor period.

To simulate the 2D RACO spectra as a function of the ψ torsion angle, the orientations of the C' CSA tensor and the C-H bond in a common molecule-fixed frame must be known. A convenient choice of the molecular frame has the z-axis perpendicular to the Cα-C'-O plane and the x-axis parallel to the Cα-C' bond (Figure 4.17). Because the most shielded axis, σ_{33}, of the C' CSA tensor is known to be perpendicular to the Cα-C'-O plane, and thus parallel to the z-axis, the Euler angles relating the molecular frame to the C' principal axes frame are (Λ, 0, 0), where Λ is a well-defined −30° for carbonyl carbons but can range from 0° to −30° for

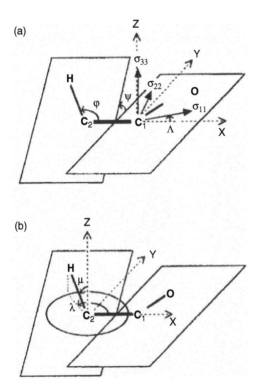

FIGURE 4.17 The orientations of C CSA (a) and Cα-Hα dipolar (b) tensors in a peptide residue. The C′ CSA orientation is defined by (Λ, 0, 0), whereas the C-H bond orientation is defined by (λ, μ, 0). The molecular frame is defined with the z-axis perpendicular to the O-C′-Cα plane, and the x-axis along the Cα-C′ bond. (Adapted from Ref. 28.)

carboxyl carbons in the case of the C-terminus residue (Figure 4.17a). The Euler angles rotating the C-H dipolar vector to the molecular frame are (λ, μ, 0) (Figure 17b). Assuming that the bond angle $\angle$Hα-Cα-C′ is 109.5°, then the angles λ and μ are related to the ψ torsion angle as

$$\lambda = \frac{\sin 109.5° \cos \psi}{|\sin 109.5° \cos \psi|} \cos^{-1}\left(\frac{\cos 109.5°}{\sqrt{\cos^2 109.5° + \sin^2 109.5° \cos^2 \psi}} \right)$$

$$= \frac{\cos \psi}{|\cos \psi|} \cos^{-1}\left(\frac{-1}{\sqrt{1 + 8 \cos^2 \psi}} \right)$$

(4.3)

$$\mu = \cos^{-1} (\sin 109.5° \sin \psi)$$

(4.4)

Therefore, the dihedral angle ψ is a unique parameter that can be determined by fitting the experimental RACO spectrum. The dependence on the C′ CSA tensor orientation Λ is minor and, moreover, Λ does not vary much among peptide carbonyl carbons. The symmetry-induced angle degeneracies for the RACO technique are (ψ, Λ), ($-\psi$, Λ), ($\pi - \psi$, $-\Lambda$), and ($\pi + \psi$, $-\Lambda$). Simulated 2D RACO spectra for a number of ψ angles and $\Lambda = -30°$ and $0°$ are shown in Figure 4.18. Demonstration of the technique on 1,2-^{13}C labeled D,L-alanine yielded a ψ angle of $137° \pm 7°$, in good agreement with the neutron diffraction result of $135.5°$.

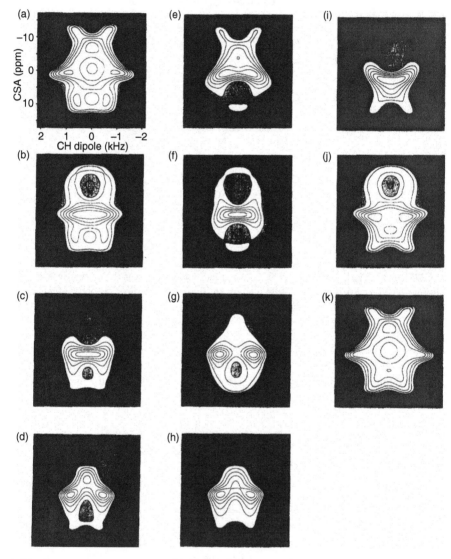

FIGURE 4.18 Simulated RACO spectra as a function of ψ angle and for $\Lambda = 0°$ and $-30°$. (a–d) top to bottom: $\Lambda = 0°$ and $\psi = 0°$, $30°$, $60°$, and $90°$. (e–k) sequentially: $\Lambda = -30°$ and $\psi = 0°$, $30°$, $60°$, $90°$, $120°$, $150°$, and $180°$. (Adapted from Ref. 28.)

The RACO technique is unique among torsion angle experiments for yielding powder-like 2D spectra under MAS. The sensitivity of the 2D patterns to the ψ angle promises high angular resolution. However, because of the complexity of the pulse sequence and the implicit requirement of 3D spectroscopy to resolve multiple sites, the technique has not been used widely.

The RACO technique is suitable under spinning speeds of less than <5 kHz. A more recent technique that uses the same spin topology but allows faster spinning was developed by Chan and Tycko.[31] The technique, ROCSA-LG, recouples $^{13}C'$ CSA in uniformly ^{13}C-labeled peptides under fast MAS (>10 kHz) by the ROCSA (recoupling of chemical shift anisotropy) sequence.[32] The Cα-Hα dipolar coupling is probed by a fixed LG-CP period. The $^{13}C'$ magnetization is dephased by the ROCSA-recoupled CSA interaction and evolves under the isotropic shift for the t_1 period. It is then transferred to Cα by RFDR and subject to the C-H dipolar interaction for a fixed time under LG-CP (Figure 4.19a). The ^{13}Cα signal is detected during t_2. In the 2D ^{13}C correlation spectrum thus obtained, the C'-Cα cross peak intensity reflects the relative orientation of the $^{13}C'$ CSA and Cα-Hα dipolar tensors. The C' CSA recoupling period is a small integer multiple ($k = 0, 1, 2$) of the LG-CP time. The cross-peak intensity as a function of k depends on the ψ angle, as shown in the simulated curves in Figure 4.19b. The main advantage of this method is that it can be applied to uniformly labeled residues, as the effect of the C-C homonuclear coupling to the CSA is minimized by the ROCSA technique.

4.2.7 $^{13}C'$ CSA CORRELATION TECHNIQUES: SIMULTANEOUS (φ, ψ) ANGLES

In analogy to the static DOQSY experiment applied to C' sites, it is possible to extract both backbone φ and ψ torsion angles simultaneously under MAS, without inducing many degeneracies, by correlating the orientation of the CSA tensors of two consecutive C' carbons. Several different pulse sequences have been designed for ^{13}C CSA correlation under MAS, all of which involve the creation of $^{13}C'$ DQ coherence. When the DQ state is prepared using the DRAWS (dipolar recovery with a windowless sequence) sequence,[33] the technique is called DQDRAWS.[34] The DRAWS sequence consists of trains of 2π pulses that compensate for chemical shifts between π/2 pulses of the original DRAMA experiment,[35] designed for homonuclear recoupling. The mutual CSA tensor orientation is contained in the spinning sideband intensities, which depend on the three Euler angles relating the principal axis systems of the two CSA tensors and two angles relating the internuclear vector to the common molecular frame. Comparison of the experimental result with numerical simulations yields the (φ, ψ) angles.[36] Figure 4.20 shows the DQDRAWS sideband spectra for the major secondary structure motifs, which are clearly distinguishable. Although the DRAWS sequence requires a relatively high radiofrequency (rf) field of $\omega_1 = 8.5\omega_p$, as the spinning speed ω_r cannot be high in the first place to generate enough sidebands, this is not a real limitation.

Tycko and coworkers developed several alternative schemes for carbonyl DQ correlation. In the DQCSA experiment,[37] an RFDR sequence is used to excite and reconvert DQ coherence, which then evolves under the CSA interaction that is

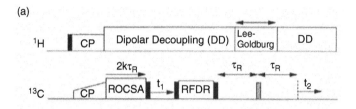

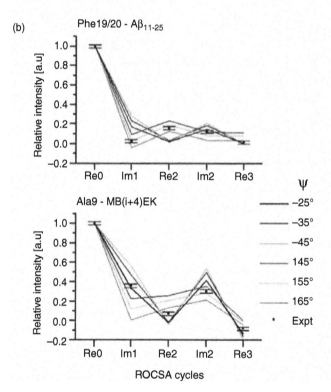

FIGURE 4.19 (a) ROCSA-LG pulse sequence. Filled and shaded rectangles indicate 90° and 180° pulses. $^{13}C'$ polarization dephases under the ROCSA-recoupled CSA for a fixed even number of rotor periods and evolves under the isotropic chemical shift during t_1. The C' polarization is then transferred to $C\alpha$ to be dephased by the $C\alpha$-$H\alpha$ dipolar coupling under LG irradiation. (b) Simulated ROCSA-LG curves for different ψ angles. The experimental data for β-sheet residues Phe19-Phe20 of $A\beta_{11-25}$ and α-helical residues Ala9 of MB(i+4)EK peptides are shown. (Adapted from Ref. 31.)

recoupled by a moving π pulse in a rotor period (Figure 4.21). The relative orientation of the two CSA tensors governs the t_1-dependence of the spinning sideband intensities. Figure 4.22 shows the simulated DQCSA t_1 signals of the center band and first-order sidebands for the main secondary structure motifs. The time oscillations differ significantly. Demonstration of the technique on the tripeptide [1-^{13}C]A[1-^{13}C]GG gave results in good agreement with the crystal-structure Gly2

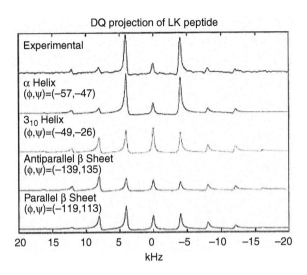

FIGURE 4.20 Experimental (top) and calculated DQDRAWS spectra for common protein secondary structures. (Adapted from Ref. 36.)

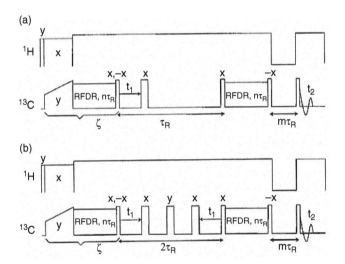

FIGURE 4.21 Pulse sequence for the DQCSA experiment under MAS. (a) With one rotor period of DQ evolution. (b) With two rotor periods of DQ evolution so that isotropic shift is refocused. (Adapted from Ref. 37.)

dihedral angles of ($\phi = 83°$, $\psi = 170°$) other than the inherent degeneracies of the technique.

Compared with the DQDRAWS technique, DQCSA places fewer demands on the ^{1}H decoupling power and rf homogeneity because only a single π pulse is applied every two rotor periods. Moreover, because the t_1 evolution is confined to a constant period of one or two rotor periods, the two RFDR blocks for DQ excitation and reconversion always maintain the same relation to the MAS rotor phase. Thus, the

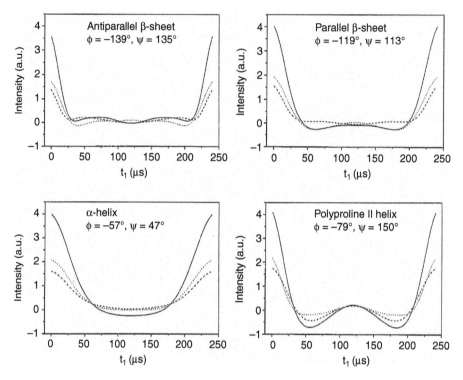

FIGURE 4.22 Simulated DQCSA curves for various secondary structures for the center band and two first-order sidebands. (Adapted from Ref. 37.)

torsion angle–dependent t_1 signal is completely independent of the two DQ mixing periods and also of the DQ line width. The sum-CSA t_1 signal is also more sensitive to the (ϕ, ψ) angles than other techniques that rely on the dependence of the DQ-filtered signals on the DQ excitation periods.

In contrast to both the DQDRAWS and the DQCSA experiments, the constant-time double-quantum filtered dipolar dephasing (CTDQFD) experiment[38] encodes torsion angle information in the excitation-time dependence of the DQ coherence. The experiment consists of three RFDR periods with durations of $L\tau_r$, $M\tau_r$, and $N\tau_r$, separated by two pairs of 90° pulses. The CP pulse, the first RFDR block, and the first 90° pulse are phase-cycled together to select for DQ coherence. The second pair of 90° pulses refocuses dipolar evolution, thus making the effective dipolar dephasing time $(M-N)\tau_r$. The total RFDR time $(L + M + N)\tau_r$ is constant to minimize the effects of transverse relaxation and residual ^{13}C-^1H dipolar coupling on the DQ dephasing curve. The CTDQFD curve is obtained by setting L to a fixed L_0, incrementing M from $M_{max}/2$ to M_{max} and setting $N = M_{max} - M$. Comparisons of the experimental curves to the calculated curves yield the (ϕ, ψ) angles between the two carbonyl carbons. The CTDQFD experiment has been applied together with the 2D MAS exchange technique to extract torsion angles more accurately, as demonstrated for the β-amyloid peptide.[39]

A different approach to simultaneous determination of the (ϕ, ψ) torsion angles in peptides is the 2D MAS exchange experiment, which correlates two $^{13}C'$ CSA tensors by detecting spin-diffusion cross peaks between the sidebands of two ^{13}C carbons.[40,41] The experiment uses the classical 2D exchange sequence, except that slow spinning speeds are used to produce sufficient $^{13}C'$ sidebands. Spin diffusion between the two consecutive carbonyl carbons occurs during a mixing time, typically 500 ms, to ensure complete exchange.

The (ϕ, ψ) torsion angles are extracted from the intensities of cross peaks, $V_{n,n'}^{i,j}$, which correlate sideband n of site i in the ω_1 dimension with sideband n' of site j in the ω_2 dimension. These interresidue cross peaks run perpendicular to the spectral diagonal and differ from the intraresidue cross peaks $V_{n,n'}^{i,i}$, which run parallel to the spectral diagonal. The intraresidue cross peaks can arise from T_1 relaxation of ^{14}N or from molecular motion. Low temperature can eliminate or significantly attenuate both processes. An example of a 2D MAS sideband spectrum is shown in Figure 4.23, acquired on the model peptide AGG with $^{13}C'$ labels at the Ala1 and Gly2 positions. The torsion angles are determined from the best fit of the experimental $V_{n,n'}^{i,j}$ for all possible (ϕ, ψ) values. The simulations use the standard C' CSA tensor orientation and incorporate the nonnegligible influence of the ^{13}C-^{14}N dipolar coupling.[41]

The main advantage of the 2D rotor-synchronized MAS exchange technique is its simplicity and low demand on probe hardware. It is most suitable for chemically synthesized peptides,[42] where the economical $^{13}C'$ labeling with respect to the final

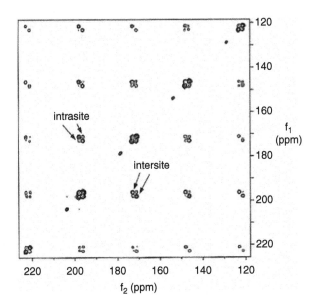

FIGURE 4.23 2D MAS exchange spectrum of 5% $^{13}C'$ doubly labeled peptide AGG, acquired at a spinning speed of 2.5 kHz. The intensities of the Ala-1-Gly-2 intersite cross peaks perpendicular to the spectral diagonal encode the (ϕ, ψ) torsion angles of Gly-2. An exchange mixing time of 500 ms was used. (Adapted from Ref. 40.)

information content is attractive. However, the experiment is not suitable for recombinant proteins, where extensive ^{13}C labeling is easier, and slow spinning and deteriorates site resolution. In applying the 2D MAS exchange technique, intermolecular spin diffusion should be avoided by diluting the labeled peptide in an excess of unlabeled peptide. It is also important to actively synchronize the t_2 and t_1 periods to make the mixing time an exact integer multiple of rotor periods. Finally, the sensitivity of the 2D MAS experiment is relatively low. It was estimated that at least five side bands are needed with a minimum signal-to-noise ratio of 5 to obtain precise torsion angles. This usually requires several days of signal averaging for each 2D spectrum.[41]

4.3   DISTANCE MEASUREMENTS FOR TORSION ANGLE DETERMINATION

In general, the distance (d) between two atoms separated by three covalent bonds depends on the torsion angle θ around the central bond sinusoidally as:

$$d^2 = c_1^2 + c_2^2 + c_3^2 - 2c_1c_2 \cos\alpha_1 - 2c_2c_3 \cos\alpha_2 +$$

$$2c_1c_3 \cos\alpha_1 \cos\alpha_2 - 2c_1c_3 \sin\alpha_1 \sin\alpha_2 \cos\theta$$

(4.5)

where α_1 and α_2 are bond angles and c_1, c_2, and c_3 are bond lengths. Thus, the distance between two nuclei separated by three bonds gives information on the dihedral angle of the central bond. Both homonuclear and heteronuclear distance experiments have been developed for this purpose recently.

4.3.1   C′-C′ DISTANCES: ϕ ANGLE DETERMINATION

The distance between two consecutive carbonyl carbons depends only on the torsion angle ϕ. The DRAWS sequence has been successfully used to measure this homonuclear dipolar coupling. It has been applied to a hexapeptide, DpSpSEEK, adsorbed to hydroxyapatite crystals.[43] The distance between two consecutive $^{13}C'$-labeled phosphoserine residues were measured. The resulting DRAWS curve indicates that the peptide has a broad distribution of conformations, with the shortest C′-C′ distance of 3.2 ± 0.1 Å, which corresponds to a ϕ angle of 90°. An α-helical ϕ angle of –60° would give a C′-C′ distance of 3.0 Å, whereas a β-sheet ϕ angle of –120° to –140° would yield a distance of 3.4 – 3.5 Å. Both these DRAWS curves disagree significantly from the experimental curve (Figure 4.24).[43]

4.3.2   ¹H-X DISTANCES: ϕ AND χ₁ ANGLE DETERMINATION

Hong and coworkers developed a ^1H-X REDOR distance technique that makes it possible to determine ϕ and χ_1 torsion angles. The ϕ angle is extracted from the intraresidue H^N-C′ distance, and the χ_1 angle is determined from the Hβ-N distance.[44] The dependence of these two distances on the respective torsion angles is shown in

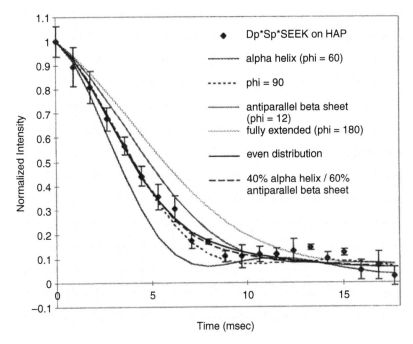

FIGURE 4.24 DRAWS dephasing curves for the hexapeptide Dp*Sp*SEEK adsorbed on hydroxyapatite (HAP). The symbol * indicates the ^{13}C-labeled residues. Simulated DRAWS curves for various conformations are also given. (Adapted from Ref. 43.)

Figure 4.25. This ^{1}H-X REDOR technique relies on the detection of a third nucleus, Y, to select the REDOR modulation of the proton of interest. The ^{1}H magnetization evolves under the REDOR-recoupled dipolar local field of the X spin. During this time ^{1}H-^{1}H homonuclear dipolar couplings are averaged by multiple-pulse sequences. The dipolar modulation of the proton is transferred selectively to the directly bonded Y spin by a short LG-CP step. The technique allows additional distance constraints to be determined using the same labeled sample as for conventional X-Y REDOR.

One advantage of the H^N-C distance experiment for determining the ϕ angle compared with the HNCH technique is that it does not require a single Hα proton. This allows Gly ϕ angles to be measured readily. Because Gly is abundant in structural proteins such as elastin, silk, and collagen, the torsion angle information would be useful for understanding their 3D structure. In principle, the T-MREV-based HNCH experiment can also be used to determine the ϕ angle in Gly residues, but the H^N-C′ distance method has the advantage that it is sensitive to both the β-sheet and α-helical conformations. The ϕ_H (H^N-N-Cα-C′) angle is related to the backbone ϕ (C′-N-Cα-C′) angle by $\phi_H = \phi + 180°$. A ϕ angle of $-60°$ (or $\phi_H = 120°$) gives a H^N-C′ distance of 3.1 Å, whereas a β-sheet ϕ angle of $-120°$ (or $\phi_H = 60°$) corresponds to a distance of ~2.8 Å. Figure 4.26 shows the H^N-C′ distance in N-t-BOC-Gly, for which two inequivalent molecules are present in the unit cell, with

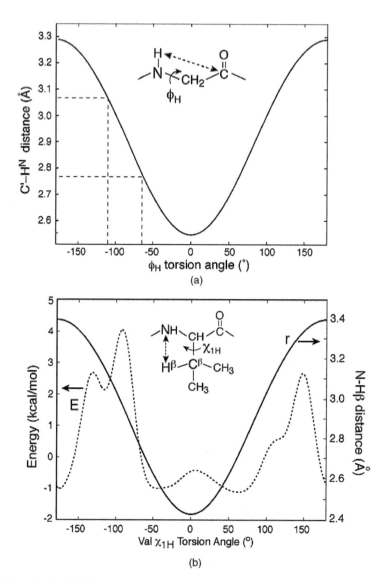

FIGURE 4.25 (a) C'-H^N distance as a function of the torsion angle ϕ_H (H^N-N-Cα-C') = ϕ (C'-N-Cα-C') + 180°. Dotted lines indicate the ϕ_H angles for two inequivalent N-t-BOC-Gly molecules in the unit cell. (b) N-β distance as a function of the χ_{1H} (N-Cα-Cβ-Hβ) = χ_1 (N-Cα-Cβ-Cγ) 120°. Dotted line: total energy of GGV as a function of Val χ_{1H} angle. (Adapted from Ref. 44.)

ϕ_H angles of –64° (2.78 Å) and –108° (3.09 Å). The measured REDOR distances, 2.75 and 3.03 Å, agree with the crystal structure well and translate to $|\phi_H|$ angles of 59° ± 5° and 100° ± 14°.

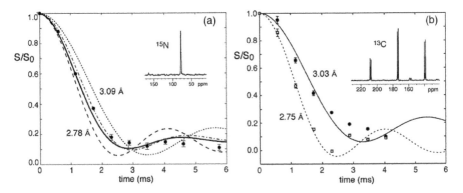

FIGURE 4.26 ^{13}C-^1H REDOR distances in N-t-BOC-Gly for determining the ϕ torsion angle. (a) ^{15}N {^{13}C-^1H} REDOR curve, not resolving the two molecules in the unit cell. (b) $^{13}C'$ detected {^{13}C-^1H} REDOR curve acquired after an additional ^{15}N to ^{13}C polarization transfer. The two molecules are now resolved. The two distances (2.75 and 3.03 Å) translate to ϕ angles of $\pm59°$ and $\pm100°$, in good agreement with the crystal structure values ($-64°$ and $-108°$). (Adapted from Ref. 44.)

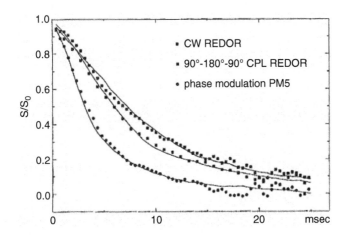

FIGURE 4.27 Experimental and simulated ^{13}C-^2H REDOR curves using CW, 90° 180° 90° composite pulses and phase-modulated PM5 pulses for ^2H. The experiment was done on singly methyl deuterated (2-d_1) Ala with detection of natural abundance $^{13}C\beta$ signal. Note the more complete dephasing of the PM5-REDOR curves. (Adapted from Ref. 46.)

4.3.3 ^2H-X Distances: ϕ, ψ Angle Determination

When the α proton in an amino acid residue is deuterated, the $^{13}C'_{i-1}$ – $^2H\alpha_i$ distance reflects the ϕ_i angle, and the $^{15}N_{i+1}$ – $^2H\alpha_i$ distance depends on the ψ_i angle. These angle-dependent three-bond distances can be measured using ^{13}C or ^{15}N observed and ^2H-dephased REDOR.[45] The main challenge is to achieve efficient inversion of the ^2H spins, which have a strong (125 kHz) quadrupolar coupling. Vega and cowork-ers showed that phase-modulated XYXYX (PM5) pulses with a total flip angle of

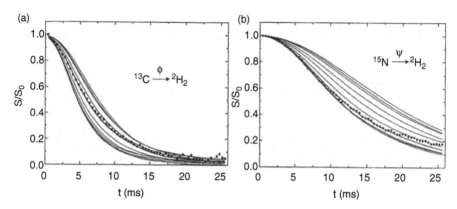

FIGURE 4.28 PM5-REDOR curves for the tripeptide LGA. (a) $^{13}C'_{Leu}$-$^{2}H\alpha_{Gly}$ REDOR curve. The experimental data is best fit with $\phi = \pm70°$. (b) $^{15}N_{Ala}$-$^{2}H\alpha_{Gly}$ REDOR curve. The experimental data is best fit with $\psi = \pm105°$. (Adapted from Ref. 45.)

180° improve the REDOR dephasing. Compared with REDOR dephasing based on 90°, 180,° 90° composite pulses or simple 180° pulses, PM5-REDOR yields the most dephasing (Figure 4.27). Its recoupling efficiency approaches that of REAP-DOR, but without the disadvantage of a fast-decaying S_0 that results from multiple 180° pulses on the observed channel. The conversion of the $^{13}C'_{i-1}$-$^{2}H\alpha_i$ and $^{15}N_{i+1}$-$^{2}H\alpha_i$ distances to torsion angles assumes standard bond lengths and bond angles and a ^{2}H quadrupolar tensor along the $H\alpha$-$C\alpha$ bond. The $^{13}C'_{i-1}$-$^{2}H\alpha_i$ PM5-REDOR experiments were demonstrated on tripeptides LGA and LAF, where the central residue was $H\alpha$ deuterated and the two flanking residues were labeled in $^{13}C'$ and ^{15}N. Figure 4.28 shows the C'_{Leu}-$^{2}H\alpha_{Gly}$ and N_{Ala}-$^{2}H\alpha_{Gly}$ PM5-REDOR curves. From, these $|\phi| = 70° \pm 10°$ and $|\psi| = 105° \pm 10°$ were determined.

4.4 CONCLUSION

Solid-state NMR spectroscopy has become a mature technique for determining torsion angles of insoluble peptides and proteins. Table 4.1 summarizes the salient features of the techniques reviewed here. The principles of these techniques are equally applicable to other biological solids such as nucleic acids and to organic compounds and have been demonstrated in several cases.[11,47,48] These torsion angle restraints are particularly useful for defining the local conformation of molecules, thus they complement long-range distance restraints, which better define the global structure of molecules.

ACKNOWLEDGMENT

M. Hong thanks Spring Smith for help in the preparation of the manuscript.

TABLE 4.1
Summary of Solid-State Nuclear Magnetic Resonance Torsion Angle Determination Methods for Peptides and Proteins

Technique	Angle	Sensitive Angle Range	S/N	Labeled Sites	Spin Interactions	ν_r Regime	Ref.
DOQSY	ψ	All	Low	$^{13}C'_i, ^{13}C'_{i+1}$	$^{13}C'$ CSA	Static	2, 9
SELFIDOQ	ψ	All	Low	$^{13}C\alpha, ^{13}C'$	C-H dipolar, ^{13}C CSA	Static	7
HCCH	χ_1	180°	High	$^{13}C\alpha, ^{13}C\beta$	C-H dipolar	<5 kHz	11
HNCH	ϕ	–120°	High	$^{15}N, ^{13}C\alpha$	N-H, C-H dipolar	<5 kHz	13, 15
T-MREV-HNCH	$\phi/\chi_1/\psi$	–120°	High	$^{15}N, ^{13}C\alpha/\beta$	N-H, C-H dipolar	5–10 kHz	18
NCCN	ψ	180°	High	$^{15}N_i, ^{13}C\alpha_i, ^{13}C'_i, ^{15}N_{i+1}$	N-C dipolar	5–15 kHz	14, 22
INADEQUATE-NCCN	ψ	180°	Medium	$^{15}N_i, ^{13}C\alpha_i, ^{13}C'_i, ^{15}N_{i+1}$	N-C dipolar	5–15 kHz	24
NCOCA-NCCN	ψ	180°	Low	$^{15}N_i, ^{13}C\alpha_i, ^{13}C'_i, ^{15}N_{i+1}$	N-C dipolar	5–15 kHz	24
HCCN	ψ	–60°	High	$^{13}C\alpha_i, ^{13}C'_i, ^{15}N_{i+1}$	C-H, C-N dipolar	>10 kHz	26
RACO	ψ	All	Low	$^{13}C\alpha_i, ^{13}C_i$	C-H dipolar, $^{13}C'$ CSA	<7 kHz	28
ROCSA-LG	ψ	All	High	$^{13}C\alpha_i, ^{13}C_i$	C-H dipolar, $^{13}C'$ CSA	>10 kHz	31
DQDRAWS	ϕ, ψ	All	High	$^{13}C'_i, ^{13}C'_{i+1}$	$^{13}C'$ CSA	<5 kHz	34
DQCSA	ϕ, ψ	All	High	$^{13}C'_i, ^{13}C'_{i+1}$	$^{13}C'$ CSA	<7 kHz	37
CTDQFD	ϕ, ψ	All	High	$^{13}C'_i, ^{13}C'_{i+1}$	$^{13}C'$ CSA	5–10 kHz	38
2D MAS exchange	ϕ, ψ	All	Medium	$^{13}C'_i, ^{13}C'_{i+1}$	$^{13}C'$ CSA	<3 kHz	40, 41
C'-C' DRAWS	ϕ	All	High	$^{13}C'_i, ^{13}C'_{i+1}$	C-C dipolar	<5 kHz	33, 43
H^N-C' REDOR	ϕ	All	High	$^{15}N, ^{13}C'$	C-H dipolar	<5 kHz	44
Hβ-N REDOR	χ_1	All	High	$^{15}N, ^{13}C\beta$	N-H dipolar	<5 kHz	44
$^{13}C'$-{2H} REDOR	ϕ	All	High	$^{13}C'_{i-1}, ^2H\alpha_i$	C-^{2}H dipolar	<5 kHz	45
^{15}N-{2H} REDOR	ψ	All	High	$^2H\alpha_i, ^{15}N_{i+1}$	N-^{2}H dipolar	5–10 kHz	45

REFERENCES

1. Dabbagh, G., Weliky, D. P. and Tycko, R., Determination of monomer conformation in noncrystalline solid polymers by two-dimensional NMR exchange spectroscopy, *Macromolecules*, 27, 6183, 1994.
2. Schmidt-Rohr, K., A double-quantum solid-state NMR technique for determining torsion angles in polymers, *Macromolecules*, 29, 3975, 1996.
3. Schmidt-Rohr, K., Complete dipolar decoupling of ^{13}C and its use in two-dimensional double-quantum solid-state NMR for determining polymer conformations, *J. Magn. Reson.*, 131, 209, 1998.
4. Schmidt-Rohr, K., Hu, W. and Zumbulyadis, N., Elucidation of the chain conformation in a glassy polyester, PET, by two-dimensional NMR, *Science*, 280, 714, 1998.
5. Havlin, R.H., Le, H., Laws, D.D., deDios, A.C. and Oldfield, E., An *ab initio* quantum chemical investigation of carbon-13C NMR shielding tensors in glycine, alanine, valine, isoleucine, serine, and threonine: comparisons between helical and sheet tensors, and the effects of chi-1 on shielding, *J. Am. Chem. Soc.*, 119, 11951, 1997.
6. Yao, X. L., Yamaguchi, S. and Hong, M., Ca chemical shift tensors in helical peptides by dipolar-modulated chemical shift recoupling NMR, *J. Biomol. NMR*, 24, 51, 2002.
7. Schmidt-Rohr, K., Torsion-angle determination in solid 13C-labeled amino acids and peptides by separated-local-field double-quantum NMR, *J. Am. Chem. Soc.*, 118, 7601, 1996.
8. Antzutkin, O.N. and Tycko, R., High-order multiple quantum excitation in 13C nuclear magnetic resonance spectroscopy of organics solids, *J. Chem. Phys.*, 110, 2749, 1999.
9. Beek, J.D.V., Beaulieu, L., Schafer, H., Demura, M., Asakura, T. and Meier, B.H., Solid-state NMR determination of the secondary structure of *Samia cynthis ricini* silk, *Nature*, 405, 1077, 2000.
10. Antzutkin, O.N. In *Solid-State NMR Spectroscopy Principles and Applications*, M.J. Duer, Ed., Blackwell Sciences, Oxford, 2002, pp. 280–390.
11. Feng, X., Lee, Y.K., Sandstroem, D., Eden, M., Maisel, H., Sebald, A. and Levitt, M.H., Direct determination of a molecular torsional angle by solid-state NMR, *Chem. Phys. Lett.*, 257, 314, 1996.
12. Lee, Y.K., Kurur, N.D., Helmle, M., Johannessen, O.G., Nielsen, N.C. and Levitt, M.H., Efficient dipolar recoupling in the NMR of rotating solids. A sevenfold symmetric radiofrequency pulse sequence, *Chem. Phys. Lett.*, 242, 304, 1995.
13. Hong, M., Gross, J.D., Rienstra, C.M., Griffin, R.G., Kumashiro, K.K. and Schmidt-Rohr, K., Coupling amplification in 2D MAS NMR and its application to torsion angle determination in peptides, *J. Magn. Reson.*, 129, 85, 1997.
14. Feng, X., Eden, M., Brinkmann, A., Luthman, H., Eriksson, L., Graslund, A., Antzutkin, O.N. and Levitt, M.H., Direct determination of a peptide torsion angle psi by double-quantum solid-state NMR, *J. Am. Chem. Soc.*, 119, 12006, 1997.
15. Hong, M., Gross, J.D. and Griffin, R.G., Site-resolved determination of peptide torsion angle phi from the relative orientations of backbone N-H and C-H bonds by solid-state NMR, *J. Phys. Chem. B*, 101, 5869, 1997.
16. Hong, M., Gross, J.D., Hu, W. and Griffin, R.G., Determination of the peptide torsion angle phi by 15N chemical shift and 13Ca-1Ha dipolar tensor correlation in the solid-state MAS NMR, *J. Magn. Reson.*, 135, 169, 1998.
17. Huster, D., Yamaguchi, S. and Hong, M., Efficient beta-sheet identification in proteins by solid-state NMR spectroscopy, *J. Am. Chem. Soc.*, 122, 11320, 2000.

18. Rienstra, C.M., Hohwy, M., Mueller, L.J., Jaroniec, C.P., Reif, B. and Griffin, R.G., Determination of multiple torsion-angle constraints in U-(13)C,(15)N-labeled peptides: 3D (1)H-(15)N-(13)C-(1)H dipolar chemical shift NMR spectroscopy in rotating solids, *J. Am. Chem. Soc.*, 124, 11908, 2002.

19. Baldus, M., Petkova, A.T., Herzfeld, J. and Griffin, R.G., Cross polarization in the tilted frame: assignment and spectral simplification in hteronuclear spin systems, *Mol. Phys.*, 95, 1197, 1998.

20. Hohwy, M., Jaroniec, C.P., Reif, B., Rienstra, C.M. and Griffin, R.G., Local structure and relaxation in solid-state NMR: accurate measurement of amide N-H bond lengths and H-N-H bond angles, *J. Am. Chem. Soc.*, 122, 3218, 2000.

21. Rienstra, C.M., Tucker-Kellogg, L., Jaroniec, C.P., Hohwy, M., Reif, B., McMahon, M.T., Tidor, B., Lozano-Perez, T. and Griffin, R.G., *De novo* determination of peptide structure with solid-state magic-angle spinning NMR spectroscopy, *Proc. Natl. Acad. Sci. USA,* 99, 10260, 2002.

22. Costa, P.R., Gross, J.D., Hong, M. and Griffin, R.G., Solid-state NMR measurement of psi in peptides: a NCCN 2Q-heteronuclear local field experiment, *Chem. Phys. Lett.*, 280, 95, 1997.

23. Sun, B.-Q., Costa, P.R., Kocisko, D., Lansbury, P.T.J. and Griffin, R.G., Internuclear distance measurements in solid state nuclear magnetic resonance: dipolar recoupling via rotor synchronized spin locking, *J. Chem. Phys.*, 102, 702, 1995.

24. Ladizhansky, V., Jaroniec, C.P., Diehl, A., Oschkinat, H. and Griffin, R.G., Measurement of multiple psi torsion angles in uniformly ^{13}C,^{15}N-labeled alpha-spectrin SH3 domain using 3D 15N-13C-13C-15N MAS dipolar-chemical shift correlation spectroscopy, *J. Am. Chem. Soc.*, 125, 6827, 2003.

25. Bennett, A.E., Ok, J.H., Griffin, R.G. and Vega, S., Chemical shift correlation spectroscopy in rotating solids: radiofrequency-driven dipolar recoupling and longitudinal exchange, *J. Chem. Phys.*, 96, 8624, 1992.

26. Ladizhansky, V., Veshtort, M. and Griffin, R.G., NMR determination of the torsion angle psi in alpha-helical peptides and proteins: the NCCN dipolar correlation experiment, *J. Magn. Reson.*, 154, 317, 2002.

27. Takegoshi, K., Nomura, K. and Terao, T., *Chem. Phys. Lett.*, 232, 424, 1995.

28. Ishii, Y., Terao, T. and Kainosho, M., Relayed anistropy correlation NMR: determination of dihedral angles in solids, *Chem. Phys. Lett.*, 256, 133, 1996.

29. Tycko, R., Dabbagh, G. and Mirau, P., Determination of chemical-shift anisotropy lineshapes in a two-dimensional magic-angle-spinning NMR experiment, *J. Magn. Reson.*, 85, 265, 1989.

30. Caravatti, P., Braunschweiler, L. and Ernst, R.R., Heteronuclear correlation spectroscopy in rotating solids, *Chem. Phys. Lett.*, 100, 305, 1983.

31. Chan, J.C.C. and Tycko, R., Solid-state NMR spectroscopy method for determination of backbone torsion angle psi in peptides with isolated uniformly labeled residues, *J. Am. Chem. Soc.*, 125, 11828, 2003.

32. Chan, J.C.C. and Tycko, R., Recoupling of chemical shift anisotropies in solid-state NMR under high-speed magic-angle spinning and in uniformly 13C-labeled systems, *J. Chem. Phys.*, 118, 8378, 2003.

33. Gregory, D., Mitchell, D.J., Stringer, J.A., Kiihne, S., Shiels, J.C., Callahan, J., Mehta, M.A. and Drobny, G.P., Windowless dipolar recoupling: the detection of weak dipolar couplings between spin 1/2 nuclei with large chemical shift anisotropies, *Chem. Phys. Lett.*, 246, 654, 1995.

34. Gregory, D.M., Mehta, M.A., Shields, J. C. and Drobny, G.P., Determination of local structure in solid nucleic acids using double quantum NMR spectroscopy, *J. Chem. Phys.*, 107, 28, 1997.

35. Tycko, R. and Dabbagh, G., Measurement of nuclear magnetic dipole-dipole couplings in magic-angle spinning NMR, *Chem. Phys. Lett.*, 173, 461, 1990.

36. Bower, P.V., Oyler, N., Mehta, M.A., Long, J.R., Stayton, P.S. and Drobny, G.P., Determination of torsion angles in proteins and peptides using solid state NMR, *J. Am. Chem. Soc.*, 121, 8373, 1999.

37. Blanco, F.J. and Tycko, R., Determination of polypeptide backbone dihedral angles in solid state NMR by double-quantum 13C chemical shift anisotropy measurements, *J. Magn. Reson.*, 149, 131, 2001.

38. Bennett, A.E., Weliky, D.P. and Tycko, R., Quantitative conformational measurements in solid-state NMR by constant-time homonuclear dipolar recoupling, *J. Am. Chem. Soc.*, 120, 4897, 1998.

39. Balbach, J.J., Ishii, Y., Antzutkin, O.N., Leapman, R.D., Rizzo, N.W., Dyda, F., Reed, J. and Tycko, R., Amyloid fibril formation by A beta 16-22, a seven-residue fragment of the Alzheimer's beta-amyloid peptide, and structural characterization by solid state NMR, *Biochemistry*, 39, 13748, 2000.

40. Weliky, D. and Tycko, R., Determination of peptide conformations by two-dimensional magic-angle spinning NMR exchange spectrocopy with rotor synchronization, *J. Am. Chem. Soc.*, 118, 8487, 1996.

41. Tycko, R., Weliky, D.P. and Berger, A.E., Investigation of molecular structure in solids by two-dimensional NMR exchange spectroscopy with magic-angle spinning, *J. Chem. Phys.*, 105, 7915, 1996.

42. Yang, J., Gabrys, C.M. and Weliky, D.P., Solid-state nuclear magnetic resonance evidence for an extended beta strand conformation of the membrane-bound HIV-1 fusion peptide, *Biochemistry*, 40, 8126, 2001.

43. Long, J.R., Dindot, J.L., Zebroski, H., Kiihne, S., Clark, R.H., Campbell, A.A., Stayton, P.S. and Drobny, G.P., A peptide that inhibits hydroxyapatite growth is in an extended conformation on the crystal surface, *Proc. Natl. Acad. Sci. USA*, 95, 12083, 1998.

44. Sinha, N. and Hong, M., X-1H rotational-echo double-resonance NMR for torsion angle determination of peptides, *Chem. Phys. Lett.*, 380, 742, 2003.

45. Sack, I., Balazs, Y. S., Rahimipour, S. and Vega, S., Efficient deuterium-carbon REDOR NMR spectroscopy, *J. Am. Chem. Soc.*, 122, 12263, 2000.

46. Sack, I. and Vega, S., *J. Magn. Reson.*, 145, 52, 2000.

47. Kaji, H. and Schmidt-Rohr, K., Conformation and dynamics of atactic poly(acrylonitrile). 2. Torsion angle distributions in meso dyads from two-dimensional solid-state double-quantum 13C NMR, *Macromolecules*, 34, 7368, 2001.

48. Harris, D.J., Bonagamba, T.J., Hong, M. and Schmidt-Rohr, K., Conformation of poly(ethylene oxide)-hydroxybenzene molecular complexes studied by solid-state NMR, *Macromolecules*, 33, 3375, 2000.

5 Structural Studies of Peptides on Biomaterial Surfaces Using Double-Quantum Solid-State Nuclear Magnetic Resonance Spectroscopy

Gary P. Drobny, Patrick S. Stayton, Joanna R. Long,
Elizabeth A. Louie, Tortny Karlsson,
Jennifer M. Popham, Nathan A. Oyler,
Peter V. Bower, and Wendy J. Shaw

CONTENTS

5.1 INTRODUCTION

Nature has evolved sophisticated strategies for engineering hard tissues through the interaction of proteins, and ultimately cells, with inorganic mineral phases. The remarkable material properties of bone and teeth thus result from the activities of proteins that function at the organic–inorganic interface. The underlying molecular mechanisms that control biomineralization are of significant interest to both medicine and dentistry, as disruption of biomineralization processes can lead to bone and tooth demineralization, artherosclerotic plaque formation, artificial heart valve calcification, kidney and gall stone build-up, dental calculus formation, and arthritis.[1,2] A better understanding of the biomolecular mechanisms used to promote or retard crystal growth could provide important design principles for the development of calcification inhibitors and promoters in orthopaedics, cardiology, urology, and dentistry.

In addition to investigating the molecular-level basis for the recognition of biomineral surfaces and the control of hard-tissue growth by proteins, the development of materials with enhanced biocompatibility is a major focus of the materials and tissue engineering communities.[3–6] There is considerable interest in the use of peptide- and protein-based coatings on materials and tissue engineering scaffolds, with the goal of recreating a natural extracellular matrix (ECM) to direct wound repair or tissue development and homeostasis.[7–11] Many of the current materials used in biomedical materials and tissue-engineering scaffolds are hydrophobic in nature. These surfaces can be modified directly with biomolecules, or the biomolecules can be conjugated through a hydrophilic layer.[12–17] In addition to the biomaterials and tissue engineering fields, there is also considerable interest in the immobilization of active peptides and proteins, often again on relatively hydrophobic material platforms, in affinity separations, diagnostics, proteomics, and cell culture technologies. A major concern in all these applications is retention of biological specificity (i.e., structure or dynamics) on adsorption.

Despite the high level of interest in elucidating and controlling the structure of proteins at material and biomineral interfaces, there is a decided lack of molecular-level structure information available for proteins at biomaterial interfaces, in general, and, in particular, for mammalian proteins that directly control calcification processes in hard tissue. The most fundamental questions regarding the secondary and tertiary structures of proteins adsorbed to material surfaces, or how proteins interact at biomaterial interfaces, remain unanswered, largely because of a lack of methods capable of providing high-resolution structural information for proteins adsorbed to material surfaces under physiologically relevant conditions (i.e., fully hydrated).

To develop a better structure–function understanding of interactions of proteins with biomaterial surfaces, we have begun to use solid-state nuclear magnetic resonance (NMR) techniques to determine the molecular structure and dynamics of proteins and peptides on inorganic crystals and on polymer surfaces. In this review, we highlight recent work that is providing insight into the structure and crystal recognition mechanisms of a salivary protein model system, but that also provides a general approach to studying protein–crystal interactions in molecular detail. In addition, the molecular insight into nature's strategy for crystal recognition has led us to the design of peptides that connect the basic studies to more applied applications in the biomaterials arena.

5.2 METHODS: DOUBLE-QUANTUM SOLID-STATE NMR AND MOLECULAR STRUCTURE

A significant complication to obtaining high-resolution information on the structural/dynamic basis of protein–surface interactions is the heterogeneous nature of the systems of interest. All high-resolution structural methods used in biology, including x-ray crystallography and solution NMR, rely on sample conditions that are homogeneous in the material sense: protein crystals with a long-range order that allow diffraction of x-rays or dilute, aqueous solutions in which macromolecules undergo rapid reorientation that removes or attenuates strong anisotropic magnetic interactions between nuclear spins, thus permitting high spectral resolution, especially at high magnetic field strengths. Recent advances in biomolecular structure determination by solid-state NMR that use uniformly isotopically enriched protein samples in combination with high-speed magic-angle spinning (MAS) and multidimensional NMR techniques rely on the preparation of protein microcrystals, which have sufficiently homogeneous magnetic properties that solid-state NMR MAS spectra achieve high-resolution spectral conditions.

The problem of determining protein structure at biomaterial interfaces is far more complex than determining the structure of a protein in a microcrystalline environment. Indeed, the biomaterial interface problem has some of the same complications and limitations encountered in NMR studies of [13]C- and ^{15}N-enriched fibrillar Alzheimer peptides,[18] bacterial cell wall components,[19] and membrane-associated peptides.[20] In the first two cases, the magnetic environment of the nuclear spins is too heterogeneous, and in the third case, molecular alignment displays mosaicity to a degree that even with high-speed MAS, typical ^{13}C NMR line widths still are in the 1–3-ppm range. For samples that display heterogeneity in excess of that typically observed in microcrystalline proteins, structural information continues to be derived from ^{13}C chemical shift data[21–26] and direct detection of internuclear distances or torsion angles that are diagnostic of protein secondary and tertiary structure, all obtained from selectively labeled samples. Before describing solid-state NMR studies of protein/peptide structure at biomaterial interfaces, we briefly review the relevant solid-state NMR methods.

As shown in Figure 5.1, the local secondary structure of a polypeptide is described by a set of Ramachandran angles ω, ϕ, and ψ. In most applications, only

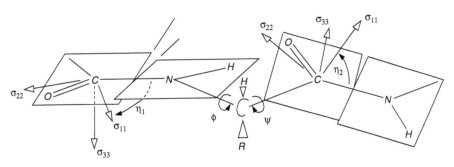

FIGURE 5.1 Peptide plane/secondary structure/Ramachandran angles.

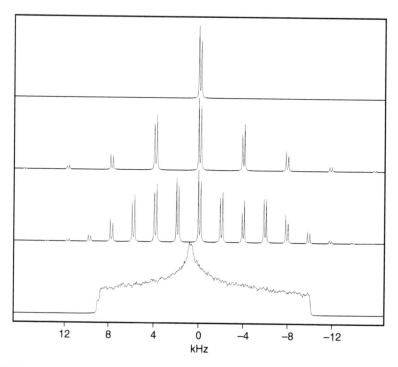

FIGURE 5.2 Solid state NMR spectra of two dipolar coupled spin 1/2 nuclei as a function of spinning rate. The static spectrum (bottom) is dominated by the chemical shift anisotropy (CSA). At spinning rates less than the CSA side bands appear at multiples of the spinning rate (middle two spectra). At spinning rates much greater than the CSA (top), side bands disappear, and two peaks at the isotropic chemical shifts of the coupled spins are observed.

the values of ϕ and ψ are at issue, so most double-quantum (DQ) NMR methods are directed toward evaluating these two angles. Torsion angles can be determined indirectly from distance measurements. The torsion angle ϕ can be obtained from a measurement of the carbonyl–carbonyl distance, whereas ψ can be measured from the amide–amide distance. However, detection of weak dipolar interactions between spins with large chemical shift anisotropies (CSAs) is problematic, as shown in the simulations in Figure 5.2.

The bottom lane of Figure 5.2 shows the static solid-state NMR spectra of two dipolar coupled carbonyl ^{13}C spins at a magnetic field strength of 9.4 T. The spectrum is dominated by the large CSA interaction of the carbonyl nuclear spins. The small dipolar coupling between the ^{13}C spins is barely discernible from the slight splitting on the spectral edges. Using MAS to remove the CSA (see lanes 1–3) has the complication of removing all anisotropic magnetic interactions at spinning rates much higher than the CSA (top lane) that transform according to the second Legendre polynomial $3\cos^2 \theta - 1$, where θ is the angle between the goniometer axis of the MAS spinner and the static magnetic field vector.[27] At spinning rates comparable to the CSA (see lanes 2 and 3), we observe the formation of side bands as a result of the modulation of all anisotropic interactions by the sample spinning. In principle,

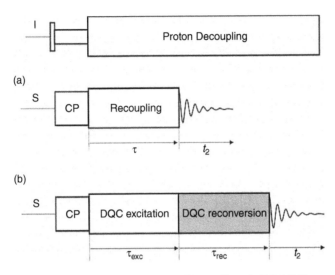

FIGURE 5.3 (a) Dipolar recoupling scheme. (b) Recoupling via DQ NMR.

the dipolar coupling strength can be determined from the spinning side-band intensities, but again the modulation of the CSA dominates the side-band intensity, and extracting the small effect of the ^{13}C-^{13}C dipolar coupling has little practical advantage over fitting a static powder pattern of the type shown in lane 1.

For reasons of accuracy and to observe very small dipolar couplings between large CSA spins, dipolar recoupling pulse sequences are used. MAS attenuates both the CSA and the dipolar interaction because both Hamiltonians transform in the same way in Cartesian space, but the CSA Hamiltonian transforms as a first-order tensor under radiofrequency (rf) irradiation, while the dipolar interaction transforms as a second-order tensor. This divergent behavior is the foundation for dipolar recoupling, defined as the selective attenuation of the CSA and recovery of the dipolar interaction in the midst of a MAS experiment by the synchronous application of resonant rf power (see Figure 5.3a for a general dipolar recoupling pulse scheme).

Many dipolar recoupling methods have appeared in the literature and have been applied with varying degrees of success.[28] Heteronuclear recoupling techniques such as rotational echo double resonance (REDOR)[29] follow the general scheme of Figure 5.3a, but homonuclear recoupling is frequently accomplished by observing the build-up of DQ coherence (DQC), as shown in Figure 5.3b. The rate of build-up of DQC depends on the magnitude of the dipolar coupling constant, which in turn is dependent on the inverse cube of the distance between the two spins, and DQ experiments performed on ^{13}C spins have the added advantage of eliminating interference from signals arising from the natural abundance background. Examples of widely used homonuclear dipolar recoupling sequences, applied according to the general scheme in Figure 5.3b, are shown in Figure 5.4. Here the intensity of single-quantum signal is monitored in t_2 after DQC is excited and subsequently reconverted to single-quantum coherence. In general, the single-quantum coherence intensity is monitored as a function of the sum of the DQC excitation time and the DQC reconversion time

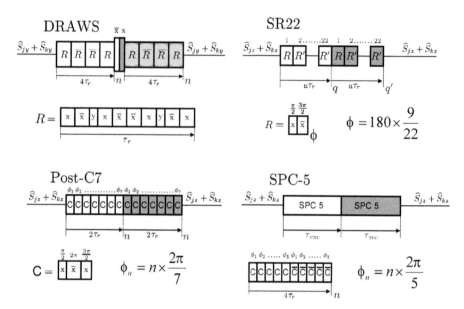

FIGURE 5.4 The sequence of pulsed irradiations used in several dipolar recoupling pulse sequences, as they are implemented in DQ experiments (see Figure 5.3b). In all cases the pulsed irradiation unit is designated C or R. A phase shift of 180 degrees is designated by an $\bar{x}$. In Post-C7 and SPC-5, pulse phases are additionally permuted by ϕ_n.

(i.e., $\tau_{exc} + \tau_{rec}$). The magnitude of the dipolar coupling constant is obtained by simulation of the DQC build-up data.

Of critical importance in a DQC experiment is the efficiency with which DQ coherence is prepared by the application of a dipolar recoupling experiment during τ_{exc} and τ_{rec}. The theoretical limit on DQ efficiency, defined as the orientationally averaged square of the DQC amplitude and understood to mean roughly the fraction of a single-quantum converted to DQ coherence, is generally a property of the structure of the dipolar recoupling pulse sequence. Suppose a crystallite is oriented relative to a goniometer-fixed frame by the solid Euler angle $\Omega^{CR} = (\alpha^{CR}, \beta^{CR}, \gamma^{CR})$. Neglecting the CSA effect, the recoupling efficiency depends on β^{CR}, γ^{CR}, in which case the recoupling is gamma-dependent, or the recoupling depends only on β^{CR}, in which case the recoupling is gamma-independent.[30–33] Three of the pulse sequences in Figure 4.4 SR22, Post-C7, and SPC-5 are γ-independent. According to the first-order average Hamiltonian theory, such sequences have a theoretical DQC efficiency of about 0.73. DRAWS, in contrast, is γ-dependent and has a theoretical efficiency of only 0.52.[28,30–33] Predictions of first-order theory are qualitatively born out by experiment, at least for coupled spin pairs with small CSAs. Figure 5.5 shows recoupling data obtained from alanine-2,3-$^{13}C_2$, and clearly DRAWS cannot match the DQC efficiency of γ-independent sequences.

For coupled spin pairs with large CSAs, the preparation of DQC is complicated by residual CSA effects. Data obtained for a coupled pair of carbonyl spins in alanylglycylglycine (AGG), see Figure 5.6, shows that the efficiency of DQC preparation in the high CSA limit is comparable for SR-22 and DRAWS.

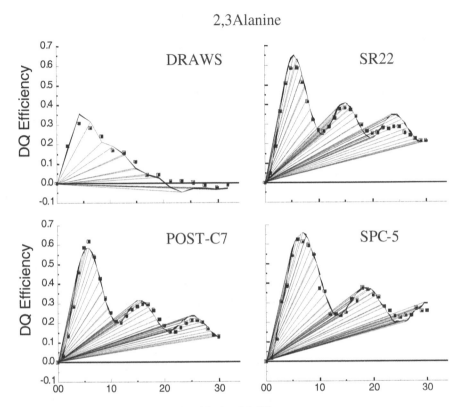

FIGURE 5.5 Dipolar recoupling data Alanine-2,3-^{13}C.

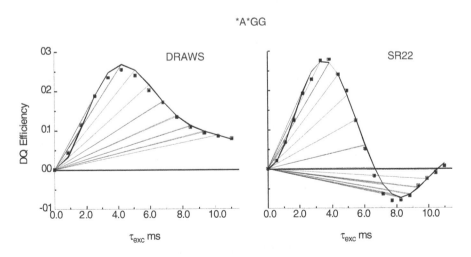

FIGURE 5.6 Dipolar recoupling data for *A*GG (High CSA limit).

We have thus far discussed solid-state NMR experiments that determine secondary structure from internuclear distances. Ramachandran angles can also be determined from the tensor sum of two magnetic interactions. Tensor sums of magnetic interactions can be obtained by observing the evolution of DQ states. A common approach is to obtain torsion angles from the sum of two CSA tensors. This general approach has been described by Bower et al.,[34] who used a DRAWS pulse sequence to create DQ coherence (i.e. DQDRAWS). Tycko and Blanco[35] used a finite-pulse RFDR sequence to accomplish the same end and called the technique DQCSA. Suppose DRAWS is used to recouple the dipolar interaction between two spins, as shown in Figure 5.3b. If at the conclusion of the preparation period, a 90° pulse is applied to produce DQ coherence, followed by a free precession period of duration t_1, and if t_1 is incremented by a time that is not an integer multiple of the rotor period, the dipolar coupling modulates the DQ signal in t_1.[36,37] The DQ spectrum will display side bands that are the result of the recoupled dipolar interaction. If, in addition, spins have CSAs that are large compared with the spinning rate, the side bands in the DQ spectrum will also have significant contributions from the tensor sum of the CSAs. The fact that the DQ NMR experiment yields the CSA tensor sum is of key importance because the sum interaction will be dependent on the set of Euler angles that define the mutual orientation of the two CSA tensors. For example, given two coupled carbonyl ^{13}C spins located in adjacent peptide bonds in a polypeptide (see Figure 5.1), the anisotropy of the CSA tensor sum can be determined by fitting the DQ interferogram or the DQ spectrum, as shown in Figure 5.7. As shown in Figure 5.8, the anisotropy of the sum CSA is sensitive to the torsion angles ϕ and ψ, which define peptide secondary structure. Independent determination of ϕ is usually performed using a DQC recoupling experiment (see Figure 5.6).

5.3 APPLICATIONS: NMR STUDIES OF PROTEINS AT MATERIAL INTERFACES

Having completed an outline of relevant solid-state NMR structure determination methods, we now review several studies of the structure of proteins and peptides at biomaterial interfaces. In Section 5.3.1 we review a study of statherin, a protein that is involved in the regulation of the primary and secondary crystallization of hydroxyapatite in the oral environment.[38–40] In Sections 5.3.2 and 5.3.3, we study the effect of immobilization methods on the structure of a peptide, first reviewing the structural effect of peptide adsorption to hydrophobic surfaces, then studying the how similar peptides are structured when covalently attached to colloidal gold nanoparticles coated with a self-assembled monolayer (SAM) composed of an alkanethiol.

5.3.1 PEPTIDES AND PROTEINS ADSORBED ONTO INORGANIC CRYSTALS

Many of the proteins directly involved in control of biomineral nucleation and growth contain acidic domains that are rich in aspartic/glutamic acid and phosphorylated serines.[41–44] To gain insight into how these acidic domains recognize hydroxyapatite,

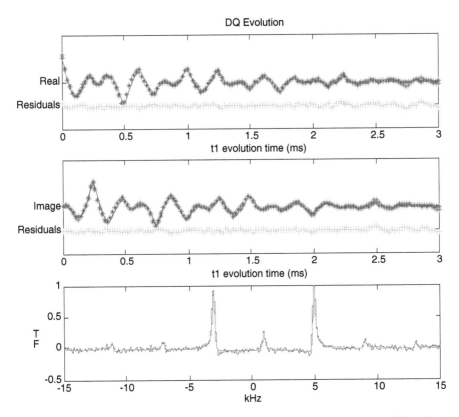

FIGURE 5.7 Experimental and simulated evolution of DQC for two coupled ^{13}C carbonyl spins. Experimental data are indicated by asterisks and simulations are solid lines.

the principal mineral phase of bone and teeth, we have been studying statherin, a particularly well studied acidic phosphoprotein found in saliva.

Salivary statherin functions biologically to inhibit the nucleation and growth of calcium phosphate minerals. Salivary statherin is a particularly well-studied acidic phosphoprotein that inhibits the nucleation and growth of calcium phosphate minerals in the oral environment and acts as a boundary lubricant.[44–55] The N terminus of statherin is highly charged, with the first five amino acids containing one aspartic acid, two phosphorylated serines, and two glutamic acids that have been shown to be important in the recognition of hydroxyapatite (HAP) (see Figure 5.9). A solution NMR study of statherin[55] acquired in a water/TFE solvent mixture found the N terminus to be alpha helical, and the C terminus was judged to be a 3_{10} helix, although the relative paucity of NOEs in the C terminus was interpreted as indicating nonrigidity. Here we briefly review recent solid-state NMR studies that have shown that the N-terminal, 15–amino acid peptide derived from statherin (called SN-15), which is believed to constitute the bulk of the HAP binding site, is partly helical on the HAP surface.[56] We also review studies of the dynamics of the SN-15 peptide adsorbed onto hydroxyapatite crystals, which shows that although the three to six

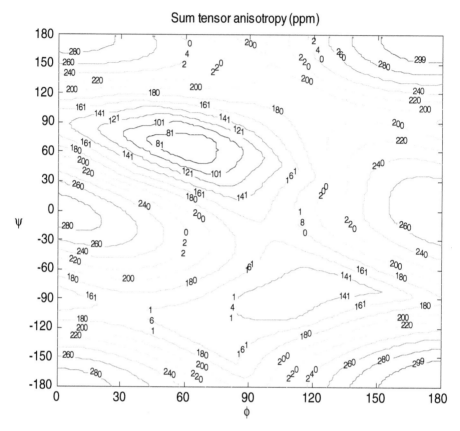

FIGURE 5.8 Dependence of the sum CSA of two coupled carbonyl ^{13}C spins on the Ramachandran torsion angles ϕ and ψ.

amino acids following the N terminus are immobilized on the surface, the remaining portion of the peptide toward the C terminus shows increased mobility.[57]

Some of the key questions driving our studies involve the interrelated aspects of protein conformation on and off the HAP surface, the binding "footprint" — or which amino acid side-chains actually contact the surface — the role of water at the protein–crystal interface, the dynamics of proteins on the surface, the orientation of the protein on the crystal surface, and finally the question of how structure and orientation are related to function. The question of how protein conformation changes on the crystal surface is connected to an important challenge faced by acidic proteins that control the nucleation or growth of biominerals. On the one hand, the structurally organized display of carboxylate side-chains could potentially be used to match the protein electrostatic surface with the complementary ionic lattice of the biomineral, leading to "capping" of growth sites. At the same time, such an organized display of ion-binding side-chains could also lead to binding stabilization of early crystal nucleation clusters and promotion of crystal growth. Such a dual activity for salivary proteins has indeed been demonstrated by Campbell et al.[39]

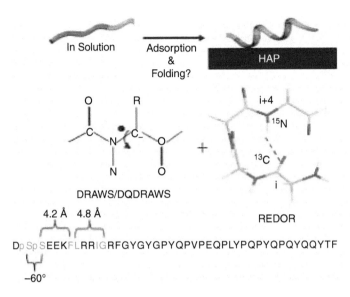

FIGURE 5.9 Binding model and NMR strategy for studying the HAP binding domain Statherin on HAP microcrystal surfaces. Amino acids labeled with ^{13}C or ^{15}N are shown in dark gray. The two phosphoserines (pS2 and pS3) were both enriched with ^{13}C at the carbonyl positions and DRAWS was used to determine ϕ angle (ca. 60°). See data and simulation in Figure 5.10. Similar measurements were repeated at F7-L8 and I11-G12. ^{13}C-^{15}N REDOR measurements were used to detect (i) to ($i + 4$) hydrogen bond lengths at pS3-F7 and L8-G12. See data in Figure 5.11.

A potential route to solving this challenge would be for the protein to be relatively unfolded in solution and to use crystal binding energy to stabilize a folded conformation in which the side-chains were optimized for interactions with the crystal surface — a model schematized at the top of Figure 5.9. We have begun to test whether statherin matches this structural profile by determining its structure on model HAP crystals. Three complementary structural techniques were used to determine the backbone structure of the N-terminal binding domain of statherin. As described in the Section 5.2, homonuclear dipolar recoupling techniques are used to measure the torsion angles ϕ and ψ, yielding model-free information on local protein secondary structure. These techniques are particularly powerful when combined with REDOR, a heteronuclear recoupling technique that provides distance measurements across putative α-helical or β-sheet hydrogen bonding interactions (i.e., (i) to ($i + 4$) positions in a helix). It should be noted that that sensitivity limitations of NMR generally require that peptides and proteins be adsorbed to high surface–area substrates: microcrystals, porous materials, nanoparticles, and so forth. In the case of statherin, the intact protein adsorbs to HAP at a level of about 1 μmol/m^2.[39] Typical HAP samples were prepared with surface areas of 80 m^2/g.

To probe the molecular backbone structure of statherin on HAP under buffered, hydrated conditions, isotopic labels were incorporated at the pS_2, pS_3, F_7, L_8, I_{11}, and G_{12} backbone carbonyl carbons or nitrogens of statherin. By way of example, the backbone ϕ angle was determined with the DQC-DRAWS techniques at the

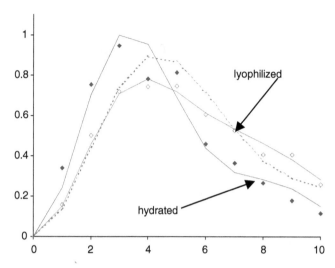

FIGURE 5.10 DQC-DRAWS data for statherin with ^{13}C labels at the carbonyl positions of pS$_2$ and pS$_3$. Vertical axis is in arbitrary units.

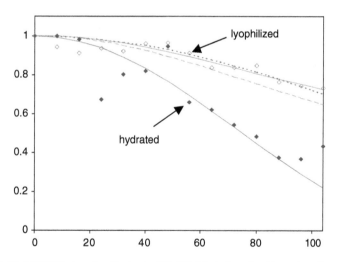

FIGURE 5.11 REDOR data for Statherin ^{13}C-^{15}N labeled at the (i) to (i + 4) positions of pS$_3$F$_7$. The vertical axis is the ratio S/S_0, where S is the intensity of the observed REDOR signal and S_0 is the reference signal intensity.

pS$_2$pS$_3$ positions (see Figure 5.9). In Figure 5.10, the build-up of DQC is plotted for statherin adsorbed onto HAP crystals and fully hydrated (solid diamonds) and for lyophilized samples (open diamonds). The line drawn through the solid diamonds is the DQC-DRAWS data simulated for an α-helical secondary structure. REDOR was used to measure the (i) to (i + 4) distance with pS$_3$F$_7$ and L$_8$G$_{12}$ labeling schemes (for sample data, see Figure 5.11). The composite results clearly define this N-terminal, 12–amino acid binding domain as α-helical on the HAP surface. The strongly acidic N-6 region displays a nearly ideal α-helical distance of 4.2 Å across the pS_3F_7 hydrogen bond, whereas the helix is more extended at 4.8 Å across the L_8G_{12} hydrogen bonding position. Low structural dispersion was observed, particularly at the acidic N terminus, but also out to the 12th amino acid, which indicates there is a relatively narrow range of molecular structures on the surface.

The same samples were lyophilized under low vacuum (100 mTorr) after the initial structural studies were completed. A significant lengthening of the pS_3F_7 distance was observed, demonstrating loss of α-helical structure in this highly acidic region. The L_8G_{12} distance remained the same, however. These results point to the importance of water in either mediating the interaction of the acidic side-chains with the HAP surface (*vide infra*) or directly stabilizing the α-helix conformation (or both). Similar results were observed in recent solid-state NMR studies of N-terminal statherin peptide fragments that were conducted in the lyophilized state both on and off the HAP surface.[56-58] The acidic N-terminus of the peptide was in an extended conformation both on and off the surface, while residues 7 through 12 were in a partly helical conformation.

It has been previously proposed that the α-helix motif might be used as a scaffolding mechanism for aligning acidic side-chain residues with HAP[59,60] by a lattice-matching mechanism, or it could align through a more general electrostatic complementarity. The initial structural characterization shows that the statherin N-terminal binding domain is indeed helical and further demonstrates the importance of water in the acidic N terminus interaction with the HAP surface. Although the comparison with the solution structure of statherin must be made with caution because the solution structure is not yet available to the same precision, it appears that the N terminus is more structured on the surface at the phosphorylated serines and carboxylate-containing aspartic and glutamic acid positions. If verified, this differential folding at the acidic domain between the unbound and bound states would be consistent with a functional role for structural disorder in crystal engineering by acidic proteins, perhaps in response to the aforementioned challenge of matching side-chain positions to inhibit crystal growth without promoting nucleation.

In addition to the structural characterization, the solid-state NMR studies provide interesting molecular dynamics information. This information is particularly useful because the dynamic properties at different atomic positions can be obtained to compare different regions of the N-terminal domain of statherin. As a result, the dynamic studies have directly provided information on the statherin "binding footprint," or which parts of the protein are in close binding contact with the HAP crystal surface. Initial dynamics studies were conducted on hydrated SN15 statherin peptides on HAP crystals. The ^{13}C isotopic backbone labels are sensitive to dynamic timescales ranging over several orders of magnitude, and complementary dynamic characterization techniques include measurement of CSAs, $T_{1\rho}$ relaxation constants, and cross-polarization efficiencies. The principal elements of the motionally preaveraged CSA tensor provide dynamic information on timescales fast compared with the CSA (i.e., >>20 kHz at a magnetic field of 11.75 T). The $T_{1\rho}$ measurements extend the dynamic information available to faster timescales, falling between 10^3 and 10^5 s.

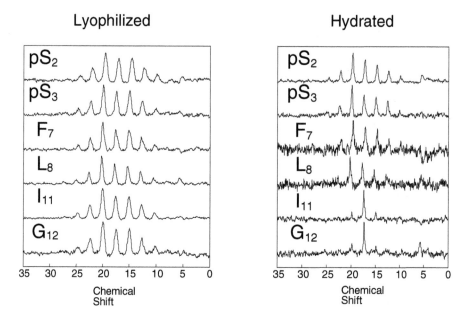

FIGURE 5.12 ^{13}C MAS spectra of carbonyl spins in the backbone of SN15 on HAP crystals as a function of position in the primary sequence X axis units are in parts per million.

The SN15 statherin peptide was characterized at six positions ranging from the N terminus to near the C terminus. Figure 5.12 compares the spectra obtained for lyophilized versus hydrated samples at these six positions. As expected, there is a large overall increase in peptide dynamics in the hydrated samples, except at the phosphoserine backbone positions, which remain strikingly immobile in both states. There is increasing motion as the label is moved toward the C terminus, with large-amplitude dynamic frequencies measured on the order of or greater than 10^5 s at the I_{11} and G_{12} positions. ^{13}C $T_{1\rho}$ measurements provided complementary information, with values for the lyophilized surface-adsorbed samples of >25 ms that demon-strated there was little motion on the kilohertz timescale. Similar values of >25 ms were obtained for the hydrated pS_2pS_3 positions, and significantly shorter values of 11 and 3 ms were measured for the F_7L_8 and $I_{11}G_{12}$ positions, respectively. These results at the middle and C-terminal ends of the peptide demonstrate increased dynamic frequencies of greater than 10^3 Hz. The peptide is thus strongly bound at the acidic N terminus but is surprisingly mobile and dynamic at the middle and C-terminal regions.

More recently, similar studies have been conducted at these same positions within the context of the full-length statherin on HAP crystals.[61,62] The N terminus is again strongly bound to the crystal surface at the phosphoserine positions, as ^{13}C $T_{1\rho}$ values were similar for the lyophilized and hydrated states, with similar cross-polarization efficiencies (Figure 5.13). In the middle and C-terminal regions of this domain of statherin, there is a significant reduction in $T_{1\rho}$ relaxation times, along with significant losses of cross-polarization efficiency at the F_7, L_8, I_{11}, and

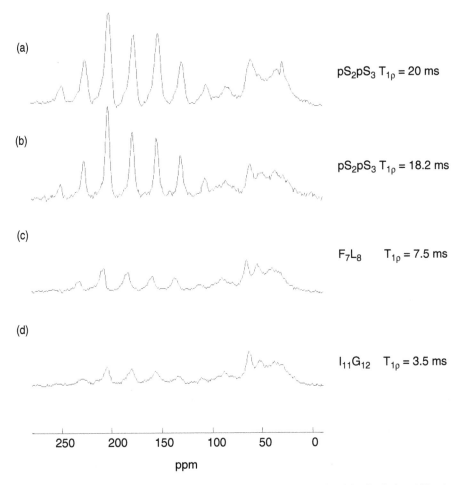

(a) pS_2pS_3 $T_{1\rho} = 20$ ms

(b) pS_2pS_3 $T_{1\rho} = 18.2$ ms

(c) F_7L_8 $T_{1\rho} = 7.5$ ms

(d) $I_{11}G_{12}$ $T_{1\rho} = 3.5$ ms

250 200 150 100 50 0

ppm

FIGURE 5.13 $T_{1\rho}$ measurements from the N terminus of Statherin: (a) pS_2pS_3 lyophilized; (b) pS_2pS_3 hydrated; (c) F_7L_8 hydrated; (d) $I_{11}G_{12}$ hydrated.

G_{12} positions in the hydrated samples. Taken together, these studies demonstrate that the regions outside the acidic N terminus display protein dynamic modes with frequencies of from 10^3 to 10^5 s, similar to those measured for the SN15 peptide. Although these results are qualitatively similar to the SN15 peptide, a comparison of the CSA at the I_{11} and G_{12} positions shows that the statherin dynamic mode exhibits a smaller amplitude, as it is nearly unchanged compared with the significantly narrowed CSA for the peptide. The smaller amplitude is consistent with the larger size of the full protein but could also include differences resulting from the stabilization of the backbone dynamics by tertiary folding interactions, protein–protein interactions on the crystal surface, or different side-chain interactions with the crystal surface. It is clear from these dynamic studies that the binding footprint is largely confined to the anionic stretch of phospho-rylated serines and acidic side-chains at the N terminus.

5.3.2 Peptides Adsorbed onto Hydrophobic Polymer Surfaces

The adsorption or covalent attachment of biological macromolecules onto polymer materials to improve their biocompatibility has been pursued using a variety of approaches, but key to understanding their efficacy is the verification of the structure and dynamics of the immobilized biomolecules. Here we present data on peptides designed to adsorb from aqueous solutions onto highly porous hydrophobic surfaces with specific helical secondary structures.[63] Small linear peptides composed of alternating leucine and lysine residues were synthesized, and their adsorption onto porous polystyrene surfaces was studied using a combination of solid-state NMR techniques. Using conventional solid-state NMR experiments and newly developed DQ techniques, the helical structure of these peptides was verified. Large-amplitude dynamics on the NMR timescale were not observed, indicating irreversible adsorption of the peptides. Their association, adsorption, and structure were examined as a function of helix length and sequence periodicity, and it was found that at higher solution concentrations, peptides as short as seven amino acids adsorb with defined secondary structures. DQCSA experiments of ^{13}C enriched peptide sequences allow high-resolution determination of secondary structure in heterogeneous environments, where the peptides are a minor component of the material. These results shed light on how polymeric surfaces may be surface-modified by structured peptides and demonstrate the level of molecular structural and dynamic information solid-state NMR can provide.

It has been shown previously in solution and at air–water interfaces that the periodicity of hydrophobic and hydrophilic residues can be more important than the helical propensity of a particular amino acid in determining peptide secondary structure, especially at concentrations high enough for aggregation to occur.[64–69] This principle should be magnified when peptides adsorb on hydrophobic surfaces where the planar environment restricts the conformational degrees of freedom.[70,71] To test this hypothesis, we synthesized peptides with varying periodicities of leucine and lysine residues, corresponding to the underlying periodicities of α- and 3_{10}-helices. We also investigated the dependence of helical structure stability on peptide length by designing short helices of 7–8 amino acids in length, and longer helices consisting of 13–14 amino acids. The peptide sequences and isotope enrichment at the backbone carbonyl carbons for NMR observation (indicated in bold) are as follows:

LK3$_{10}$7:	Ac-LL**K**LL**L**KL-NH$_2$
LKα8:	Ac-LL**KK**LL**K**L-NH$_2$
LK3$_{10}$13:	Ac-LL**K**LL**K**LL**K**LL**K**L-NH$_2$
LKα14:	Ac-L**K**KLL**K**LL**KK**LL**K**L-NH$_2$

Qualitative information about peptide secondary structure on and off the polystyrene surface may be determined by measuring isotropic chemical shifts and the CSA tensor.[21–26] A chemical shift analysis of peptide structure, both lyophilized and on the polystyrene surface, was initially performed as a function of peptide concentration in solution. It was observed that relatively high concentrations of peptide in solution

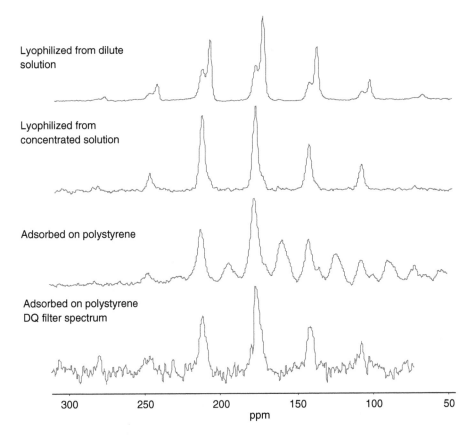

Lyophilized from dilute
solution

Lyophilized from
concentrated solution

Adsorbed on polystyrene

Adsorbed on polystyrene
DQ filter spectrum

300   250   200   150   100   50
ppm

FIGURE 5.14 ^{13}C MAS spectra of LK3$_{10}$7: lyophilized from dilute and concentrated solutions, adsorbed onto polystyrene beads. A DQ–filtered spectrum is shown for comparison. Note elimination of the natural abundance background arising from the polystyrene. X-axis units are in parts per million.

were necessary to drive appreciable adsorption onto the high–surface area polystyrene. At all concentrations, the peptides remained in solution, indicating that any aggregates forming were limited in size. This behavior was previously seen in solutions of similar peptides.[72] Figure 5.14 shows the ^{13}C chemical shift spectra of the LK3$_{10}$7 peptide when it was lyophilized from dilute solution, lyophilized from concentrated solution, and adsorbed on polystyrene from a concentrated solution. A DQ filtered spectrum of the peptide adsorbed on polystyrene is also shown for comparison. Although only 30% of the signal from the isotope-enriched carbonyl carbons is recovered in the DQ experiment, sufficient signal is obtained and the interpretation of the spectrum is greatly simplified because all contributions from natural abundance carbon in the peptide and polystyrene are removed.

In both the spectra of the peptide on polystyrene and the peptide lyophilized from a concentrated solution, the isotope-enriched carbonyl carbons have a chemical shift of 177 ppm and a CSA of 0.6. Previously, it has been shown that these values correlate to helical structure in peptides.[72] In contrast, the spectrum of the peptide lyophilized from a dilute solution shows two distinct resonances — a sharp peak at 177 ppm ($\eta = 0.6$) and a broader peak at 172 ppm ($\eta = 0.9$). These peaks correspond to helical and random coil conformations, respectively.

These results indicate that aggregation and formation of secondary structure in solution precedes adsorption onto the polystyrene surface. Given the relatively small sizes of the peptides, this concentration dependence is not unexpected. Similarly, the other three peptides were found to have little secondary structure or affinity for polystyrene at low concentrations (<1 mmol). At higher concentrations (>1 mmol), the peptides aggregated with well-defined secondary structures and subsequently adsorbed to polystyrene at a much higher level. This behavior is reminiscent of other amphipathic molecules such as lipids, where surface adsorption from micelles has been extensively studied. However, the formation of hydrogen bonding networks in these peptide aggregates yields an added degree of control/complexity in the self-assembly characteristics.[64–67] It should be noted that at even higher concentrations (10 mmol) the helical peptides remain soluble, indicating the formation of helical bundles with a limited size distribution. Similar systems making use of β-sheet-type structures have exhibited aggregation out of solution.[62,65] This may hinder their adsorption in systems making use of porous, high–surface area polymers.

The CSA values can also provide information about the presence of dynamics. As mobility increases, the anisotropy will become increasingly averaged until it approaches the liquid state spectrum — the completely averaged or isotropic state. As is evident in Figure 5.15, the CSAs indicate that there is little or no peptide motion on the timescale (<1 kHz) of the NMR experiments even for the shorter peptides. Measurement of the ^{13}C $T_{1\rho}$ relaxation constants[73] and the build-up of DQ efficiency in the DQDRAWS experiment also indicate there is no detectable motion of the adsorbed peptides on the timescales of the experiments. To determine whether this was true for the entire lengths of the peptides, a second LKα8 peptide was synthesized with the carbonyl carbon labels placed at the N terminus. CPMAS spectra of this peptide adsorbed on polystyrene were indistinguishable from those of the LKα8 peptide isotopically enriched in the middle of the sequence. Taken together, these findings demonstrate that the peptides are tightly bound to the surface along their sequence. This is in contrast to what we have observed previously for peptides adsorbed on ionic surfaces; see Section 5.3.1.[74]

High-resolution determination of the secondary structure of these peptides was accomplished using the DRAWS-based DQC build-up experiment to measure φ and the two-dimensional DRAWS-based DQC experiment for simultaneous measurement of (φ,ψ). When the peptides are adsorbed on polystyrene, a significant natural abundance contribution from the aromatic groups in polystyrene complicates any quantitative analysis of single-quantum spectra. Thus, the DQ buildup experiment was used to measure the distances between adjacent carbonyl carbons in the peptides backbones. Figure 5.15 displays the build-up of DQ coherence for the four peptides adsorbed on polystyrene. Two-dimensional DQC projections for the longer peptides, LKα14 and LK3$_{10}$13, are shown in Figure 5.16. The *S/N* ratios for the shorter peptides were significantly poorer as a result of less adsorption to the surface. As a consequence, only the DQ buildup was monitored.

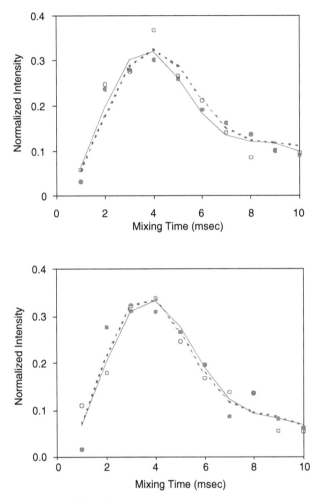

FIGURE 5.15 DQC-DRAWS buildup curves for hydrated, surface-adsorbed peptides on polystyrene. (Top) Longer helices LKα14 (●) and LK3$_{10}$13 (O) and simulations for φ torsion angles of –62° and –67°. (Bottom) shorter helices LKα8 (●) and LK3$_{10}$7 (O) and simulations for φ torsion angles of –64° and –70°.

The results obtained for the four peptides adsorbed to polystyrene are fairly similar to those for the peptide aggregates and demonstrate that all the peptides are helical when adsorbed to the surface. However, the torsion angles measured in the peptides adsorbed on polystyrene differ slightly from those measured for the aggregates. For the LK3$_{10}$13 peptide, φ increased from –65° ± 1° to –67° ± 7° and ψ decreased from –36° ± 1° to –33° ± 2°. For the LKα14 peptide, φ decreased from –63° ± 1° to –62° ± 6°, and ψ decreased from –39° ± 1° to –34° ± 2°. For the shorter peptides, φ decreased in both cases — from –72° ± 3° to –64° ± 11° for LKα8 and from –78° ± 2° to –70° ± 11° for LK3$_{10}$7.

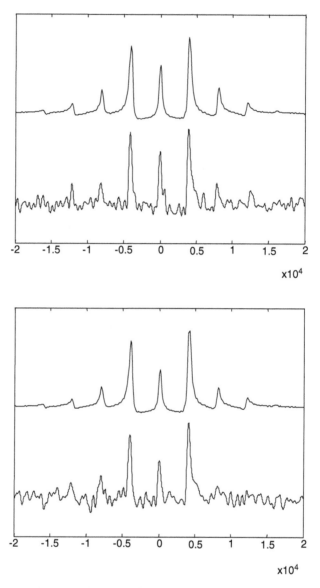

FIGURE 5.16 Projections of two-dimensional DQDRAWS spectra onto the double-quantum axis for hydrated peptides adsorbed on polystyrene. (Top) LKα14 and (bottom) LK3$_{10}$13 along with simulations for (−62, −34) and (−67, −33), respectively. *X*-axis units are in Hertz.

These changes indicate that adsorption on polystyrene may induce more helical structure in the shorter peptides and that slight changes in torsion angles are needed for the longer peptides to adsorb on the polystyrene surface.

5.3.3 Peptides Covalently Attached to Monolayer-Protected Gold Nanoparticles

A number of surface-modification techniques can be used to produce a biocompatible protein coating over a material surface. Simple physical adsorption onto a polymer bead exploiting the hydrophobic effect and helical periodicity of the peptide structure was described in the previous section. Another approach is to functionalize the surface with reactive groups that may be used to tether the biomolecule. Key to this approach is to retain secondary structure once the biomolecule is tethered and immobilized.

A versatile approach for varying surface chemistry is based on the use of colloidal gold nanoparticles and alkanethiols. The synthetic preparation of colloidal gold has been extended to include self-protection of the gold cluster by self-assembled monolayers (SAMs) of functionalized alkanethiols and other sulfur-containing capping molecules[75] into what are commonly called monolayer-protected clusters (MPCs). Gold coatings are easily constructed on many underlying materials, and the coverage with a SAM provides a versatile interface for subsequent masking of the material with peptide coatings. Following the results of our solid-state NMR study of the structures of peptides with varying periodicities of leucine and lysine residues adsorbed on polystyrene beads, described earlier, we synthesized and used in this study a peptide displaying a periodic alternation of leucine and lysine amino acids: Ac-LKKLLKLLKKLLKL-NH$_2$ (LKα14, see Section 5.3.2). This peptide assumes a helical structure when adsorbed onto polystyrene beads. In this study, LKα14 was synthesized, adsorbed, and covalently tethered to gold nanoparticles coated with an alkanethiolate SAM and studied using a combination of solid-state NMR dipolar recoupling techniques.

To enable covalent attachment of LKα14, a water-soluble, functionalized MPC gold bead is required. MPC's consisting of ω-carboxylic acid functionalized alkanethiolates can be made water soluble by minimizing the length of the methylene spacer chain and by introducing polar functional groups between the thiol group and the ω-carboxyl group. An approach developed by Murray et al.[76] for the preparation of water-soluble MPC gold particles uses thiol-containing biomolecules like tiopronin (N-2-mercaptopropionylglycine; see Figure 5.17), a drug used in the treatment of cystinuria and arthritis. LKα14 was covalently attached to the MPC through the creation of amide bonds between the primary amine groups of the lysine side chains and the terminal carboxyl groups of the tiopronin-MPC.[77,78]

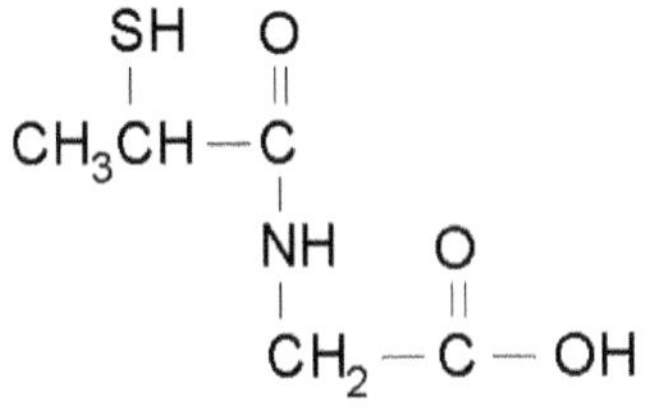

$$
\begin{array}{c}
\overset{\displaystyle SH}{\underset{\displaystyle |}{}} \quad \overset{\displaystyle O}{\underset{\displaystyle ||}{}} \\
CH_3CH-C \\
| \\
\overset{\displaystyle NH}{\underset{\displaystyle |}{}} \quad \overset{\displaystyle O}{\underset{\displaystyle ||}{}} \\
CH_2-C-OH
\end{array}
$$

FIGURE 5.17 Tiopronin.

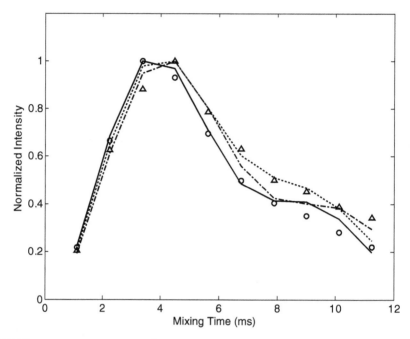

FIGURE 5.18 The build-up of ^{13}C double-quantum coherence accomplished with the DRAWS pulse sequence. Open circle: $LK\alpha14$ ^{13}C enriched at the carbonyl positions of L4/L5 and adsorbed onto tiopronin-protected gold nanoparticles; open triangle: $LK\alpha14$ ^{13}C enriched at the carbonyl positions of L11/L12 and adsorbed onto tiopronin-protected gold nanoparticles; dashed line: simulated data assuming model consisting of 20% beta sheet, 80% alpha helix; dash-dot: carbonyl–carbonyl distance = 3.10 Å; solid line: carbonyl–carbonyl distance = 3.02 Å.

To probe structure at selected positions in the peptide, samples of $LK\alpha14$ were pair-wise enriched at the carbonyls sites of L4/L5 and L11/L12. Figure 5.18 shows the DRAWS-based DQC build-up data for the $LK\alpha14$ covalently immobilized on MPC particles. In the former sample, the carbonyl–carbonyl DQF-DRAWS data were fitted to a model in which the internuclear distance is 3.10 Å, corresponding to $\phi = 78°$, which deviates somewhat from values typical of α-helical peptides and approaches the range of 3_{10} helices.[79] The fit can be improved, as shown in Figure 5.8, by adding small populations of nonhelical conformers, but in all models, the major conformer has a ϕ value in the α helical region. The L11/L12 sample yields DQF-DRAWS data with a similar interpretation. The carbonyl–carbonyl distance is 3.02 Å, corresponding to $\phi = 64°$, again in the α-helical region of the Ramachandran map.

A more thorough definition of the secondary structure of the MPC-bound peptide is obtained from the projection of the two-dimensional DQDRAWS spectrum. For the L11/L12 doubly carbonyl enriched sample, the DQ interferogram is shown in Figure 5.19. As shown in Figure 5.19, the real and imaginary DQ interferograms are actually fitted and indicated a value of $\psi = 30° \pm 5°$. The simulated DQ interferograms were produced assuming a ϕ value derived from the DQF-DRAWS experiment.

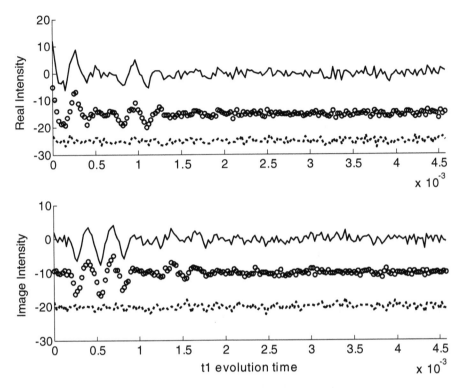

FIGURE 5.19 Real (top) and imaginary (bottom) components of the DQ interferogram, obtained from L11/L12 carbonyl-labeled LKα14 adsorbed onto tiopronin-protected gold nanoparticles. In both plots the line is a simulation, produced assuming a carbonyl–carbonyl distance of 3.02 Å, corresponding to $\phi = 64°$ and $\psi = 30° \pm 5°$. Open circles correspond to experimental data, and the dashed line is the residual.

Combining the two data sets, the L11/L12 labels yield torsion angle values of $(\phi, \psi) = (-64° \pm, -30° \pm 5°)$, values typical of well-ordered helical peptides.

5.4 SUMMARY

Our long-term goal is to understand the structural and dynamical basis for the control of hard-tissue formation by proteins, to understand the influence that surface chemistry and surface preparation exerts on protein structure and dynamics, and to exploit information derived from such studies in the design of well-structured and oriented peptide or protein surface coatings on biomaterial, tissue engineering, and device surfaces (e.g., diagnostic, chip, microfluidic channels). Being able to control the structure and orientation of these devices on the surface will enhance their ability to display signal sequences that interact with target cell receptors, to display larger proteins on chip surfaces, and to open avenues for templating off of specific side-chain residues that have defined periodicity. One particularly important application is in the surface modification of three-dimensional, porous scaffolds for tissue engineering.[80,81] Immobilization chemistry is difficult in these complex structures, but peptides that bind tightly in an ordered and oriented fashion from solution may allow better penetration and coverage.

A key component of the approach described in this review is the verification of structure on the relevant biomaterial surfaces using solid-state NMR and, particularly, DQ solid-state NMR. Vibrational and optical spectroscopies have provided important structural information, primarily with regard to secondary structure content and protein unfolding on surfaces. However, such methods can only detect averaged secondary structures, and they provide little dynamic information. Many vibrational and optical spectroscopies are also limited by their inability to probe rough or porous structures. Solid-state NMR alone can provide high-resolution structural and dynamic information about peptides on hydrated surfaces. In this review we have shown that dipolar recoupling techniques such as DRAWS are well suited to measuring distances and torsion angles in peptides and proteins adsorbed onto biomaterial and biomineral surfaces. These studies also demonstrate the feasibility of using solid-state NMR to obtain high-resolution protein structures on material surfaces.

ACKNOWLEDGMENTS

We gratefully acknowledge the support provided by the National Science Foundation (grant numbers EEC-9529161 and DMR-0110505) and by the National Dental Institute (DE 12554). P. V. Bower acknowledges support from the Danish Research Academy. Helpful and informative discussions with Professor David G. Castner are gratefully acknowledged.

REFERENCES

1. Coe, F.L., Parks, J. H. and Asplin, J.R., The pathogenesis and treatment of kidney stones, *N. Engl. J. Med.*, 327, 1141, 1992.
2. McCarty, D.J., Crystals and arthritis, *Dis. Month.* June, 258, 1994.
3. Healy, K.E., *Curr. Opin. Solid State Mat. Sci.*, 4, 381, 1999.
4. Angelova, N. and Hunkeler, D., *Trends Biotech.*, 17, 409, 1999.
5. Griffith, L.G., *Acta Mater.*, 48, 263–277, 2000.
6. Langer, R., *Acc. Chem. Res.*, 33, 94–101, 2000.
7. Borkenhagen, M., Clemence, J.F., Sigrist, H. and Aebischer, P.J. *Biomed. Mat. Res.*, 40, 392–400, 1998.
8. Belcheva, N., Baldwin, S.P., Saltzman, W.M., *J. Biomat. Sci. – Polymer Ed.*, 9, 207–226, 1998.
9. Rezania, A. and Healy, K.E., *Biotech. Progress*, 15, 19–32, 1999.
10. Gilbert, M., Shaw, W.J., Long, J.R., Nelson, K., Drobny, G.P., Giachelli, C.M. and Stayton, P.S., *J. Biol. Chem.*, 275, 16213, 2000.
11. Whang, K., Goldstick, T.K. and Healy, K.E., *Biomaterials*, 21, 2545, 2000.
12. Drumheller, P.D., Elbert, D.L. and Hubbell, J.A., *Biotech. Bioeng.*, 43, 772, 1994.
13. Walluscheck, K.P., Steinhoff, G., Kelm, S. and Haverich, A., *Eur. J. Vasc. Endovac. Surg.*, 12, 321, 1996.

14. Neff, J.A., Caldwell, K.D. and Tresco, P.A., *J. Biomed. Mat. Res.*, 40, 511, 1998.
15. Tong, Y.W. and Shoichet, M.S., *J. Biomat. Sci. – Polymer Ed.*, 9, 713, 1998.
16. Pakalns, T., Haverstick, K.L., Fields, G.B., McCarthy, J.B., Mooradian, D.L. and Tirrell, M., *Biomaterials*, 20, 2265, 1999.
17. Bhadriraju, K. and Hansen, L.K., *Biomaterials*, 21, 267, 2000.
18. Tycko, R., *Annu. Rev. Phys. Chem.*, 52, 575, 2001.
19. Kim, S.J., Cegelski, L., Studelska, D., et al.. *Biophys. J.*, 82(1), 2289, 2002.
20. Warshawski, D., Trailkia, M., Devaux, P. and Bodenhausen, G., *Biochimie*, 80, 437, 1998.
21. Asakawa, N., Kurosu, H., Ando, I., Shoji, A. and Ozaki, T., *J. Mol. Struct.*, 317, 119, 1994.
22. Tsuchiya, K., Takahashi, A., Takeda, N., Asakawa, N., Kuroki, S., Ando, I., Shoji, A. and Ozaki, T., *J. Mol. Struct.*, 350, 233, 1995.
23. Kameda, T., Takeda, N., Kuorki, S., Kurosu, H., Ando, S., Ando, I., Shoji, A. and Ozaki, T., *J. Mol. Struct.*, 384, 17, 1996.
24. Ando, I. and Kuroki, S., *The Encyclopedia of NMR*, John Wiley, New York, 1996, p. 4458.
25. Kameda, T. and Ando, I., *J. Mol. Struct.*, 412, 197, 1997.
26. Ando, I., Kameda, T., Asakawa, N., Kuroki, S. and Kurosu, H., *J. Mol. Struct.*, 441, 213, 1998.
27. Mehring, M., *Principles of High Resolution NMR in Solids*, Springer, Berlin, 1976, p. 40.
28. Karlsson, T., Popham, J.M., Long, J.R. and Drobny, G.P., *J. Amer. Chem. Soc.*, 125(24): 7394, and references cited therein, 2003.
29. Gullion, T. and Schaefer, J., *J. Magn. Reson.*, 196, 1989.
30. Hohwy, M., Jakobsen, H.J., Eden, M., Levitt, M.H. and Nielsen, N.C. *J. Chem. Phys.*, 108(7), 2686, 1998.
31. Hohwy, M., Rienstra, C.M., Jaroniec, C.P. and Griffin, R.G., *J. Chem. Phys.*, 110(16), 7983, 1999.
32. Carravetta, M., Eden, M., Johannessen, O.G., Luthman, H., Verdegem, P.J.E., Lugtenburg, J., Sebald, A. and Levitt, M.H., *J. Amer. Chem. Soc.*, 123, 10628, 2001.
33. Nielsen, N.C., Bildsoe, H., Jacobsen, H.J. and Levitt, M.H., *J. Chem. Phys.*, 101(3), 1805, 1994.
34. Bower, P.V., Oyler, N.A., Mehta, M.A., Long, J.R., Stayton, P.S. and Drobny, G.P., *J. Amer. Chem. Soc.*, 121, 8373, 1999.
35. Blanco, F.J. and Tycko, R., *J. Magn. Reson.*, 149, 131, 2001.
36. Gregory, D.M., Mehta, M.A., Shiels, J.C. and Drobny, G.P., *J. Chem. Phys.*, 107, 28, 1997.
37. Gottwald, J., Demco, D.E., Graf, R. and Spiess, H.W., *Chem. Phys. Lett.*, 243, 314, 1995.
38. Johnson, M., Richardson, C.F., Bergey, D.J., Levine, M.J. and Nancollas, G.H., *Arch. Oral Biol.*, 36, 631, 1991.
39. Campbell, A.A., Ebrahimpour, A., Perez, L., Smesko, S.A. and Nancollas, G.H. *Calcif. Tissue Int.*, 45, 122, 1989.
40. Schwartz, S.S., Hay, D.I. and Schluckebier, S.K., *Calc. Tissue Int.*, 50, 511, 1992.
41. Hunter, G.K. and Goldberg, H.A., *Biochem. J.*, 302, 175, 1994.
42. Hunter, G.K., Hauschka, P.V., Poole, A.R., Rosenberg, L.C. and Goldberg, H.A., *Biochem. J.*, 317, 59, 1996.
43. Fuijisawa, R. and Kuboki, Y., *Eur. J. Oral Sci.*, 106, 249, 1998.
44. Schlesinger, D.H. and Hay, D.I., *J. Biol. Chem.*, 252, 1689, 1977.

45. Raj, P.A., Johnsson, M., Levine, M.J. and Nancollas, G.H., *J. Biol. Chem.*, 267, 5968, 1992.
46. Schwartz, S.S., Hay, D.I. and Schluckebier, S.K., *Calc. Tissue Int.,* 50, 511, 1992.
47. Ramasubbu, N., Thomas, L.M., Bhandary, K.K. and Levine, M.J., *Crit. Rev. Oral Biol. Med.*, 4, 363, 1993.
48. Gururaja, T.L. and Levine, M.J., *Peptide Res.,* 9, 283, 1996.
49. Naganagowda, G.A., Gururaja, T.L. and Levine, M.J. *J. Biomol. Struct. Dyn.,* 16, 91, 1998.
50. Schlesinger, D.H. and Hay, D.I., *J. Biol. Chem.,* 252, 1689, 1977.
51. Raj, P.A., Johnsson, M., Levine, M.J. and Nancollas, G.H., *J. Biol. Chem.,* 267, 5968, 1992.
52. Schwartz, S.S., Hay, D.I. and Schluckebier, S.K., *Calc. Tissue Int.,* 50, 511, 1992.
53. Ramasubbu, N., Thomas, L.M., Bhandary, K.K. and Levine, M.J., *Crit. Rev. Oral Biol. Med.*, 4, 363, 1993.
54. Gururaja, T.L. and Levine, M.J., *Peptide Res.,* 9, 283, 1996.
55. Naganagowda, G.A., Gururaja, T.L. and Levine, M.J., *J. Biomol. Struct. Dyn.,* 16, 91, 1998.
56. Shaw, W.J., Long, J.R., Dindot, J.L., Campbell, A.A., Stayton, P.S. and Drobny, G.P., *J. Amer. Chem. Soc.*, 122, 1709., 2000.
57. Shaw, W.J., Long, J.R., Stayton, P.S. and Drobny, G.P., *J. Amer. Chem. Soc.,* 122, 7118, 2000.
58. Long, J.R., Dindot, J.L., Zebroski, H., Kiihne, S., Clark, R.H., Campbell, A.A., Stayton, P.S. and Drobny, G.P., *Proc. Natl. Acad. Sci. USA*, 95, 12083, 1998.
59. Hauschka, P.V. and Carr, S.A., *Biochemistry*, 21, 2538–2547, 1982.
60. DeOliviera, D.B. and Laursen, R.A., *J. Am. Chem. Soc.* 119, 10627, 1997.
61. Long, J.R., Shaw, W.J., Stayton, P.S. and Drobny, G.P., *Biochemistry*, 40, 154151, 2001.
62. Moreno, E.C., Varughese, K. and Hay, D.I., Effect of human salivary proteins on the precipitation kinetics of calcium phosphate. *Calc. Tissue Int.,* 28, 7, 1979.
63. Long, J.R., Oyler, N., Drobny, G.P. and Stayton, P.S. Assembly of alpha-helical peptide coatings on hydrophobic surfaces, *J. Am. Chem. Soc.,* 124(22), 6297, 2002.
64. Perez-Paya, E., Houghten, R.A. and Blondelle, S.E., *J. Biol. Chem.*, 271, 4120, 1996.
65. Aggeli, A., Bell, M., Boden, N., Keen, J.N., Knowles, P.F., McLeish, T.C.B., Pitkeathly, M. and Radford, S.E., *Nature*, 386, 259, 1997.
66. Clark, T.D., Buriak, J.M., Kobayashi, K., Isler, M.P., McRee, D.E. and Ghadiri, M.R,. *J. Am. Chem. Soc.*, 120, 8949, 1998.
67. Fujita, K., Kimura, S. and Imanishi, Y., *Langmuir*, 15, 4377, 1999.
68. Maget-Dana R., Lelievre, D. and Brack, A., *Biopolymers*, 49, 415, 1999.
69. Powers, E.T. and Kelly, J.W., *J. Am. Chem. Soc.*, 123, 775, 2001.
70. Wattenbarger, M.R., Chan, H.S., Evans, D.F., Bloomfield, V.A. and Dill, K.A., *J. Chem. Phys.*, 93, 8343, 1990.
71. Chan, H.S., Wattenbarger, M.R., Evans, D.F., Bloomfield, V.A. and Dill, K.A., *J Chem. Phys.*, 94, 8542, 1991.
72. Degrado, W.F. and Lear, J.D. *J. Am. Chem. Soc.*, 107, 7684, 1985.
73. Schaefer, J., Stejskal, E.O. and Buchdahl, R., *Macromolecules*, 10, 384, 1977.
74. Shaw, W.J., Long, J.R., Campbell, A.A., Stayton, P.S. and Drobny, G.P., *J. Am. Chem. Soc.*, 122, 7118, 2000.
75. Healy, K.E., *Curr. Opin. Solid State Mater. Sci.*, 4, 38, 1991.
76. Templeton, A.C., Chen S., Gross S. M., Murray R. W., *Langmuir*, 15, 66, 1999.
77. Hermanson, G.T., *Bioconjugate Techniques*, Academic Press, New York, 1996.

78. Bower, P., Long, J.R., Louie, E., Stayton, P.S. and Drobny, G.P., *Langmuir*, 2004.
79. Millhauser, G.L., *Biochemistry*, 34, 3873, 1995.
80. Brauker, J.H., Carr-Brendel, V.E., Martinson, L.A., Crudele, J., Johnston, W.D. and Johnson, R.C., *J. Biomed. Materials Res.*, 29, 1517, 1995.
81. Sieminski, A.L. and Gooch, K.J., *Biomaterials*, 21, 2233, 2000.

6 Sensitivity Enhancement by Inverse Detection in Solids

Kay Saalwächter and Ayyalusamy Ramamoorthy

CONTENTS

ABSTRACT The most disappointing aspect of solid-state NMR spectroscopy is the requirement of a large amount of sample and/or a long acquisition time. Some of the most interesting biomolecules, such as membrane proteins and RNA, are not abundantly available and also do not withstand long high power solid-state NMR experiments. Therefore, there is a considerable interest in developing methods to increase the sensitivity of this powerful tool. To accomplish this goal, a number of proton-detection techniques have recently been reported. In this chapter, we address the need for higher sensitivity experiments, discuss the difficulties in designing such techniques, and explain the merits of different types of proton-detection methods.

KEY WORDS: *cross-polarization, coherence transfer, REDOR, SEMA, TANSEMA, fast-MAS, PRIDE*

6.1   INTRODUCTION

Solid-state nuclear magnetic resonance (NMR) spectroscopy has proven to be a powerful method to obtain atomistic-level information from a variety of crystalline, noncrystalline, and amorphous samples.[1] Recent technique and instrumental advancements further strengthened the scope of the technique. For example, the advent of higher magnetic fields, the art of sample preparation and fast magic-angle spinning (MAS) capabilities dramatically increased the sensitivity and resolution of spectra even from biological complexes such as membranes and amorphous materials.[2–4] Although spectral resolution rendered by solid-state NMR experiments is on par with that of high-resolution solution NMR experiments, the sensitivity is still a major concern. The techniques demand a rather large (milligrams) quantity of biological solids in spite of their capability to provide impressive "solution-like" spectra from most solids. However, the mandatory requirement of long experiments to enhance the sensitivity suffers because of instabilities of the spectrometer and the sample. For example, typically, signal acquisition for more than a day is necessary to obtain a two-dimensional PISEMA (polarization inversion spin exchange at the magic-angle)[3,5–11] spectrum of a mechanically aligned lipid bilayer sample containing a few milligrams of ^{15}N-labeled membrane-associated protein.[12–16] Thus, the development of techniques that further enhance the sensitivity of solid-state NMR experiments is of great importance.

Highly abundant nuclei with large gyromagnetic ratios (such as ^{1}H and ^{19}F) are very NMR sensitive and therefore should be the first choice for detection in NMR experiments. Although this has become true for solution NMR experiments in which ^{1}H-detected (or inversely detected) multidimensional techniques are routinely applied,[17–20] it is still under development for solid-state studies. The key factors that have been preventing the application of ^{1}H-detected solid-state NMR are the large ^{1}H-^{1}H dipolar interaction, short relaxation (T_2 and $T_{1\rho}$) times, and need for high radio frequency (rf) power. Some of these difficulties have been overcome by the recent developments in the field. As a result, several reports have demonstrated the feasibility of ^{1}H-detected solid-state NMR experiments as well as significant gain in sensitivity.[3,21–31] Because this remarkable methodology, based on coherence transfer (CT) between coupled nuclear spins, can in principle be applied to any heteronuclear spin system, in this chapter we refer to it as "inverse detection" instead of "^{1}H-detection." Although most of these inverse-detected techniques are efficient under fast MAS conditions,[21,22,26,30,31] some of the techniques for static or slow spinning speeds are also reported.[3,23–25,27,29] Most of these techniques employ a reverse polarization transfer step based on ramp[32,33] or Lee-Goldburg (LG) cross polarization (CP),[34] as well as other coherence transfers, possibly involving some type of recoupling.[26,35]

This happy circumstance is naturally a major advancement in the field and certainly will have a major effect on NMR studies of biological solids. In addition, the development of a plethora of new multidimensional solid-state NMR techniques based on inverse detection is also in progress in several laboratories. In this chapter, ^{1}H-detected methods reported in the literature are reviewed with a special emphasis on the basic concepts that led to the development of ^{1}H-detected solid-state NMR experiments under static and MAS conditions. Advantages, disadvantages, and hardware requirements for each method are briefly discussed with some examples from the literature. In addition, possible applications of these remarkable techniques to investigate small molecules and large proteins are highlighted.

6.2 BASIC CONCEPTS AND TECHNIQUES

The sensitivity of an NMR experiment depends on the gyromagnetic ratio (γ) of the nuclei that are being prepared and the γ of the detected nuclei. This sensitivity is further increased if the nuclei under preparation and detection are highly abundant. The potential to significantly enhance the sensitivity of heteronuclear correlation experiments via inverse detection, particularly of low-γ nuclei such as ^{15}N and ^{13}C, was recognized at the early stages of the development of two-dimensional NMR.[17–19] The basic idea is to make use of the large γ, which, first of all, provides a large magnetic moment that leads to a large induction voltage in the receiver coil. Second, most solution-state inverse-detected experiments take advantage of the high equilibrium magnetization of protons by coherent transfer of polarization.[18,19] Because the transverse relaxation has negligible effects on the timescale needed for CT via isotropic scalar couplings, transfer efficiencies close to 100% are possible in solution samples.[36] Therefore, the magnitude of the acquired signal in an optimized inverse detected experiment will not depend on the low gyromagnetic ratio (γ_S) of the heteronucleus, and thus inverse detection may lead to a gain proportional to $(\gamma_H/\gamma_S)^{3/2}$.

A serious limitation of this approach in ^{1}H-detected solution NMR spectroscopy arises when the heteronucleus is not highly abundant. This results in receiver saturation by unwanted (uncoupled and also solvent) proton signal. However, this problem can be overcome by employing a suitable phase cycle to suppress the solvent signal. As the wanted signal is relatively weak and appears as a difference of large numbers, significant amounts of noise are introduced into the spectrum, thus often spoiling the sensitivity gain because of inverse detection. This difficulty was solved by the use of pulsed field gradients for coherence selection as well as by dephasing unwanted coherences,[37] which allow the receiver to be optimized for the wanted signal.

In solid-state NMR spectroscopy, however, inverse detection was not regarded as a useful procedure for a long time. First, the high filling factor of solid rather than solution samples, along with polarization enhancement by CP, line narrowing by moderate MAS, high-power dipolar decoupling, and possibly isotopic labeling, makes low-abundance heteronucleus spectroscopy quite feasible in many relevant cases. Second, the signal-to-noise ratio (S/N) gained by an inverse detection[21] depends on the quality factor of the detection circuits, Q; the effective line width, W, of the two heteronuclei; and the efficiency of the additional CT step, f_{XH}, as given by the following equation

$$\xi = \frac{(S/N)_{id}}{(S/N)_{dd}} \propto f_{XH} \left(\frac{\gamma_H}{\gamma_X} \right)^{3/2} \left(\frac{W_X}{W_H} \right)^{1/2} \left(\frac{Q_H}{Q_X} \right)^{1/2} \tag{6.1}$$

Although Q_H/Q_X is usually greater than unity even in most commonly used commercial double-resonance probes (and can certainly be further improved), the proton line width was traditionally considered prohibitively large (i.e., on the order of 40 kHz even at moderate MAS). In addition, CT via scalar couplings in solids is not as efficient as in solution. Although full CT via J couplings is theoretically possible in solids and has experimentally been demonstrated,[38] it is subject to large losses resulting from strong T_2 relaxation. All other alternatives rely on the use of orientation-dependent dipolar couplings.[39,40] Apart from experiments in ordered samples or single (liquid) crystals, the inevitable random orientation of polycrystallites will always limit the efficiency of CT to values on the order of 50%.[39,41] As an alternative, adiabatic transfer schemes may be considered,[42] but these often suffer from the need to have short and selective single-bond transfers.

Even though first reports of ^{1}H-detected double-resonance experiments date back much further than the above-mentioned milestones in solution-state NMR,[43] few applications were reported, mainly in the field of CP, involving quadrupolar nuclei.[44] A first notable gain in sensitivity was described for the indirect detection of rare-spin resonances such as ^{113}Cd, ^{77}Se, and ^{29}Si via their J coupling to ^{31}P in MAS NMR of inorganic solids,[45] using a method that is essentially a variant of the original heteronuclear multiple-quantum coherence (HMQC) experiment.[18] ^{31}P represents a favorable case of a comparatively sensitive nucleus with rather weak T_2^* relaxation times and, thus, narrow lines in MAS spectra.

Renewed interest in sensitivity enhancement of solid-state experiments has mainly been spawned by the growing interest in studying biological solids, where even when the heteronuclei of interest are isotopically labeled and very low S/N are common because of the high dilution of specific nuclei in a given macromolecule. Also, in the case of membrane-associated proteins, labeled proteins often need to be reconstituted in artificial membranes, possibly stacked between thin glass plates, which further lowers the effective filling factor. Significant technological advances, particularly the possibility to do fast MAS (spinning speed > 20 kHz) and the development of other efficient proton line-narrowing techniques have finally opened avenues to successful inverse detection via protons in solids.

A schematic representation of various types of inverse-detection experiment is depicted in Figure 6.1. During the first stage, initial ^{1}H polarization is transferred to the heteronucleus (S) using either CP or some other CT scheme. In the second step, spectral information on S, most commonly its chemical shift but also dipolar couplings or quadrupole coupling, is encoded in an indirect dimension (t_1). In addition, when significant polarization is on the S spin (and possibly stored along the z-axis to avoid loss resulting from T_2), specific measures can be taken to remove any unwanted proton magnetization. This is particularly important when isotopically dilute systems are investigated, in which leakage of such signals into the detection stage is often deleterious. Finally, the S spin polarization needs to be transferred to protons for detection on the proton channel, possibly under heteronuclear decoupling.

Most approaches discussed in the next three subsections follow this scheme. Sections 6.2.1 and 6.2.2 comprise techniques designed to work under fast MAS,[21,22,26,28,30,31] where the CP-based techniques (given in Section 6.2.1)[23,27,29] are mostly used to establish chemical-shift correlations and semiquantitative distance constraints. Although the REDOR-based[46] heteronuclear CT (Section 6.2.2) gives high-precision information on heteronuclear dipolar couplings,[26,28] it is also useful for the investigation of local structure and dynamics.[35] More specialized approaches applicable in static samples are summarized in Section 6.2.3.

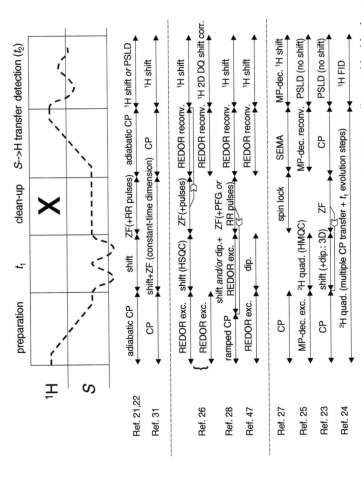

FIGURE 6.1 Basic steps of a typical inverse-detected 2D experiment and a survey of techniques discussed in this chapter. Spectral information included in t_1 can be chemical shifts or dipolar or quadrupolar coupling. The three different schemes are discussed in Sections 6.2.1–6.2.3. In some cases, two or more stages are combined into one, and in three cases, the location of the clean-up stage differs, as indicated by the arrows. The abbreviations are: CP, cross polarization; ZF, z-filter; RR, rotary resonance; REDOR, rotational-echo double-resonance; HSQC, heteronuclear single-quantum coherence; DQ, double-quantum; PFG, pulsed field gradient; SEMA, spin exchange at the magic-angle; MP-dec, multiple-pulse decoupled; HMQC, heteronuclear multiple-quantum coherence; PSLD, pulsed spin-lock detection.

6.2.1 TECHNIQUES BASED ON CPMAS

A sensitivity gain in a solid-state inverse ^{15}N-^{1}H shift correlation experiment was successfully demonstrated under fast MAS conditions.[21] For both heteronuclear polarization transfer steps in the pulse sequence (Figure 6.1), a standard CP sequence was used. An optimized polarization transfer can be achieved by the use of adiabatic spin-lock pulses during CP on one of the rf channels.[42] This is particularly important at fast MAS frequencies of around 30 kHz, which are crucial, as the observed gain is only possible with rather narrow proton lines (Equation 6.1). Although an adiabatic transfer is certainly recommendable for the initial enhancement of ^{15}N polarization, actual applications such as the resonance assignment in proteins might require a more specific transfer at the second stage; in such applications, polarization transfer among directly dipolar coupled spin pairs is desired, as relayed polarization transfer or spin diffusion defeat the purpose of heteronuclear correlation experiments. In many cases, a short, ramped CP[32] has proven to be suitable at fast MAS. However, an off-resonance spin-lock[33,34] can be used for effective polarization transfer among directly dipolar coupled spin pairs under static or slow spinning conditions.

When this technique is to be applied at low isotopic abundance, surplus proton signals from uncoupled protons that are bound or close to ^{14}N or ^{12}C must be removed. A simple and also rather robust approach was presented later for the case of ^{13}C-^{1}H correlation at natural abundance,[22] where two half-millisecond pulses at the rotary resonance recoupling (RRR) condition ($v_1 = nv_R$) were applied during a ^{13}C z-filter delay to convert the dipolar-coupled ^{1}H spin bath into nonobservable higher-quantum coherences. A sample spectrum acquired with this technique is given in Figure 6.2, in comparison with a conventional CP-based HETCOR. Another interesting option was also demonstrated: When only one-dimensional spectra of the heteronucleus are sought, it is possible to employ traditional stroboscopic observation while the transverse ^{1}H magnetization is locked with short pulses (i.e., pulsed

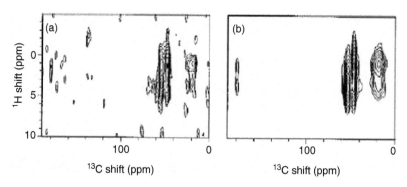

FIGURE 6.2 Directly detected (a) and inverse-detected (b) ^{1}H-^{13}C correlation spectra of poly(methyl methacrylate) with ^{13}C in natural abundance, measured with a double CP technique at 31 kHz MAS, demonstrating sensitivity gains for the different signals of about 2–3. Reproduced from reference 22, with permission from the *Journal of the American Chemical Society*.

spin-lock detection, PSLD). In this procedure, even though the ^1H chemical shift information is lost, the length of the FID is increased and the effective line width is consequently small. The main disadvantage of this method is that a two-dimensional acquisition is mandatory to construct the S nucleus spectrum. Nevertheless, a twofold signal gain per unit time could be obtained.[22]

RRR pulses to remove unbound protons are only efficient in rather immobile solids. In samples with significant molecular motions such as semisolids, such as hydrated lipid bilayers, pulsed field gradients applied during a z-filter are one method of choice, although it requires special hardware. This method was demonstrated on a uniformly labeled microcrystalline SH3 protein sample at only 10 kHz MAS.[30] This condition offers additional narrowing of ^1H spectral lines. Another important strategy is to use isotopic dilution of protons that can be achieved by perdeuteration of a protein, followed by the back-exchange of N-Ds to N-Hs.

A recent study demonstrated that pulsed field gradients are not even necessary to remove excess water signal.[31] Introduction of a constant-time interval with a prolonged heteronuclear decoupling in the pulse sequence (Figure 6.1) effectively dephases the water signal. The constant-time interval comprises the t_1 evolution time and the z-filter, where simply the filter store pulse is moved from $t_1 = 0$ to $t_{1,max}$. As all pulses on the proton channel stay exactly the same during this procedure, any residual water signal is suppressed by a simple two-step phase cycle on the S-channel pulse. In fact, this study showed that it is possible to obtain a two-dimensional ^{15}N-^1H chemical shift correlation spectrum of a uniformly ^{15}N labeled protein in about 10 min with no additional phase cycling. At 20 kHz MAS, proton line widths as narrow as 0.2 ppm were reported.

6.2.2 REDOR-Based Techniques

A second family of inverse-detection schemes was devised using a quantifiable dipolar transfer step at the very fast spinning speeds necessary to achieve high proton resolution. Earlier work has shown that rotational-echo double-resonance (REDOR), originally designed to recouple dipolar interactions[46] and transfer polarization[35] between isolated pairs of heteronuclei, is surprisingly efficient even for ^1H-S systems.[48,49] REDOR-type CT among ^1H and S nuclei was made possible mainly because fast MAS on the order of 30 kHz effectively suppresses ^1H-^1H dipolar couplings. In this limit, the ^1H-S spin system can, to a very good approximation, be analyzed, using simple expressions derived using product operator formalism in conjunction with the simple REDOR average Hamiltonian and a complete neglect of ^1H homonuclear couplings.

Apart from the replacement of free scalar-coupling evolution periods by REDOR π pulses spaced by half the rotor period, this whole family of techniques closely resembles the structure of solution-state inverse-detected experiments. In liquid-state, CT via the scalar coupling is usually restricted to protons directly bound to the S nucleus, such that $H_n \rightarrow S$ and $S \rightarrow H_n$ transfer periods are considered equivalent and are optimized for a specific J value and the multiplet type (or a compromise when different kinds of multiplets are measured at the same time). However, more subtleties arise in the solid-state: although the evolution of S-spin transverse magnetization is influenced by the joint dipolar field of differently positioned ^{1}H nuclei that are close to the S nucleus, an evolving ^{1}H transverse coherence typically feels the dipolar field of only one S nucleus. The former situation is commonly referred to as separated local field (SLF), and the latter is termed proton-detected local field (PDLF) and has the advantage that the theoretical description embodies a simple summation over spin pairs.[50] For this reason, the details of the CT process in a solid-state REDOR experiment depend sensitively on the nucleus that represents the transverse part of the coherence evolving under REDOR recoupling. The modulation of the final signal intensities observed as a function of the recoupling time or t_1 rotor encoding (see below) can be very sensitive to the coupling constants as well as the local coupling topology, and the choice of transfer pathway should not be made solely on the basis of sensitivity considerations. In fact, when directly detected and inverse-detected experiments are combined, valuable information on the local coupling topology ("spin triangulation") becomes accessible.[47]

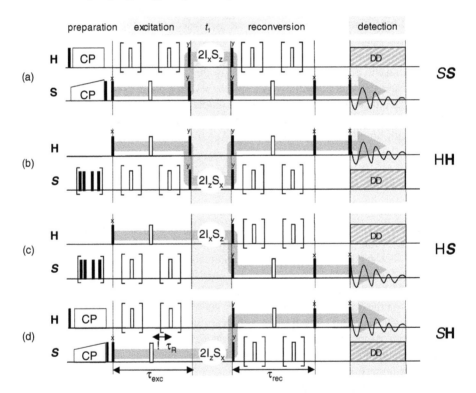

FIGURE 6.3 Variants of REDOR-based heteronuclear single-quantum shift correlation (HSQC) experiments that differ in their SH coherence transfer pathways. Panels (a) and (b) are "symmetric" with respect to the spin which is transverse during the two recoupling periods, and (c) and (d) are "asymmetric" and embody a net polarization transfer. Panel (a) represents the classic SLF configuration, (b) is a PDLF experiment, and (c) and (d) feature both types of spin configuration in separate recoupling periods. Panels (b) and (d) are inverse-detected experiments.

There are essentially four possible permutations of transfer pathways, which are shown in Figure 6.3 for the specific case of heteronuclear single-quantum shift correlation (HSQC) experiments, in which evolution of a heteronuclear antiphase coherence is probed in t_1. An HSQC experiment derived from the traditional REDOR scheme is shown in Figure 6.3a. The intensities of the cross signals in such an experiment can be analyzed in terms of the strongest couplings of S-spin to its surrounding protons,[49] and as the S-spin coherence is always transverse during this experiment, it has the lowest T_2^* losses during recoupling. When the channels are switched, the initial CP can be omitted, however, at the expense of larger losses by T_2^* of protons during recoupling (Figure 6.3b). This experiment is conceptually identical to one of the first inverse HSQC experiments in solution,[19] and its use in solid-state shift correlation has been demonstrated.[26] It has been proven to provide 5–10-fold sensitivity enhancements over a directly detected version (Figure 6.3c), which also does not require an initial CP. It was further demonstrated that one could omit t_1 and use the sequence only as a heteronuclear editing filter in front of a ^{1}H homonuclear DQ shift correlation experiment.

The experiment in Figure 6.3c is very similar to the transferred-echo, double-resonance (TEDOR experiment),[35] the only difference being that HSQC coherence is monitored in the middle of the transfer process as opposed to observing single ^{1}H-spin coherence before excitation and reconversion. It combines PDLF and SLF coupling topologies, and calculations show that this type of CT is only efficient for single S-H moieties at the shortest possible recoupling times (one or two rotor periods at 30 kHz MAS). Under these conditions, its total performance is still somewhat inferior to a well-optimized CP, but this disadvantage is compensated for by the possibility to determine the actual heteronuclear coupling with high precision.[48]

Finally, the experiment in Figure 6.3d is the only one that was successfully used to obtain ^{15}N-^{1}H correlation spectra and determine their distances in small molecules with ^{15}N natural abundance.[28,47] A more detailed scheme can be inspected in Figure 6.4. The initial CP is essential in that it creates ^{15}N polarization, which can be stored along the z-axis during the time needed for the removal of the overhead ^{1}H magnetization. There are two important differences: first, the chemical shift modulation occurs before the REDOR excitation period, and second, the coherence present during a second indirect dimension t_1 is heteronuclear dipolar order, which is rotor-encoded when t_1 is incremented in fractions of the rotor period. Moving the chemical shift dimension up front has the advantage that the t_1 increment can be chosen freely to suit the required spectral width. This is not possible in the HSQC variants, in which increments are restricted to integer rotor periods. This is because noninteger rotor period increments between REDOR excitation and reconversion lead to the appearance of a special kind of sideband spectra with spinning sidebands separated by $2\nu_R$. These increments depend sensitively on the heteronuclear dipolar coupling constant.[47] Their large frequency separation is not compatible with the simultaneous encoding of chemical shifts, which would require an extremely high number of slices in the indirect dimension when considering the concept in the case of HSQC. This problem is overcome by encoding heteronuclear dipolar order (which does not evolve) during t_1 simultaneously with the initial t_1, but with different increments. In this way, the sideband pattern is folded into the chemical shift range, and its apparent frequency spread is scaled by the ratio of $\Delta t_1/\Delta t_1$.

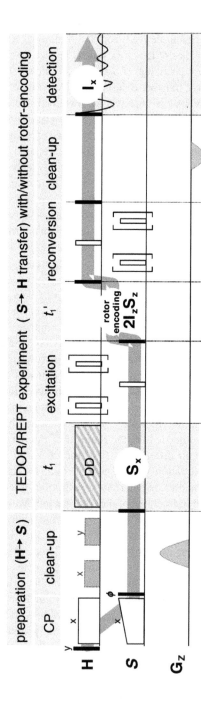

FIGURE 6.4 An inverse-detected SH correlation experiment to study samples without isotopic labeling. Unwanted magnetization is purged either by a gradient or by a pair of RRR pulses. There are two separate t_1 evolution periods, which can be incremented simultaneously (with different step sizes). The first leads to S chemical shift encoding, and the second splits the signal into a spinning sideband manifold that depends on the S-H dipolar coupling.

Results obtained using this experiment are shown in Figure 6.5. Note that heteronuclear dipolar order rotor encoding (HDOR) can also be performed without the additional chemical shift dimension (Figure 6.5d). Efficient processing strategies help to further minimize the acquisition time and yield precise coupling information within reasonable experimental times.[48] Because the combined shift–side band spectrum of Figure 6.5b takes about a day of acquisition time on a 700-MHz spectrometer, this technique, when applied at natural abundance, is clearly limited to the study of small molecules. Experiments on labeled proteins are certainly feasible and promise rich structural insights via the exact measurement of amide-NH distances, which provides information on hydrogen bonds.

Finally, we note that probably the most promising inverse experiment of Figure 6.3b has not been applied at high isotopic dilution. It does not require an initial CP and features an efficient symmetric REDOR transfer pathway, which is expected to more than compensate its increased losses resulting from proton T_2^*. The HSQC dimension is easily split into a separate HSQC t_1 and an HDOR t_1. The only major requirement would be the application of pulsed field gradients for coherence pathways selection, as ^{15}N is never present in a pure magnetization state, and a simple z-period for clean up cannot be implemented.

6.2.3 INVERSE DETECTION IN STATIC SAMPLES

The foregoing sections have presented widely useful experiments that yield high-resolution shift and dipolar coupling spectra under fast MAS. In this section, we review the inverse-detection techniques that are specifically designed for studies under static or slow spinning conditions. In one of the techniques, an off-resonance spin-lock[34] (i.e., using LG[51,52] or flip-flop Lee-Goldburg [FFLG][53,54] pulse sequence) is used to transfer the amide-^{15}N transverse magnetization to its dipolar coupled ^{1}H spin. This polarization transfer step via SEMA (spin exchange at magic-angle) avoids the relay of polarization transfer as well as suppresses spin diffusion via ^{1}H-^{1}H dipolar couplings.[34,54] Employing this step in an inverse-detection experiment (Figure 6.1), the selective acquisition of amide ^{1}H chemical shifts under multipulse decoupling (or CRAMPS-type detection[1]) has been demonstrated.[27] A spin-lock pulse in the S-spin channel after the t_1 period is used to suppress the ^{1}H channel before the second polarization transfer step. This sequence can also be used to obtain S nuclei chemical shift and heteronuclear dipolar coupling in the t_1 dimension. Because this method involves CRAMPS-type acquisition of the ^{1}H signal,[1] it is applicable to solids under slow spinning condition and to studying static aligned samples such as mechanically/magnetically aligned bilayers and liquid crystalline materials.[7]

A recent study has demonstrated that the PISEMA sequence can be modified using the inverse-detection procedure.[7] Because transverse magnetization is exchanged between I and S nuclei during the t_1 period (or SEMA sequence) of the PISEMA sequence, a stroboscopic observation of the ^{1}H signal can be used to observe the heteronuclear dipolar coupling. This method has been successfully demonstrated on site-specifically ^{15}N-labeled single-crystalline and polycrystalline peptide samples. The main disadvantage of this sequence is the requirement of the sampling window during a multiple-pulse sequence such as FFLG. Because FFLG

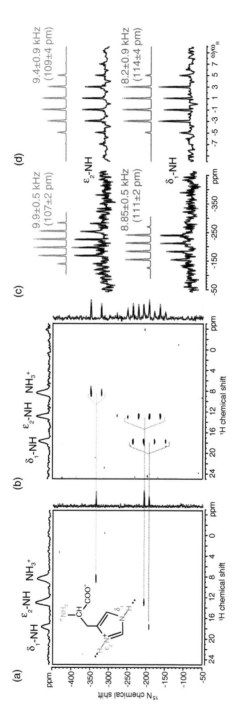

FIGURE 6.5 Inverse-detected ^{1}H-^{15}N correlation spectra of L-histidine·HCl·H$_2$O at 700 MHz. (a) Shift correlation with incremented t_1', (b) additional rotor encoding by also incrementing t_1', (c) sideband spectra extracted from (b), and (d) sideband spectra obtained by pure rotor ending in t_1'. The extracted distance compare favorably with results from other methods. Reproduced from reference 28, with permission from the *Journal of the American Chemical Society.*

is a windowless sequence, the insertion of rf-free delays reduces the efficiency of the sequence. In addition, the bulk magnetization from residual solvent (mainly water) or from protons that are not involved in the polarization transfer process results in a zero-frequency peak in the IS dipolar coupling spectrum. To avoid this difficulty, it is possible to use a two-dimensional version of this technique that will suppress the nonparticipating ^1H magnetization as well as use the spin-lock detection in the acquisition dimension. Such sequences will be useful in structural studies of membrane proteins.

A proton inverse-detected deuteron (PRIDE) NMR experiment is demonstrated for the measurement of ^2H wide line spectra from a small amount of sample (on the milligram scale) within 2 h.[25] This technique is expected to be useful in studying molecular motions in complex organic solids. Evolution under heteronuclear dipolar coupling although suppressing the homonuclear dipolar couplings, is used to create and reconvert HMQC coherences, which are modulated by the ^2H quadrupolar coupling in an indirect dimension. Proton magnetization is detected using PSLD, which is essential for the observed enhancements on the order of 10–20. An important feature is that the shape of the ^2H spectra is free of artifacts and independent of the HMQC excitation/reconversion times. This would be expected for a powder-average dependent transfer via single ^1H-^2H dipolar couplings but is not observed because the ^2H obtains its polarization from many surrounding protons. No specific clean-up stage was introduced, yet good suppression of large amounts of mobile signal components was also demonstrated. Its suitability to do ^2H spectroscopy in natural abundance will need to be examined.

Two-dimensional correlation of ^1H-^{15}N dipolar coupling with ^{15}N chemical shift using inverse detection was demonstrated for a static sample.[23] It employs a simple double CP, between which a conventional SLF-type ^{15}N-^1H dipolar coupling dimension featuring a partially multipulse-decoupled Hahn echo is combined with a second isotropic ^{15}N chemical shift dimension. PSLD is the preferred choice for ^1H detection, resulting in a three-dimensional measurement protocol, in which, of course, only two dimensions are of interest. Proton inverse-detected nitrogen static (PRINS) NMR has been demonstrated on a labeled peptide and a membrane channel domain protein, with about twofold sensitivity enhancements. Successful applications in the field of aligned biological samples are anticipated.

Another recent study[24] has revived the early experiments of Grannell et al.,[43] in which the whole process of cross-polarization to an S nucleus, its chemical shift evolution, and the back transfer to protons is combined within a single extended proton spin lock period, during which an S spin lock is interleaved with multiple t_1 evolution periods. A phase cycling on the S spin lock pulses is apparently sufficient to remove large overheads of uncoupled proton magnetization, such that it provides static ^2H wide line spectra in natural abundance (0.015%). This is remarkable considering that the broad ^1H signal was detected directly, without taking advantage of the large potential gain of PSLD used in other cases when ^1H chemical shift information is of minor importance. Line shape distortions are apparent in the resultant spectra, but it is proposed that use of modern hardware would eliminate such artifacts. The main drawback of this sequence is the use of long spin-lock periods, which restricts the applications to samples with a long $T_{1\rho}$.

6.2.4 WHICH EXPERIMENT TO USE?

The above sections provided an overview of published approaches to inverse detection in different samples, in which different types of polarization transfer steps have been employed. Apart from the examples concerning static samples, all other techniques benefit from fast MAS conditions in excess of 20 kHz, for which special equipment is mandatory. The choice between CP- and REDOR-based techniques is simple: Given a reasonable spectrometer stability, CP approaches appear to be the more efficient ones and are the methods of choice when qualitative distance constraints are sought. LG-CP certainly represents a very promising alternative for a quantifiable transfer,[34,43,56] but a slight modification to meet the Hartmann–Hahn condition under MAS will be useful for studies under fast spinning frequencies (Figure 6.6).[7,57] Because FFLG is efficient in suppressing ^{1}H-^{1}H dipolar couplings, SEMA-type sequence is preferred to selectively transfer coherence from an S nucleus to its dipolar coupled ^{1}H spins.[34] The TANSEMA (time-averaged mutation SEMA) sequence (Figure 6.6) has been shown to dramatically reduce the rf power needed in the polarization transfer process.[55] Although CRAMPS-type homonuclear dipolar decoupling sequences under slower spinning conditions might also provide sufficient proton line narrowing to make inverse detection feasible, it seems that the simplicity of fast MAS will favor this approach. This method will particularly be useful to study wet biological solids that cannot be spun faster. However, the use of small rotors for fast MAS is often not a serious restriction, as most biological macromolecules are often not available in large amounts.

Pulsed field gradients have been used in some cases to remove mobile water signals and surplus uncoupled ^{1}H magnetization in samples with a low abundant S nucleus. However, both problems can be solved with conventional probes and cleanup based on rf irradiation (i.e., prolonged ^{1}H decoupling during a z-filter in a constant time protocol and RRR pulses, respectively). Gradients might, however, prove indispensable when recoupling pulse sequences, such as the symmetric REDOR-HSQC, do not allow for the introduction of S nucleus z-filter delays, during which cleanup rf schemes can be applied.

The REDOR-based schemes are at present the best choice when precise dipolar couplings between the heteronuclei are to be determined. Also, in this case, we stress the fact that such studies do not depend on a scaling factor, as REDOR is rather robust and forgiving of experimental errors resulting from nonideality of π pulses. This holds in particular when the rotor-encoding spinning sidebands, instead of spectral intensities, are used for the determinations. In the following section, we present a model study in which not only precise distances but also a full coupling topology was elucidated by these methods.

6.3 APPLICATIONS

In general, most of the techniques discussed herein have just recently been developed, and not many actual applications have emerged until now. The sensitivity gain is the key factor that will enable many studies to be completed within a fraction of the time needed to do the experiments in the conventional way. We here append a

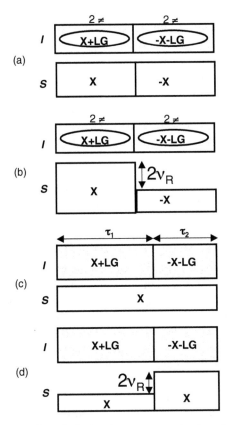

FIGURE 6.6 Off-resonance ^{1}H spin-lock pulse sequences that have been used for coherence transfer among heteronuclei via dipolar coupling in inverse-detected experiments. (a) Spin exchange at magic-angle (SEMA) for static condition,[28] (b) SEMA under MAS with a spinning speed v_R,[3,51] (c) time-averaged nutation spin exchange at magic-angle (TANSEMA) for static conditions, and (d) TANSEMA under MAS. In (b), the S-spin-lock power ($B_{eff,S}$) is set to equal to $B_{eff,I} \pm v_R$. In TANSEMA under static condition, $B_{eff,S} = (\tau_1 \; \tau_2)B_{eff,I}/(\tau_1 + \tau_2)$, whereas under MAS, $B_{eff,S} = (\tau_1 \; \tau_2)B_{eff,I}/(\tau_1 + \tau_2) \pm v_R$. Half the duration of the sequences in (a) and (b) can also be used for coherence transfer.[3,28,50] If needed, supercycles and ramp spin-lock pulses can be used to overcome effects caused by offset and mismatch in rf power.[48]

few further examples that highlight the great potential of the new methods, which we expect to have a significant effect on the routine toolbox of the solid-state NMR spectroscopists.

6.3.1 Applications to Study Small Molecules and Materials

In Section 6.2.2, it was indicated that the REDOR-based approaches not only hold promise for remarkable sensitivity enhancements by inverse detection but also serve to elucidate local coupling geometries by combining inverse- and directly detected experiments and making use of the different influence of SLF- and PDLF-type coupling topologies on the measured signals. We highlight this aspect here by reviewing a study concerned with the structure of N-H...N hydrogen bonds in the enol form of *N*-butylaminocarbonyl-6-tridecyl-isocytosine. A dimer structure (Figure 6.7a) is stabilized by four hydrogen bonds, and we focus on the two equivalent central ones between a urea-N and a pyrimidine-N. Figure 6.7b shows a directly detected sideband spectrum using the pulse scheme of Figure 6.3a, in which HDOR rather than HSQC was selected and rotor-encoded during t_1 and was then Fourier-transformed. Each urea-N has a relatively tightly bound ^1H with secondary couplings being comparably weak; thus, the bond length is easily determined from the sideband intensities using a simple spin-pair solution for the fit. When analogous sideband spectra are taken with the inverse-detected version of the same experiment (Figures 6.7c and 6.7d), the local coupling topology around the evolving and detected ^1H coherence is strongly influenced by both nitrogens in the hydrogen bond. The sideband modulation then depends not only on the two distances but also on the angle between the two coupling tensors. This unique correlation between distances and angles can be further amplified by using different REDOR recoupling times for excitation and reconversion. This leads to some intensity loss (as this is not the condition for optimum transfer), but still the obtained sideband spectra are considerably less noisy than the one in Figure 6.7b as a result of inverse detection.

The missing longer distance and the hydrogen bond angle can by deduced by comparison with simulated sideband spectra plotted in Figure 6.8. The accuracy of the determination can be increased by analyzing further sideband spectra taken under different recoupling conditions and fitting them to the same set of angle and distance. Obviously, this approach bears a large potential for applications in the biological and material sciences, in which hydrogen bonds are probably the most important supramolecular structure–directing interaction. NMR thus provides access to structure–function relationships in materials that do not need to be crystalline.

6.3.2 Biomolecular Applications

Most of the presently used solid-state NMR experiments require a large quantity of sample or a long acquisition time to obtain a reasonable *S/N* multidimensional spectrum from biological solids. Because most of the interesting biological molecules, such as membrane-associated proteins and RNA, are not available in large quantities, applications of existing NMR techniques to such systems are highly restricted. However, in many cases, even when obtaining large quantities of the system of interest is feasible, the use of a large quantity of such a system does not provide biologically relevant information. For example, in the structural studies of membrane-permeating peptides (such as antimicrobial peptides, toxins, fusion peptides, and channel forming peptides), there is considerable interest in obtaining data from biologically relevant concentrations of the peptide in lipid bilayers, as the increase in the peptide concentration can lead to the disruption of the lipid bilayer structure.[58,59] Ligand-binding studies are another example in which the concentration of the ligand cannot be increased to acquire signal from the bound ligand. In addition, the mandatory sample size restriction is a heavy price to pay for static as well as MAS experiments with a higher magnetic field. Long signal acquisitions used to

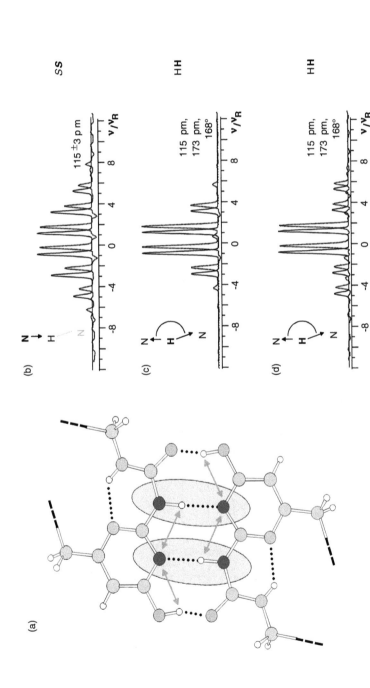

FIGURE 6.7 (a) Hydrogen bonds in *N*-butylaminocarbonyl-6-tridecyl-isocytosine. (b–d) Sideband patterns obtained from HDOR experiments measured in a ^{15}N-enriched sample at 30 kHz MAS following the experimental schemes in Figure 6.3. The REDOR excitation and reconversion times are (b) τ_{exc} = τ_{rec} = 8τ_R, (c) τ_{exc} = 3τ_R and τ_{rec} = 6τ_R, (d) τ_{exc} = 4τ_R and τ_{rec} = 8τ_R. Reproduced from reference 47, with permission from *Solid State Nuclear Magnetic Resonance*.

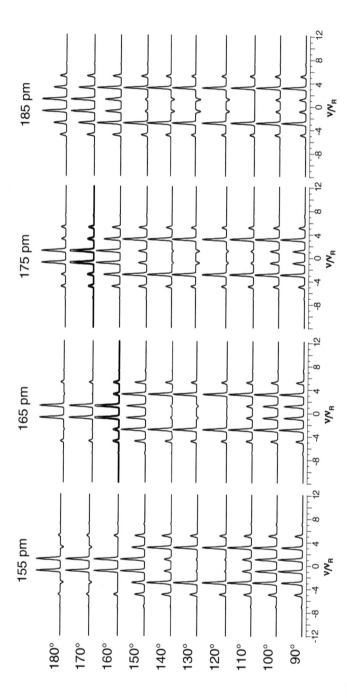

FIGURE 6.8 Sample sideband patterns calculated for different coupling topologies of an N-H-N hydrogen bond, based on the experiment in Figure 6.3b with $\tau_{exc} = 4\tau_R$ and $\tau_{rec} = 8\tau_R$. The short NH distance was taken to be 115 ppm. Reproduced from reference 47, with permission from *Solid State Nuclear Magnetic Resonance*.

avoid these concerns suffer from spectrometer instability and sample denaturing resulting from rf heat dissipation. Therefore, the remarkable sensitivity gained through an inverse-detected experiment could be the long-awaited "magic wand" for NMR applications on biological solids.

Inverse-detected NMR experiments for studies on biosolids fall under two categories: aligned or unaligned crystalline samples. Experiments on aligned samples are usually performed under static condition, whereas unaligned samples are studied under MAS. Experiments need to be chosen on the basis of the nature of the sample under study. The advantages of using fast sample spinning and REDOR-based coherence transfer cannot be used for experiments on aligned samples or on samples that cannot be spun faster. However, the rate of CT among heteronuclear spins and the extent of line broadening caused by homo- and heteronuclear dipolar couplings depend on the rigidity of the sample. For example, because of molecular motions in wet samples, dipolar couplings are partially averaged out, and therefore the contact time for optimum cross polarization needs to be carefully chosen. In addition, the use of constant time period is not possible if T_2 is not favorable.

Among the methods proposed for inverse detection under MAS, the most recently demonstrated ^{1}H-detected 2D ^{1}H/^{15}N HSQC MAS experiment[31] is straightforward to implement in any solid-state NMR spectrometer. In this experiment, a two-step phase cycle is used to suppress water. This experiment is expected to be applicable to any crystalline proteins. It provides a sevenfold increase in sensitivity. 2D spectra that correlate the chemical shifts of amide-^{1}H and amide-^{15}N in ubiquitin nanocrystals are given in Figure 6.9. Experiments on a perdeuterated protein under fast MAS with ^{15}N decoupling using WALTZ-16 provided a ^{1}H line width of about 0.2 ppm. This experimental plan can be used to further design a family of multidimensional MAS experiments to determine the three-dimensional structure of crystalline proteins.

Inverse-detection experiments designed based on the SEMA concept[7,27] are well-suited to studies on aligned samples. In these experiments, in the absence of MAS, homonuclear dipolar couplings are suppressed using multiple pulse sequences. For this purpose, FFLG sequence is chosen, as it can be used to accomplish coherence transfer among dipolar coupled heteronuclear spins selectively. This technique was used to obtain an amide-^{1}H spectrum of a single-site ^{15}N-labeled magainin2 peptide.[27] Therefore, this experiment will also be useful in determining chemical shift tensors of amide-^{15}N, [13]C$_\alpha$, and amide-^{1}H that can be used in the dynamic studies.

A recent study demonstrated the application of the PRIDE technique to determine the slant angle of helices in membrane proteins.[29] This is so far mainly done using the (directly detected) PISEMA experiment,[7,15] which is used to measure [1]H-^{15}N dipolar frequencies in oriented samples. Because the coupling constant is known, one can directly convert the dipolar frequency into the tilt angle of the NH axis with respect to the membrane normal and derive the helix orientation from this. By exchanging the amide protons by deuterium (which is done by merely adding heavy water), one can take advantage of the large quadrupolar interaction, which has exactly the same symmetry axis as the heteronuclear dipolar tensor. The spectra in Figures 6.10a and 6.10b show that directly detected ^{2}H spectroscopy is not at all suited to measuring the quadrupolar splitting associated with the amide groups of the peptide,

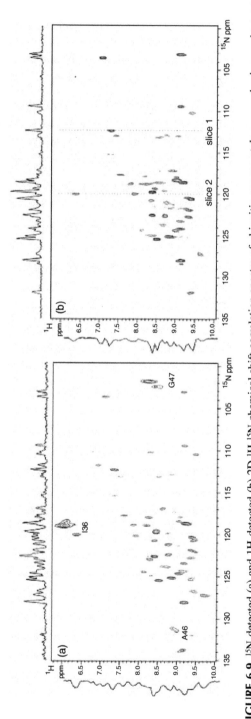

FIGURE 6.9 ^{15}N-detected (a) and 1H-detected (b) 2D [1]H-^{15}N chemical shift correlation spectra of ubiquitin nanocrystals prepared using perdeuterated 2-methyl-2,4-pentanediol (MPD) under 25 kHz MAS. The directly detected spectrum in (a) was acquired with eight times as many scans. Reproduced from reference 31, with permission from the *Journal of the American Chemical Society.*

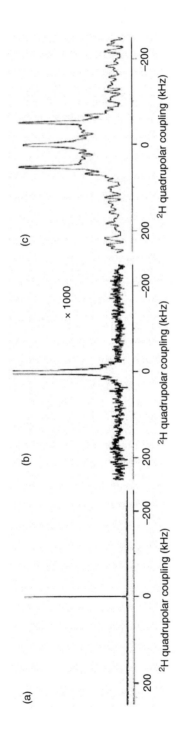

FIGURE 6.10 Static ^2H spectra of 2.5 mg back-exchanged ovispirin, constituted in membrane lipids (mole ratio 1:27). (a) Directly detected spectrum, (b) the same, magnified 1000 times vertically, and (c) proton inverse-detected spectrum, experimental time: 6 h. Reproduced from reference 29, with permission from the *Journal of Magnetic Resonance*.

as the signal is dominated by residual D_2O. PRIDE efficiently removes this signal and separates out the amide-^2H signals. The spectrum is consistent with the in-plane orientation of the helix reported earlier. Notably, the central peak is not an artifact but is associated with N-D axes that are oriented close to the magic-angle with respect to the membrane normal. This orientation represents a "blind spot" for PISEMA, in which the polarization is transferred via the N-H coupling and is therefore zero. PRIDE transfers magnetization of surrounding protons to the deuteron in question, therefore circumventing this type of problem.

REFERENCES

1. Schmidt-Rohr, K. and Spiess, H.W., *Multidimensional Solid-State NMR and Polymers*, Academic Press, New York, 1994.
2. Castellani, F., van Rossum, B., Diehl, A., Schubert, M., Rehbein, K. and Oschkinat, H., Structure of a protein determined by sold-state magic-angle-spinning NMR spectroscopy, *Nature,* 420, 98, 2002.
3. Baldus, M., Correlation experiments for assignment and structure elucidation of immobilized polypeptides under magic-angle spinning, *Prog. NMR Spectrosc.*, 41, 1, 2002.
4. Saito, H., Tuzi, S., Tanio, M. and Naito, K., Dynamic aspect of membrane proteins and membrane associated peptides as revealed by ^{13}C NMR: Lessons from bacteriorhodopsin as an intact protein, *Ann. Rep. NMR Spectrosc.*, 47, 41, 2002.
5. Saito, H., Yamaguchi, S., Ogawa, K., Tuzi, S., Marquez, M. and Sanz, C., Glutamic acid residues of bacteriorhodopsin at the extracellular surface as determinants for conformation and dynamics as revealed by site-directed solid-state C-13 NMR, *Biophys. J.*, 86, 1673, 2004.
6. Fujiwara, T., Todokoro, Y., Yanagishita, H., Tawarayama, M., Kohno, T., Wakamatsu, K. and Akutsu, H., Signal assignments and chemical-shift structural analysis of uniformly C-13, N-15-labeled peptide, mastoparan-X, by multidimensional solid-state NMR under magic-angle spinning, *J. Biomol. NMR*, 28, 311, 2004.
7. Ramamoorthy, A., Wei, Y. and Lee, D.K., PISEMA solid-state NMR spectroscopy, *Ann. Rep. NMR Spectrosc.*, 52, 1, 2004.
8. Marassi, F.M. and Opella, S.J., Structure determination of membrane proteins by NMR spectroscopy, *Chem. Rev.*, 104, 3587, 2004.
9. Wu, C.H., Ramamoorthy, A. and Opella, S.J.J., High resolution heteronuclear dipolar solid-state NMR spectroscopy, *Magn. Reson.*, A109, 270, 1994.
10. Ramamoorthy, A. and Opella, S.J., Two-dimensional chemical shift/heteronuclear dipolar coupling spectra obtained with polarization inversion spin exchange at the magic-angle and magic-angle sample spinning (PISEMAMAS), *Solid State NMR Spectrosc.*, 4, 387, 1995.
11. Ramamoorthy, A., Marassi, F. and Opella, S.J., Applications of multidimensional solid-state NMR spectroscopy to membrane proteins, in *Proceedings of the International School of Biological Magnetic Resonance, 2nd course, Dynamics and the Problem of Recognition in Biological Macromolecules,* Jardetsky O. and Lefeure, J., Eds., Plenum, New York, 1996, p. 238.
12. Jelinek, R., Ramamoorthy, A. and Opella, S.J., High-resolution three-dimensional solid-state NMR spectroscopy of a uniformly ^{15}N-labeled protein, *J. Am. Chem. Soc.*, 117, 12348, 1995.

13. Marassi, F.M., Ramamoorthy, A. and Opella, S.J., Solid-state NMR spectroscopy of uniformly ^{15}N-labeled protein oriented in lipid bilayers, *Proc. Natl. Acad. Sci. USA*, 94, 8551, 1997.
14. Wang, J., Denny, J., Tian, C., Kim, S., Mo, Y., Kovacs, F., Song, Z., Nishimura, K., Gan, Z., Fu, R., Quine, J.R. and Cross, T.A., Imaging membrane protein helical wheels, *J. Magn. Reson.*, 144, 162, 2000.
15. Marassi, F.M. and Opella, S.J., Simultaneous assignment and structure determination of a membrane protein from NMR orientational restraints, *Protein Sci.*, 12, 403, 2003.
16. Thiriot, D.S., Nevzorov, A.A., Zagyanskiy, L., Wu, C. and Opella, S.J., Structure of the coat protein in Pf1 bacteriophage determined by solid-state NMR spectroscopy, *J. Mol. Biol.*, 341, 869, 2004.
17. Maudsley, A.A., Müller, L. and Ernst, R.R., Cross-correlation of spin-decoupled NMR-spectra by heteronuclear 2-dimentional spectroscopy, *J. Magn. Reson.*, 28, 463, 1977.
18. Müller, L., Sensitivity enhanced detection of weak nuclei using heteronuclear multiple quantum coherence, *J. Am. Chem. Soc.*, 101, 4481, 1979.
19. Bodenhausen, G. and Ruben, D.J., Natural abundance N-15 NMR by enhanced heteronuclear spectroscopy, *Chem. Phys. Lett.*, 69, 185, 1980.
20. Bax, A., Multidimensional nuclear-magnetic-resonance methods for protein studies, *Curr. Opin. Struct. Biol.*, 4, 738, 1994.
21. Ishii, Y. and Tycko, R., Sensitivity enhancement in solid state-N-15 NMR by indirect detection with high-speed magic-angle spinning, *J. Magn. Reson.*, 142, 199, 2000.
22. Ishii, Y., Yesinowski, J.P. and Tycko, R., Sensitivity enhancement in solid-state C-13 NMR of synthetic polymers and biopolymers by H-1 NMR detection with high-speed magic-angle spinning, *J. Am. Chem. Soc.*, 123, 2921, 2001.
23. Hong, M. and Yamaguchi, S., Sensitivity-enhanced static N-15 NMR of solids by H-1 indirect detection, *J. Magn. Reson.*, 150, 43, 2001.
24. Khitrin, A.K. and Fung, B.M., Indirect NMR detection in solids with multiple cross-polarization periods, *J. Magn. Reson.*, 152, 185, 2001.
25. Schmidt-Rohr, K., Saalwächter, K., Liu, S.-F. and Hong, M., High-sensitivity H-2 NMR in solids by H-1 detection, *J. Am. Chem. Soc.*, 123, 7168, 2001.
26. Schnell, I., Langer, B., Sontjens, S.H.M., van Genderen, M.H.P., Sijbesma, R.P. and Spiess, H.W., Inverse detection and heteronuclear editing in H-1-N-15 correlation and H-1-H-1 double-quantum NMR spectroscopy in the solid state under fast MAS, *J. Magn. Reson.*, 150, 57, 2001.
27. Wei, Y., Lee, D.-K., Hallock, K.J. and Ramamoorthy, A., One-dimensional ^{1}H-detected solid-state NMR experiment to determine amide-^{1}H chemical shifts in peptides, *Chem. Phys. Lett.*, 351, 42, 2002.
28. Schnell, I. and Saalwächter, K., N-15-H-1 bond length determination in natural abundance by inverse detection in fast-MAS solid-state NMR spectroscopy, *J. Am. Chem. Soc.*, 124, 10938, 2002.
29. Yamaguchi, S. and Hong, M., Determination of membrane peptide orientation by H-1-detected H-2 NMR spectroscopy, *J. Magn. Reson.*, 155, 244, 2002.
30. Chevelkov, V., van Rossum, B.J., Castellani, F., Rehbein, K., Diel, A., Hohwy, M.H., Steuernagel, S., Engelke, F., Oschkinat, H. and Reif, B., H-1 detection in MAS solid-state NMR spectroscopy of biomacromolecules employing pulsed field gradients for residual solvent suppression, *J. Am. Chem. Soc.*, 125, 7788, 2003.
31. Paulson, E.K., Morcombe, C.R., Gaponenko, V., Dancheck, B., Byrd, R.A. and Zilm, K.W., Ramped-*J. Am. Chem. Soc.*, 125, 15831, 2003.

32. Metz, G., Wu, X. and Smith, S.O., Ramped-amplitude cross-polarization in magic-angle-spinning NMR, *J. Magn. Reson. A*, 110, 219, 1994.
33. Shekar, S.C., Lee, D.K. and Ramamoorthy, A., Chemical shift anisotropy and offset effects in cross polarization solid-state NMR spectroscopy, *J. Magn. Reson.*, 157, 223, 2002.
34. Ramamoorthy, A., Wu, C.H. and Opella, S.J., Experimental aspects of multidimensional solid-state NMR correlation spectroscopy, *J. Magn. Reson.*, 140, 131, 1999.
35. Hing, A.W., Vega, S. and Schaefer, J., Transferred-echo, double-resonance NMR, *J. Magn. Reson.*, 96, 205, 1992.
36. Ramamoorthy, A. and Chandrakumar, N., Comparison of the coherence-transfer efficiences of laboratory- and rotating-frame experiments, *J. Magn. Reson.*, 100, 60, 1992
37. Sattler, M., Schleucher, J. and Griesinger, C., Heteronuclear multidimensional NMR experiments for the structure determination of proteins in solution employing pulsed field gradients, *Progr. Nucl. Magn. Reson. Spectrosc.*, 34, 93, 1999.
38. Lesage, A., Steuernagel, S. and Emsley, L., Carbon-13 spectral editing in solid-state NMR using heteronuclear scalar couplings, *J. Am. Chem. Soc.*, 120, 7095, 1998.
39. Taylor, D.A. and Ramamoorthy, A., Analysis of dipolar-coupling mediated coherence transfer in a homonuclear two spin 1/2 solid-state system, *J. Magn. Reson.*, 141, 18, 1999.
40. Shekar S.C. and Ramamoorthy, A., The unitary evolution operator for cross-polarization schemes in NMR, *Chem. Phys. Lett.*, 342, 127, 2001.
41. Taylor, D.A. and Ramamoorthy, A., Coherence transfer through homonuclear dipolar coupling in an unoriented two spin-1/2 solid-state system, *J. Mol. Struct.*, 602, 115, 2001.
42. Hediger, S., Meier, B.H. and Ernst, R.R., Adiabatic passage Hartmann–Hahn cross-polarization in MNMR under magic-angle sample-spinning, *Chem. Phys. Lett.*, 240, 449, 1995.
43. Grannell, P.K., Mansfield, P. and Whitaker, M.A.B., C-13 double-resonance Fourier-transform spectroscopy in solids, *Phys. Rev. B*, 8, 4149, 1973.
44. Vega, S., Shattuck, T.W. and Pines, A. Double-quantum cross-polarization NMR in solids, *Phys. Rev. A*, 22, 638, 1980.
45. Franke, D., Hudalla, C. and Eckert, H., Heteronuclear X-Y double quantum MAS NMR in crystalline inorganic solids. Applications for indirect detection and spectral editing of rare-spin resonances, *Solid State Nucl. Magn. Reson.*, 1, 33, 1992.
46. Gullion, T. and Schaefer, J., Rotational-echo double-resonance NMR, *J. Magn. Reson.*, 81, 196, 1989.
47. Saalwächter, K. and Schnell, I., REDOR-based heteronuclear dipolar correlation experiments in multi-spin systems: rotor-encoding, directing, and multiple distance and angle determination, *Solid State Nucl. Magn. Reson.*, 22, 154, 2002.
48. Saalwächter, K., Graf, R. and Spiess, H.W., Recoupled polarization-transfer methods for solid-state H-1-C-13 heteronuclear correlation in the limit of fast MAS, *J. Magn. Reson.*, 148, 398, 2001.
49. Saalwächter, K. and Spiess, H.W., Heteronuclear H-1-C-13 multiple-spin correlation in solid-state nuclear magnetic resonance: combining rotational-echo double-resonance recoupling and multiple-quantum spectroscopy, *J. Chem. Phys.*, 114, 5707, 2001.
50. Schmidt-Rohr, K., Nanz, D., Emsley, L. and Pines, A., NMR measurement of resolved heteronuclear dipole couplings in liquid-crystals and lipids, *J. Phys. Chem.*, 98, 6668, 1994.

51. Lee, M. and Goldburg, W.J., Nuclear magnetic resonance line narrowing by a rotating rf field, *Phys. Rev. A*, 140, 1261, 1965.

52. Ravikumar, M. and Ramamoorthy, A., Exact evaluation of the line-narrowing effect of the Lee–Goldburg pulse sequence in solid-state NMR spectroscopy, *Chem. Phys. Lett.*, 286, 199, 1998.

53. Mehring, M. and Waugh, J.S., Magic-angle NMR experiements in solids, *Phys. Rev. B*, 5, 3459, 1972.

54. Bielecki, A., Kolbert, A.C., de Groot, H.S.M., Griffin, R.G. and Levitt, M.H., Frequency-switched Lee–Goldburg sequences in solids, *Adv. Magn. Reson.*, 14, 111, 1990.

55. Lee, D.K., Narasimhaswamy, T. and Ramamoorthy, A., *Chem. Phys. Lett.*, 399, 359, 2004

56. van Rossum, B.-J., de Groot, C.P., Ladizhanski, V., Vega, S. and de Groot, H.J.M., A method for measuring heteronuclear (H-1-C-13) distances in high speed MAS NMR, *J. Am. Chem. Soc.*, 122, 3465, 2000.

57. Dvinskikh, S.V., Zimmermann, H., Maliniak, A. and Sandstrom, D., Heteronuclear dipolar recoupling in liquid crystals and solids by PIDEMA-type pulse sequences, *J. Magn. Reson.*, 164, 165, 2003.

58. Lee, D.K., Henzler-Wildman, K.A. and Ramamoorthy, A., Solid-state NMR spectroscopy of aligned lipid bilayers at low temperatures, *J. Am. Chem. Soc.*, 126, 2318, 2004.

59. Henzler-Wildman, K.A., Perturbation of the hydrophobic core of lipid bilayers by the human antimicrobial peptide LL-37, *Biochemistry*, 43, 8459, 2004.

7 Uniaxial Motional Averaging of the Chemical Shift Anisotropy of Membrane Proteins in Bilayer Environments

*Alexander A. Nevzorov, Anna A. De Angelis,
Sang Ho Park, and Stanley J. Opella*

CONTENTS

ABSTRACT Magnetically aligned bicelles are an attractive system for structure studies of membrane proteins. They provide a planar liquid crystalline bilayer environment that both immobilizes and aligns the proteins. Bicelles align with the bilayer normal perpendicular to the direction of the magnetic field; therefore, the proteins must undergo rapid rotational diffusion about the normal. We describe a quantitative dynamic model based on the stochastic Liouville equation including uniaxial rotational diffusion. The results show that bicelles modeled as disks undergo rotational diffusion fast enough to average the cylindrical distribution about the bicelle axis. Experimental spectra of a 15N-labeled protein in bicelles are consistent with these calculations. Experimental spectra of mechanically aligned bilayers indicate that the protein itself undergoes rotational diffusion fast enough for the requisite averaging.

KEY WORDS: bicelles, membrane proteins, motional narrowing, uniaxial diffusion, uniaxial distribution, stochastic Liouville equation

7.1 INTRODUCTION

Membrane proteins are important targets for structure determination by nuclear magnetic resonance (NMR) spectroscopy. Phospholipid bilayers are particularly attractive samples for structure determination because they provide a planar liquid-crystalline environment that closely mimics that found in biological membranes under physiological conditions. Moreover, protein-containing bilayer samples can be prepared so that they fulfill the two principal requirements for structure determination by solid-state NMR of aligned samples: the proteins are immobilized and uniaxially aligned by their interactions with the lipids. Phospholipid bilayers can be mechanically aligned with their bilayer normals either parallel or perpendicular to the direction of the applied magnetic field. Bilayer disks (bicelles) also provide a planar bilayer environment for membrane proteins with the advantage that they are aligned by the magnetic field and do not require the use of glass plates.[1] In the presence of lanthanide ions, bicelles are "flipped," such that the bilayer normal is parallel to the applied magnetic field.[2] However, "unflipped" bicelles, which have their bilayer normals perpendicular to the field, are of greater interest because the spectroscopic and chemical effects of the lanthanides can be avoided. In this case, rotational diffusion about the bilayer normal is essential for the motional averaging of the chemical shift anisotropy to obtain high-resolution, single-line ^{15}N NMR spectra of proteins.[3]

Liquid crystalline bicelles have been described as disks,[4-7] perforated lamellae ("swiss cheese"), and worm-like micelles,[8] all of which can be effectively modeled[9] as a "virtual disk" in which the proteins undergo fast rotational diffusion about the bilayer normal, which is aligned perpendicular to the applied magnetic field, as well as rapid "wobble" motion of that axis characterized by an order parameter. In this chapter, we show that under physiological conditions, it is possible to obtain single-line resonances from labeled sites in membrane proteins in bicelles aligned with their normals perpendicular to the magnetic field. This, in turn, provides a starting point for obtaining high-resolution NMR spectra of membrane proteins labeled in multiple sites for structural studies.[3]

7.2 ROTATIONAL DIFFUSION OF MEMBRANE PROTEINS IN BICELLES

Protein-containing bicelles are prepared by solubilizing the purified protein in short-chain lipids (e.g., dihexanoylphosphatidylcholine (DHPC)) and then mixing with long-chain lipids (e.g., dimyristoylphosphatidylcholine (DMPC)). The resulting bicelles are characterized by q, the ratio of the long-chain lipid to the short-chain lipid. Within a relatively broad range of lipid ratios ($q > 2.5$) and concentrations, the bilayers align with their normals perpendicular to the magnetic field, as illustrated in Figure 7.1.[1,5] The small-amplitude wobbling of the bicelle axis that slightly reduces

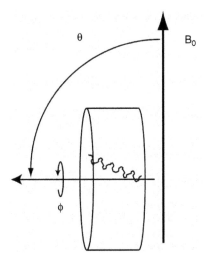

FIGURE 7.1 Diagram showing the angles described in the text for a trans-membrane α-helix in a large bicelle. When bicelles are placed in a high magnetic field, they orient so that their axes are, on average, perpendicular to the direction of the applied magnetic field. At the same time, both the protein molecules and the bicelles undergo rotational diffusion about their axes. There is also a wobbling motion fast enough to be characterized by an order parameter.

the span of the chemical shift frequencies is characterized by an order parameter, S_{bicelle}. Notably, the disks or the proteins undergo rotational diffusion about the bicelle axis. If the rate of diffusion of the bicelles is fast enough, regardless of whether it arises from motions of the protein molecules within the bilayer environment or the bicelles themselves, then the cylindrical disorder in a static uniaxial distribution of protein molecules is averaged out and it is possible to obtain high-resolution NMR spectra of membrane proteins in bilayer environments, even if the proteins themselves are otherwise immobilized by the lipids. In this context, an outstanding question has been whether protein-containing bicelles large enough to be aligned by the magnetic field undergo rotational diffusion fast enough to effect complete averaging of the chemical shift anisotropy.

An estimate of the rotational diffusion rates of bicelles in water can be obtained by applying the well-known Stokes–Einstein relation. Approximating the bicelle as an oblate ellipsoid of revolution, we use Equation (7.1)[10,11]:

$$D_{\parallel} = \frac{3\rho(\rho - \sigma)}{2(\rho^2 - 1)} \frac{kT}{8\pi\eta ab^2} \tag{7.1}$$

where $\rho = a/b$, the ratio of the ellipsoid semiaxes, and

$$\sigma = (1 - \rho^2)^{-1/2} \arctan[(1 - \rho^2)^{1/2}/\rho] \tag{7.2}$$

For a typical ratio of long-chain to short-chain lipids, $q = 3$, we substitute: $a = 20$ Å, $b = 100$ Å (6), $\eta = 1$ cPoise for pure water and obtain

$$D_{\parallel} \approx 3 \times 10^5 \, s^{-1} \tag{7.3}$$

Because the frequency span for a ^{15}N amide chemical shift tensor in a high-field magnet is around 10^4 Hz, this rate of rotational diffusion would not be fast enough to achieve the motional narrowing from the solution NMR point of view, even if it occurred isotropically. To evaluate the situation for solid-state NMR spectroscopy, instead of using the classical Redfield theory,[12] we invoke the Stochastic Liouville Equation (SLE),[13] which couples the quantum spin transitions to diffusional reorientations. For calculating one-dimensional solid-state ^{15}N NMR spectra, we assume complete ^{1}H-^{15}N heteronuclear decoupling from ^{1}H irradiation. We further assume that wobbling about the bicelle axis is fast enough that the final results can be further scaled by an order parameter. The value for the order parameter can be obtained from the phase diagram for bicelles.[1] The order parameter is $S_{bicelle} = 0.8$ for DMPC/DHPC bicelles with $q = 3$.

7.3 SLE FORMULATION FOR A TWO-SPIN SYSTEM INCORPORATED INTO A BICELLE

The SLE for the density matrix $\rho(\phi, t)$ of a spin system incorporated in a bicelle undergoing rotational diffusion about its axis can be written as

$$\frac{\partial \rho(\phi, t)}{\partial t} = -i\left[-\frac{1}{2}H_0 + H(\phi), \rho(\phi, t)\right] + \Gamma \rho(\phi, t) \tag{7.4}$$

Here H_0 is the static spin Hamiltonian in which the chemical shifts are scaled by a factor of $-1/2$ relative to the isotropic values because of the perpendicular orientation of the bicelle relative to the magnetic field, $H(\phi)$ is the part of the Hamiltonian that is explicitly dependent on the uniaxial rotation ϕ of the bicelle. The operator Γ is the rotational diffusional operator given by

$$\Gamma = D_{\parallel} \frac{\partial^2}{\partial \phi^2}$$

If fast enough averaging of the operator $H(\phi)$ takes place, then the system effectively evolves under the static Hamiltonian, $-1/2H_0$.

The NMR signal averaged over all instances of ϕ in time is obtained by taking the trace of the product of the detection operator, S_-, and the density matrix, ρ, followed by integrating from 0 to 2π over the ϕ–space

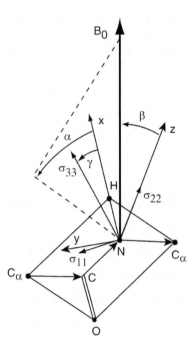

FIGURE 7.2 The orientation of the principal axis system for the ^{15}N amide chemical shift tensor in the molecular frame of the peptide plane. The orientation of the magnetic field $\mathbf{B_0}$ relative to the molecular frame is given by the angles α and β. The angle between the σ_{33} axis of the ^{15}N chemical shift tensor and the N-H bond axis is given by $\gamma = 17$–$19°$.

$$G(t) = Tr\left[S_-\rho(t)\right] = \int_0^{2\pi} Tr\left[S_-\rho(\phi,t)\right]d\phi \qquad (7.5)$$

Consider the chemical shift Hamiltonian for ^{15}N nuclei. The decomposition of the Hamiltonian as in Equation (7.4) can be carried out in the following manner. As shown in Figure 7.2, the molecular frame (MF) is chosen so that its x-axis is along the N-H bond, and its z-axis is perpendicular to the peptide plane.

In terms of the orientation (α, β) of the magnetic field relative to the peptide plane (as given by MF), the chemical shift anisotropy (CSA) can be rewritten in the spherical form as

$$H_{CSA} = S_z\left(\sigma_{iso} + \Delta\sigma\left[\frac{3\cos^2\beta - 1}{2} + \frac{\eta}{2}\sin^2\beta\cos 2(\alpha - \gamma)\right]\right) = (\sigma_{iso} + \sigma)S_z$$

$$(7.6)$$

where σ_{iso} is the isotropic chemical shift, $\Delta\sigma = 2\sigma_{22} - \sigma_{iso}$; $\eta = (\sigma_{33} - \sigma_{11})/(2\sigma_{22} - \sigma_{iso})$; and $\gamma = 17°$–$19°$ is the angle between the σ_{33} shielding axis and the N-H bond vector.

Next, the decomposition of the CSA into the static and motional parts is achieved by introducing a series of transformations using the Wigner rotation matrices of rank 2, acting on the basis of spherical harmonics. For the bicelle tilt angle of $\theta = \pi/2$, we have

$$Y_m^{(2)}(\beta,\alpha) = \sum_{m'=-2}^{2} Y_{m'}^{(2)}(\theta,\phi) D_{m'm}^{(2)}(0,\beta_0,\pi-\alpha_0) \tag{7.7}$$

Here α_0 and β_0 refer to the orientation of the magnetic field relative to the molecular frame of the peptide plane in the absence of the bicelle tilt ($\theta = 0$), and ϕ is the instantaneous rotation of the bicelle about its axis. Performing the algebra, we obtain

$$H_{CSA} = -\frac{1}{2} H_{CSA}^{(0)} + \Delta\sigma S_z \left(A_{CSA} \cos 2\phi + B_{CSA} \sin 2\phi \right) \tag{7.8}$$

where the ϕ-independent part of the Hamiltonian is given by:

$$H_{CSA}^{(0)} = S_z \left(\sigma_{iso} - \frac{\Delta\sigma}{2} \left[\frac{3\cos^2\beta_0 - 1}{2} + \frac{\eta}{2} \sin^2\beta_0 \cos 2\left(\alpha_0 - \gamma\right) \right] \right)$$

$$= \left(\sigma_{iso} - \frac{\Delta_0}{2} \right) S_z$$

The coefficients of the ϕ-dependent part of the CSA Hamiltonian are given by

$$A_{CSA} = \frac{3}{4} \sin^2\beta_0 + \frac{\eta}{4}(1+\cos^2\beta_0)\cos 2(\alpha_0 - \gamma) \;; \quad B_{CSA} = -\frac{\eta}{2}\cos\beta_0 \sin 2(\alpha_0 - \gamma)$$

$$\tag{7.9}$$

To solve the SLE, we employ the direct-product formalism in the trace metric space of the spin-1/2 operators to construct the relevant superoperators and the spin-density vectors.[14] This is equivalent to replacing the spin operator S_z by

$$S_z \rightarrow [(S_z \otimes e) + (e \otimes S_z)] \equiv \mathbf{S}_z \tag{7.10}$$

where e is a 2×2 unit matrix, and the symbol $\otimes$ denotes the direct or Kronecker product.

In the superoperator notation, the solution for the Stochastic Liouville Equation free-induction decay (FID) is then given by

$$G(t) = \int_{0}^{2\pi} \mathbf{g}_1^T \exp\left[(-i\mathbf{L} + \Gamma)t\right] \mathbf{g}_0 d\phi \tag{7.11}$$

where $\mathbf{L}$ is the Liouvillian superoperator, $\mathbf{g}_0$ is the starting vector, $\mathbf{g}_1$ is the projection vector, and T stands for the matrix transpose. The time evolution of the chemical shift is governed by the Liouvillian

$$\mathbf{L} = -\frac{1}{2} \mathbf{H}_{CSA}^{(0)} + S_z \Delta\sigma \left(A_{CSA} \cos 2\phi + B_{CSA} \sin 2\phi \right) \tag{7.12}$$

The starting spin-state vector is chosen so that it describes the initial density operator S_x:

$$\mathbf{g}_0 = \mathbf{i}_+ + \mathbf{i}_- \tag{7.13}$$

where

$$\mathbf{i}_+ = \begin{pmatrix} 1 \\ 0 \\ 0 \\ 0 \end{pmatrix}, \quad \mathbf{i}_\alpha = \begin{pmatrix} 0 \\ 1 \\ 0 \\ 0 \end{pmatrix}, \quad \mathbf{i}_\beta = \begin{pmatrix} 0 \\ 0 \\ 1 \\ 0 \end{pmatrix}, \quad \mathbf{i}_- = \begin{pmatrix} 0 \\ 0 \\ 0 \\ 1 \end{pmatrix} \tag{7.14}$$

The basis vectors $\mathbf{i}_\pm$ and $\mathbf{i}_{\alpha,\beta}$ have been described in detail.[14] The projection vector corresponds to the $S_- = S_+^T$ operator and is given by

$$\mathbf{g}_1 = \mathbf{i}_+ \tag{7.15}$$

To integrate the SLE over ϕ, we use a Fourier expansion for its solution, $\mathbf{g}(t)$,

$$\mathbf{g}(t) = \sum_{n=0}^{N_{max}-1} \mathbf{g}_n^c(t) \cos n\phi + \sum_{n=1}^{N_{max}} \mathbf{g}_n^s(t) \sin n\phi \tag{7.16}$$

We renumber the vector-coefficients $\mathbf{g}_n$ of the expansion so that the first N_{max} vectors correspond to the cosines, and the vectors numbered from $N_{max} + 1$ to $2N_{max}$ would correspond to the sines. This leads to two $(2N_{max}) \times (2N_{max})$ matrices of the coupling coefficients C and S with the following nonzero entries:

$$C_{nn'} = \dfrac{\displaystyle\int_0^{2\pi} \cos n\phi \cos 2\phi \cos n'\phi \, d\phi}{\displaystyle\int_0^{2\pi} \cos^2 n\phi} \qquad (7.17a)$$

$$C_{nn'} = \dfrac{\displaystyle\int_0^{2\pi} \sin n\phi \cos 2\phi \sin n'\phi \, d\phi}{\displaystyle\int_0^{2\pi} \sin^2 n\phi} \qquad (7.17b)$$

$$S_{nn'} = \dfrac{\displaystyle\int_0^{2\pi} \cos n\phi \sin 2\phi \sin n'\phi \, d\phi}{\displaystyle\int_0^{2\pi} \cos^2 n\phi} \qquad (7.17c)$$

$$S_{nn'} = \dfrac{\displaystyle\int_0^{2\pi} \sin n\phi \sin 2\phi \cos n'\phi \, d\phi}{\displaystyle\int_0^{2\pi} \sin^2 n\phi} \qquad (7.17d)$$

These integrals can be readily evaluated analytically. It is clear that the even indexes n are coupled to even n', and odd n are coupled to odd n'. Only the functions having even indices are of interest. In this basis, the diffusion operator Γ defined in Equation (7.4) is of a simple diagonal form:

$$\Gamma_{nm} = -D_\| n^2 \qquad (7.18)$$

After performing the necessary matrix-vector multiplications in the spin space, and integrating over the ϕ-space Equation (7.11), the FID solution in uniaxially tumbling bicelles can be reduced to a single quadratic form:

$$G(t) = \exp\left[-i\left(\sigma_{iso} - \frac{\sigma_0}{2}\right)t\right]\left\langle \mathbf{x}^{\mathrm{T}} \exp\left\{[-i\Delta\sigma(A_{CSA}C + B_{CSA}S) + \Gamma]\,t\right\}\,\mathbf{x}\right\rangle$$

$$(7.19)$$

where $\mathbf{x}^T = (1,0,...,0)$ denotes a row of length $2N_{max}$ having unity in the first entry. The line shape is centered at the frequency $\sigma_{iso} - \sigma_0/2$, modulated by the motional part, as given by the quadratic form in angular brackets.

7.4 NUMERICAL SIMULATIONS

Equation (7.19) can be solved numerically by explicit diagonalization of the matrix exponential. In this work, the size of the matrices is less than 1000×1000, so the standard MATLAB diagonalization routine can be employed. Simulations for a tilted ideal α-helix were performed over a range of values of the diffusion coefficient, $D_\parallel$. As illustrated in Figure 7.1, the helix was assumed to have 18 amino acid residues and be tilted 30° relative to the normal of the bilayer. The values for the static peptide plane orientations (α_0, β_0) for each residue have been calculated by using the polypeptide chain propagation method.[15] The term N_{max} was set from 10 to 40, depending on the value for $D_\parallel$.

Simulated spectra are presented in Figure 7.3. Figure 7.3A shows the spectral lines that would arise in the case of complete alignment for three different peptide

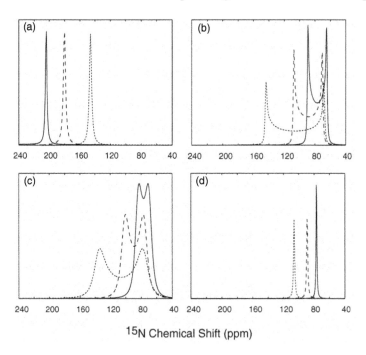

15N Chemical Shift (ppm)

FIGURE 7.3 Simulations of ^{15}N solid-state NMR spectra for three different peptide plane orientations in an α-helix shown by solid, dashed, and dotted lines. (a) Completely aligned bilayers. (b–d) Bicelles oriented perpendicular to the magnetic field, as in Figure 7.1. (b) Slow uniaxial rotational diffusion, $D_\parallel = 10$ s^{-1}; (c) intermediate rotational diffusion, $D_\parallel = 10^3$ s^{-1}; (d) fast motional averaging, $D_\parallel = 5 \times 10^5$ s^{-1}. With fast uniaxial diffusion, the resonances have narrow line widths and the spectra are scaled by a factor of 0.5.

plane orientations with no motional averaging. Figure 7.3B shows one-dimensional ^{15}N NMR spectra for the same orientations in the case of bicelles aligned perpendicular to the magnetic field and undergoing rotational diffusion with $D_\parallel = 10^1$ s^{-1} ($N_{max} = 40$). As can be seen in Figure 7.3B, the spectra essentially represent cylindrical powder patterns under these ultraslow motional averaging conditions. Figure 7.3C shows the same orientations at an intermediate uniaxial diffusion rate of $D_\parallel = 10^3$ s^{-1}, which is insufficient to provide complete motional averaging. However, at the diffusion rate of $D_\parallel = 5 \times 10^5$ s^{-1}, the powder distributions are completely averaged for all three orientations. This results in near-Lorentzian lines with positions scaled by a factor of –0.5 relative to the isotropic chemical shift (120 ppm) (Figure 7.3D). There are notable variations in averaging the uniaxial distribution for different orientations of the corresponding peptide planes relative to the bicelle axis.

7.5 EXPERIMENTAL RESULTS

The samples of membrane proteins in phospholipid bicelles used in the NMR experiments were prepared by dissolving ^{15}N-labeled polypeptides in aqueous solution containing 6-O-PC and then adding this solution to a dispersion of 14-O-PC in water.[3] Bicelles were obtained after a few cycles of freezing and heating above 45°C. Although it is possible to optimize bicelle samples for specific proteins by employing different aqueous buffers, a simple mixture of 6-O-PC/14-O-PC/water has proven effective in incorporating a variety of hydrophobic membrane proteins.

In detail, the pure, lyophilized polypeptide is dissolved in trifluoroethanol (TFE) and sonicated for 5–30 min in a bath sonicator. The organic solvent is then evaporated under a stream of N$_2$ gas to obtain a thin, transparent protein film, which is placed under high vacuum overnight. This dry protein film can be stored at –20°C if needed. The lipids (Avanti Polar Lipids) are obtained dissolved in chloroform. The solvent is evaporated under a stream of N$_2$ gas, and the lipids are placed under high vacuum overnight. An aqueous solution of 6-O-PC is added to the dry protein film. A dispersion of 14-O-PC in H$_2$O is prepared by adding water to the lipid, followed by extensive vortexing and three freeze–heating cycles (liquid nitrogen/40°C). The protein-containing 6-O-PC solution is then added to the 14-O-PC dispersion, previously warmed above 40°C. The resulting solution is briefly vortexed, freeze–heated a few times, and then allowed to equilibrate to room temperature. On bicelle formation, the previously opaque dispersion of 14-O-PC becomes a clear solution between 0° and 10°C. A typical sample has $q = 3.2$ and 28% w/v, and contains 46.5 mg of 14-O-PC, 9.5 mg of 6-O-PC, and 3 mg of protein in a 200-μl volume. The bicelle solution is prechilled in an ice-water bath for a few seconds before being transferred to the NMR tube. A small, flat-bottomed NMR tube with a 5-mm outer diameter (New Era Enterprises) is filled with about 160 μl of the bicelle solution, which is transferred to the tube with a glass Pasteur pipette. The tube is sealed with a tight-fitting plastic cap that is then pierced with a thin syringe to remove excess air from the sample and create a tight seal.

Samples of protein-containing bicelles yield NMR spectra consistent with uniaxial orientation and motional averaging about the bilayer normal at 43°C. The lipid alignment can be verified experimentally by ^{31}P NMR, ^{13}C NMR, or ^{2}H NMR after adding small amounts of deuterated lipids to the bicelles.[3] The one-dimensional spin-lock cross-polarization experiments are optimized by varying the contact-time and [1]H-carrier frequency and compensated for mismatch by using CP-MOIST (cross polarization mismatch optimized IS transfer).[16] The composite-pulse heteronuclear decoupling scheme SPINAL-16 (small phase incremental alteration) is used to perform complete heteronuclear decoupling in high field spectrometers.[17] If bicelles aligned perpendicular to the direction of the magnetic field undergo fast enough rotational diffusion about their normals, the static uniaxial distribution of orientations is averaged to yield a single-line resonance for each ^{15}N site. One-dimensional NMR spectra of membrane proteins in bilayers mechanically aligned on glass plates have resonances with line widths similar to those observed in single crystals of model peptides, indicating a high degree of alignment. In bicelles, the spectra also exhibit narrow single-line resonances, indicating complete motional averaging and a high degree of alignment about the bicelle axis.

Solid-state ^{15}N NMR spectra of the transmembrane domain of virus protein "u" (Vpu) from HIV-1 have been thoroughly characterized and the resonances assigned to specific residues.[18] In the case of alignment of bilayers on glass plates, the one-dimensional spectrum spans the frequency region from 185 to 225 ppm. By contrast, in the case of bicelles oriented perpendicular to the magnetic field, this range has to be inverted with respect to the isotropic value of 120 ppm and then scaled by a factor of 0.5.

Experimental NMR spectra of the transmembrane domain of Vpu ^{15}N labeled at Leu 11 in mechanically aligned bilayer and magnetically aligned bicelle samples are compared in Figure 7.4.[18] The resonance frequencies vary with the alignment of the bilayers. When the bilayer normals are parallel to the magnetic field (Figures 7.4a and 7.4b), the ^{15}N amide resonance is near 200 ppm, as expected for the N-H bond being approximately parallel to the field in a transmembrane helix. The resonance for the bicelle samples has its frequency shift from the isotropic position reduced by the 0.8 order parameter that results from the wobble present in bicelles but not bilayers. With the bilayer normals parallel to the field, single-line resonances are present regardless of the presence or absence of motional averaging.

In contrast, rotational diffusion is essential to obtain single-line resonances when the bilayer normals are perpendicular to the field (Figures 7.4b and 7.4d). As described above, bicelles modeled by disks undergo rotational diffusion that is rapid enough to perform this averaging. Because planar bilayers are extended in the two dimensions, they cannot undergo overall diffusional motions, and the single line resonance in Figure 7.4B must result from the proteins themselves undergoing rotational diffusion within the bilayers. Therefore, these spectra demonstrate that the protein molecules undergo rapid rotational diffusion.

7.6 CONCLUSIONS

The experimental results presented in this chapter demonstrate that it is possible to obtain high-resolution NMR spectra from ^{15}N-labeled proteins in mechanically aligned bilayers and magnetically aligned bicelles with their normals parallel and perpendicular to the direction of the applied magnetic field. For samples with uniaxial

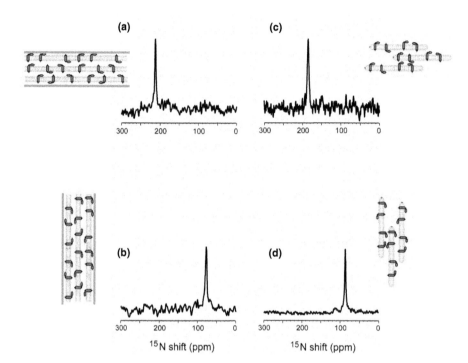

FIGURE 7.4 ^{15}N spectra of a trans-membrane α-helix labeled in a single backbone amide site with ^{15}N. The 36-residue polypeptide has a sequence corresponding to the trans-membrane domain of the protein Vpu from HIV-1. It is labeled at Leu 11. (a,b) Experimental NMR spectra obtained from a sample of oriented lipid bilayers on glass plates. (c,d) Experimental spectra obtained from a sample of bicelles in solution. The bicelles in (c) are "flipped" by the addition of lanthanide ions. As illustrated, the bilayer normals are parallel in (a) and (c), and perpendicular in (b) and (d), to the direction of the applied magnetic field.

alignment parallel to the direction of the magnetic field, this is expected even in the complete absence of rotational diffusion. However, when the alignment is perpendicular to the field, rotational diffusion about the direction of the normal is essential. The calculations described in this chapter show that bicelles modeled as disks undergo rotational diffusion that is fast enough for this to occur. However, the experimental results for the mechanically aligned bilayer sample show that the protein molecules themselves undergo diffusion in the plane of the bilayer that is fast enough to accomplish this averaging.

Two sources of motional averaging in the plane of bilayers must be taken into account in understanding why membrane proteins can be studied effectively by solid-state NMR of aligned samples. The orientational information about the peptide plane is carried by the residual chemical shift Hamiltonian. Experimental results for a selectively labeled transmembrane α-helix are in agreement with the above theoretical predictions. The same considerations hold for two-dimensional solid-state NMR experiments, including PISEMA (polarization inversion with spin exchange at the magic-angle)[19] or SAMMY,[20] which yield high-resolution separated local field spectra. As a result, protein-containing bicelles have the potential to serve as samples for structure determination of membrane proteins by solid-state NMR spectroscopy.[21]

ACKNOWLEDGMENTS

We thank A. Mrse, C. V. Grant, and C. H. Wu for assistance with the instrumentation and advice. This research was supported by grants EB002169, GM64676, and GM66978, used the Resource for NMR Molecular Imaging of Proteins supported by grant EB002031, and A.D. was supported by postdoctoral fellowship GM65833, all from the National Institutes of Health.

REFERENCES

1. Sanders, C.R., Hare, B.J., Howard, K.P. and Prestegard, J.H., Magnetically oriented phospholipid micelles as a tool for the study of membrane-associated molecules, *Prog. NMR Spectrosc.*, 26, 421, 1994.
2. Prosser, R.S., Hunt, S.A., DiNatale, J.A. and Vold, R.R. Magnetically aligned membrane model systems with positive order parameter: switching the sign of S_{zz} with paramagnetic ions. *J. Am. Chem. Soc.*, 118, 269, 1996.
3. De Angelis, A.A., Nevzorov, A.A., Park, S.H., Howell, S.C., Mrse, A.A. and Opella, S.J. High-resolution NMR spectroscopy of membrane proteins in aligned bicelles. *J. Am. Chem. Soc.,* 126, 15340, 2004.
4. Sanders, C.R. and Schwonek, J.P. Characterization of magnetically orientable bilayers in mixtures of DHPC and DMPC by solid state NMR, *Biochemistry*, 31, 8898, 1992.
5. Vold, R.R. and Prosser, R.S., Magnetically oriented phospholipid bilayered micelles for structural studies of polypeptides. Does the ideal bicelle exist? *J. Magn. Reson. B*, 113, 267, 1996.
6. Struppe, J. and Vold, R.R., Dilute bicellar solutions for structural NMR work, *J. Magn. Reson.*, 135, 541, 1998.
7. Glover, K.J., Whiles, J.A., Wu, G., Yu, N.-J., Deems, R., Struppe, J.O., Stark, R.E., Komives, E.A. and Vold, R.R., Structural evaluation of phospholipid bicelles for solution-state studies of membrane-associated biomolecules, *Biophys. J.*, 81, 2163, 2001.
8. Gaemers, S. and Bax, A. Morpology of three lyotropic liquid crystalline biological NMR media studies by translational diffusion anistropy. *J. Am. Chem. Soc.*, 123, 12343, 2001.
9. Nieh, M.P., Raghunathan, V.A., Glinka, C.J., Harroun, T.A., Pabst, G. and Katsaras, J. Magnetically alignable phase of phospholipid "bicelle" mixtures is a chiral nematic made up of wormlike micelles. *Langmuir*, 20, 7893, 2004.
10. Perrin, F., Mouvement Brownian d'un ellipsoide. I. Dispersion dielectrique pour des molecules ellipsoidales, *J. Phys. Radium*, 5, 497, 1934.
11. Perrin, F., Mouvement Brownian d'un ellipsoide. II. Rotation libre et depolarisation des fluorescences, *J. Phys. Radium*, 7, 1, 1936.
12. Abragam, A., *The Principles of Nuclear Magnetism*, Oxford University Press, London, 1961.
13. Schneider, D.J. and Freed, J.H., Spin relaxation and motional dynamics, *Adv. Chem. Phys.*, 73, 387, 1989.

14. Nevzorov, A.A. and Freed, J.H., Direct-product formalism for calculating magnetic resonance signals in many-body systems of interacting spins, *J. Chem. Phys.*, 115, 2401, 2001.

15. Nevzorov, A.A. and Opella, S.J., Structural fitting of PISEMA spectra of aligned proteins, *J. Magn. Reson.*, 160, 33, 2003.

16. Levitt, M., Heteronuclear cross polarization in liquid-state nuclear magnetic resonance: mismatch compensation and relaxation behavior, *J. Chem. Phys.*, 94, 30, 1986.

17. Fung, B.M., Khitrin, A.K. and Ermolaev, K., An improved broadband decoupling sequence for liquid crystals and solids, *J. Magn. Reson.*, 142, 97, 2000.

18. Park, S.H., Mrse, A.A., Nevzorov, A.A., Mesleh, M.F., Oblatt-Montal, M., Montal, M. and Opella, S.J., Three-dimensional structure of the channel-forming trans-membrane domain of virus protein "u" (Vpu) from HIV-1, *J. Mol. Biol.*, 333, 409, 2003.

19. Wu, C.H., Ramamoorthy, A. and Opella, S.J., High-resolution heteronuclear dipolar solid-state NMR spectroscopy, *J. Magn. Reson. A*, 109, 270, 1994.

20. Nevzorov, A.A. and Opella, S.J., A "magic sandwich" pulse sequence with reduced offset dependence for high-resolution separated local field spectroscopy, *J. Magn. Reson.*, 164, 182, 2003.

21. Opella, S.J. and Marassi, F.M., Structure determination of membrane proteins by NMR spectroscopy, *Chem. Rev.*, 104, 3587, 2004.

8 Nuclear Magnetic Resonance Structural Studies of the FXYD Family Membrane Proteins in Lipid Bilayers

Carla M. Franzin, Jinghua Yu, and Francesca M. Marassi

CONTENTS

ABSTRACT Solid-state NMR spectroscopy enables the structures of membrane peptides and proteins to be determined in lipid bilayers. Regardless of fold, all membrane proteins adopt three-dimensional structures with a unique direction in space defined by the membrane environment. Because this directionality is an intrinsic characteristic of membrane protein structure and function, it is highly desirable to carry out structure determination within the context of bilayer lipid membranes. Solid-state NMR spectroscopy of uniaxially oriented planar bilayer lipid samples is ideally suited for this purpose. This chapter outlines the methodology for membrane protein structure determination using solid-state NMR of oriented planar lipid bilayer samples. Recent developments in sample preparation, recombinant bacterial expression systems for the preparation of isotopically labeled membrane proteins, pulse sequences for high-resolution spectroscopy, and structural indices that guide the structure assembly process, have greatly extended the capabilities of the technique, and are described. The methods are illustrated with examples from the FXYD proteins, a family of auxiliary regulatory subunits of the Na,K-ATPase.

KEY WORDS: *FXYD, lipid bilyars, membrane proteins, orientation, PISA wheel, PISEMA, solid-state NMR*

8.1 INTRODUCTION

Membrane proteins regulate some of the most basic cellular functions, and therefore it is not surprising that they constitute approximately 30% of all expressed genes and that they are major targets for drug discovery initiatives. However, despite their importance, only 140 structures of membrane proteins have been deposited in the PDB (Protein Data bank), compared with the approximately 23,000 coordinates deposited for globular proteins to date (http://www.rcsb.org/pdb/). This shortage is a result of the lipophilic character of membrane proteins, which makes them difficult to overexpress and purify, complicates crystallization for x-ray analysis, and results in broadened and highly overlapped solution nuclear magnetic resonance (NMR) spectral lines. NMR spectroscopy offers two approaches to membrane protein structure determination. Solid-state NMR methods can be applied to samples of membrane proteins in lipid bilayers, enabling structures to be determined in a native-like environment, whereas solution NMR methods can be used on samples of proteins dissolved in lipid micelles. The two approaches are complementary and can be used in combination as a unified approach to membrane protein structure.

High-quality solution NMR spectra can be obtained for some fairly large helical membrane proteins in micelles, but it is very difficult to measure and assign a sufficient number of long-range nuclear Overhauser effects (NOEs) to determine protein folds. This limitation can be overcome by preparing weakly aligned micelle samples for the measurement of residual dipolar couplings (RDCs) and residual chemical shift anisotropies (RCSAs). High-resolution solid-state NMR spectra can be obtained for membrane proteins that are expressed, isotopically labeled, and reconstituted in uniaxially oriented planar lipid bilayers. The spectra have characteristic resonance patterns that directly reflect protein structure and topology, and this direct relationship between spectrum and structure provides the basis for methods that enable the simultaneous sequential assignment of resonances and the measurement of orientation restraints for protein structure determination. Recent developments in sample preparation, recombinant bacterial expression systems for the preparation of isotopically labeled membrane proteins, pulse sequences for high-resolution spectroscopy, and structural indices that guide the structure assembly process have greatly extended the capabilities of the technique. The structures of a variety of membrane peptides and proteins have been investigated using this approach, and several atomic-resolution structures have been determined and deposited in the PDB.[1-7] In this chapter, the methods are illustrated with examples from the FXYD family proteins, auxiliary regulatory subunits of the Na,K-ATPase.

8.2 THE FXYD FAMILY PROTEINS

The FXYD family proteins are expressed in tissues that perform fluid and solute transport (breast/mammary gland, kidney, colon, pancreas, prostate, liver, lung, and placenta) or that are electrically excitable (muscle, nervous system), where they function to regulate the flux of transmembrane ions, osmolytes, and fluids.[8] The protein sequences are highly conserved through evolution and are characterized by a 35–amino acid FXYD homology (FH) domain, which includes the transmembrane (TM) domain (Figure 8.1). The short motif PFXYD (Pro, Phe, X, Tyr, Asp), preceding the transmembrane domain, is invariant in all known mammalian examples and is identical in other vertebrates, except for the proline. Residue X is usually Tyr but can also be Thr, Glu, or His. In all these proteins, conserved basic residues flank the TM domain, the extracellular N-termini are acidic, and the cytoplasmic C-termini are basic.

PLM (phospholemman) is one of the best characterized members of the FXYD family and is the major substrate of hormone-stimulated phosphorylation by cAMP-dependent protein kinase A and C in the heart.[9] CHIF (corticosteroid-hormone-induced factor) is upregulated by aldosterone and corticosteroids in mammalian kidney and intestinal tracks, where it regulates Na+ and K+ homeostasis.[10] Mat8 (mammary tumor protein 8 kDa) is expressed in breast, prostate, lung, stomach, and colon, as well as in human breast tumors, breast tumor cell lines, and prostate cancer cell lines, after malignant transformation by oncogenes,[11,12] and other FXYD proteins are also induced by oncogenic transformation. All three proteins, PLM, Mat8, and CHIF, induce ionic currents in *Xenopus* oocytes, and PLM also forms ion channels in phospholipid bilayers.[10,11,13] The identification of several FXYD family members, including PLM and CHIF, as regulators of the Na,K-ATPase, points to a mechanism for regulation of the pump that involves the expression of an auxiliary subunit.[8]

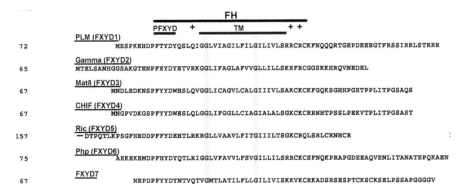

FIGURE 8.1 Amino acid sequences of mammalian FXYD membrane proteins. The FXYD homology (FH) domain encompasses the FXYD consensus sequence, and the transmembrane (TM) domain is flanked by conserved positively charged residues. Conserved Gly residues in the TM domain are highlighted in gray.

Recently, we described the recombinant expression, purification, and sample preparation in lipid micelles and bilayers for three members of the FXYD family: PLM, Mat8, and CHIF.[14] The solid-state NMR spectra in lipid bilayers provide the first view of their structures and topologies and are described later.

8.3 PROTEIN EXPRESSION AND PURIFICATION

NMR structural studies typically require samples containing milligram quantities of isotopically labeled proteins. Smaller peptides can be prepared by solid-phase peptide synthesis; however, this is impractical for larger proteins and for the preparation of uniformly labeled samples, in which efficient expression systems are essential. The ability to express membrane proteins in bacteria enables a wide variety of isotopic labeling schemes to be incorporated in the NMR experimental strategy. Selective labeling, by amino acid type, is accomplished by growing the bacteria transformed with the protein gene on media in which only one type of amino acid is labeled and all others are not. Uniform labeling, where all the nuclei of one or several types (^{15}N, ^{13}C, ^{2}H) are labeled, is accomplished by growing the bacteria on media containing ^{15}N-labeled ammonium sulfate, ^{13}C labeled glucose, D$_2$O, or any combination of these.

Several *Escherichia coli* expression systems and mutant cell strains have been developed for membrane protein expression and purification.[15–19] Many involve the use of fusion proteins, the formation of inclusion bodies, and the incorporation of engineered N-terminal His-tags for metal affinity chromatography to greatly simplify protein isolation and purification. After inclusion body isolation, fusion protein purification, and cleavage, the target membrane protein is finally purified and then reconstituted into lipid bilayers for NMR studies.

To produce PLM, CHIF, and Mat8, we used the *E. coli* pMMHa fusion protein expression vector, which directs the synthesis of the fusion protein His$_9$-TrpΔLE-FXYD.[14,18] This vector has been used successfully for the production of other membrane or toxic proteins ranging in size from 80 to 150 amino acids.[20] The TrpΔLE fusion partner, from the Trp leader amino acid sequence, is very effective at forming inclusion bodies and is thus protected from proteolysis. The fusion protein is not toxic to the *E. coli* host cells and is expressed at levels up to 20% of total cellular protein in *E. coli* strain C41(DE3),[15] grown on M9 minimal media for isotopic labeling. Intact FXYD proteins are liberated from the fusion partner using CNBr (cyanogen bromide), which cleaves specifically after Met residues.[21] The use of chemical cleavage eliminates difficulties such as poor specificity and enzyme inactivation that are often encountered with protease treatment of membrane proteins in detergents.

The protein content of cells isolated before and after IPTG induction is shown in Figure 8.2a (lanes 1 and 2), where fusion protein overexpression is marked by the appearance of an intense band near 21 kDa. After protein expression, the inclusion bodies enriched in fusion protein were separated from the *E. coli* lysate by a series of wash and centrifugation steps (lane 3), the fusion protein was isolated by nickel affinity chromatography (lane 4), and each FXYD protein was cleaved from the fusion partner using CNBr (lane 5). This yields a fragment nearly 14 kDa,

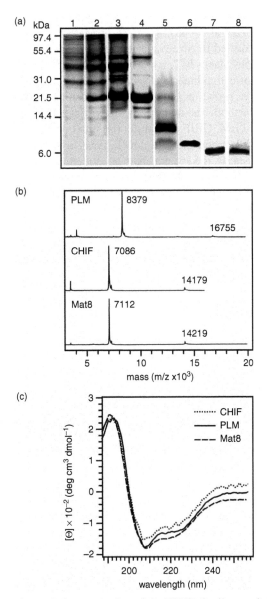

FIGURE 8.2 Expression and characterization of the FXYD family proteins PLM, CHIF, and Mat8. (a) Protein fractions at different stages of the purification protocol were analyzed on Coomassie-stained 16% SDS-PAGE. Lane 1: entire cell before IPTG induction. Lane 2: entire cell after IPTG induction with the fusion protein band at 20 kDa. Lane 3: isolated inclusion bodies enriched in fusion protein. Lane 4: fusion protein isolated by Ni affinity chromatography. Lane 5: His$_9$-TrpLE fusion partner and PLM resulting from CNBr cleavage of the fusion protein. Lane 6: purified PLM. Lane 7: purified CHIF. Lane 8 purified Mat8. (b) MALDI-TOF mass spectra of purified PLM (8379 Da), CHIF (7086 Da), and Mat8 (7,112 Da). (c) CD spectra of purified PLM, CHIF, and Mat8 in SDS micelles. The samples contained 20 μM FXYD protein, 500 mM SDS, 20 mM sodium phosphate, 1 mM sodium azide, pH 5.[14]

corresponding to His$_9$-TrpΔLE, and a smaller fragment corresponding to PLM at 8.4 kDa (lane 6), CHIF at 7.1 kDa (lane 7), or Mat8 at 7.1 kDa (lane 8). Finally, the proteins were purified by size-exclusion chromatography, followed by reverse-phase chromatography (lanes 6, 7, and 8). Typically, 2.5 mg of purified protein are obtained from 1 L of cell culture in ^{15}N-labeled minimal media.

The mass spectra shown in Figure 8.2b demonstrate the high degree of purity obtained with this method. The major peaks have masses that correspond exactly to the FXYD proteins PLM, CHIF, and Mat8; the small peaks at half mass arise from doubly charged species; and those at double mass arise from a small fraction of FXYD dimer. In all cases, the spectra show no evidence of degradation or chemical modifications. The CD spectra obtained for the proteins in SDS (sodium-n-dodecyl sulfate) micelles are shown in Figure 8.2c. For all three FXYD family members, the two minima at 208 and 222 nm are characteristic of α-helical structures, and the helical content, estimated from the CD spectra with the k2d program (http://www.embl-heidelberg.de/~andrade/k2d/),[22] is approximately 40%. These purified recombinant proteins were used for NMR structural studies in lipid micelles and lipid bilayers.

8.4 NMR IN MICELLES

Solution NMR methods rely on rapid molecular reorientation for line narrowing and can be successfully applied to membrane proteins in micelles.[14,23–35] The size limitation is substantially more severe than for globular proteins because the many lipid molecules associated with each polypeptide slow its overall reorientation rate. Micelles afford rapid and effectively isotropic reorientation of the protein, and their amphipathic nature simulates that of membranes, offering a realistic alternative to organic solvents for studying membrane proteins. Moreover, for the proteins examined by both solution and solid-state NMR, similar structural features have been found in micelle and bilayer samples.[36–38]

The first step in solution NMR studies of proteins is the preparation of folded, homogeneous, and well-behaved samples, and several lipids are available for membrane protein solubilization.[35] For membrane-bound proteins, small micelles containing approximately 60 lipids and one protein provide a generally effective model membrane environment without the damaging effects of organic solvents. The primary goal in micelle preparation is to reduce the effective rotational correlation time of the protein so that resonances will have the narrowest possible line widths. Careful handling of the protein throughout the purification is essential, as subtle changes in the protocol can have a significant effect on the quality of the resulting spectra. It is essential to optimize the protein concentration, lipid nature and concentration, counter ions, pH, and temperature to obtain well-resolved NMR spectra, with narrow ^{1}H and ^{15}N resonance line widths. These parameters also influence the ability of protein–micelle solutions to soak into polyacrylamide gels that are either stretched or compressed for weak sample alignment.

High-quality solution NMR spectra can be obtained for some fairly large helical membrane proteins in micelles,[35,39] but there are only a few cases in which it has been possible to measure and assign sufficient long-range NOEs to determine protein folds. This limitation can be overcome by preparing weakly aligned micelle samples for the measurement of RDCs and RCSAs[40,41] in all backbone amide sites and the analysis of these orientation restraints in terms of dipolar waves, PISA wheels, and other indices of protein structure.[36,37,42,43] Stressed polyacrylamide gels[44–47] provide an ideal orientable medium because they do not suffer from the drawbacks of bicelles, which bind tightly to membrane proteins, and phage particles, which are destroyed by micelles. Several laboratories have reported the use of stressed poly-acrylamide gels for NMR of proteins, including membrane proteins in lipid micelles.[36,37,42,48–51] Another useful approach to compensate for insufficient NOEs involves the combination of site-directed spin labeling and NMR,[52] in which distances derived from paramagnetic broadening of NMR resonances are used to determine global fold. In addition, spin label probes can be incorporated within the micelles to probe protein insertion.[53–57]

The two-dimensional $^1H/^{15}N$ HSQC (heteronuclear single quantum correlation) spectra of the uniformly ^{15}N-labeled FXYD proteins, PLM, CHIF, and Mat8, in SDS micelles are shown in Figure 8.3. Each resonance represents a single ^{15}N-labeled site of the protein and is characterized by ^1H and ^{15}N chemical shift frequencies that reflect the local environment. The samples were optimized to obtain narrow resonance line widths and high resolution, and the presence of one well-defined resonance for each amide site in the protein is indicative of a high-quality micelle sample. Resonances from the Gly (^{15}N shift = 105–110 ppm), Trp indole (^1H shift = 10 ppm), and Gln and Asn side chain (^1H shift = 6.5–7.5 ppm) nitrogens are clearly resolved. The limited chemical shift dispersion reflects the helical structures of these proteins. The resonances in the spectrum of Mat8 were assigned using uniformly and selectively labeled samples with additional ($^1H/^{15}N$) double- and ($^1H/^{15}N/^{13}C$) triple-resonance NMR experiments.[58–61]

The measurements of as many homonuclear $^1H/^1H$ NOEs as possible among the assigned resonances provide the short-range and long-range distance restraints required for structure determination.[62–64] These restraints are supplemented by other structural restraints, such as spin–spin coupling constants, chemical shift correlations, deuterium exchange data, and RDCs to assign resonances and to characterize the secondary structure of the protein. The HSQC spectra of samples in D_2O solutions identify the most stable helical residues and give useful information on the topology of membrane proteins in micelles.[65] In addition, hydrogen–deuterium fractionation experiments extend the range of exchange rates that can be monitored to identify more subtle structural features.[66] For the FXYD protein Mat8 in SDS micelles, D_2O exchange clearly identifies a single stable transmembrane helix (Figure 8.4b), spanning about 20 hydrophobic amino acids from Val19 to Ala39 (Figure 8.4a).

The two-dimensional HSQC spectra also serve as the basis for the measurement of the ^1H and ^{15}N relaxation parameters of protein backbone amide site, which are useful for describing protein dynamics. The heteronuclear ^1H-^{15}N NOEs of the backbone amide sites provide remarkably direct and sensitive information on local protein dynamics.[67–69] They can be measured with and without ^1H irradiation to saturate the ^1H magnetization.[70] The ^1H-^{15}N NOEs (Figure 8.4c), as well as measurements of $^1H/^{15}N$ resonance line intensities (Figure 8.4d), strongly discriminate

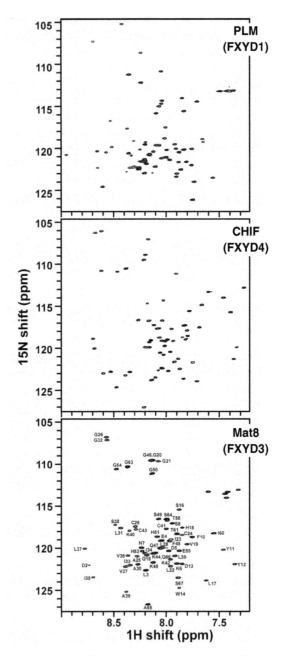

FIGURE 8.3 Two-dimensional $^1H/^{15}N$ HSQC spectra of uniformly ^{15}N-labeled PLM, CHIF, and Mat8 in SDS micelles. The spectrum of Mat8 has been completely assigned. The samples contained 1–2 mM FXYD protein, 500 mM SDS, 20 mM sodium citrate, 10 mM DTT, 1 mM sodium azide, pH 5. The ^{15}N and ^1H chemical shifts are referenced to 0 ppm for liquid ammonia and tetramethylsilane.[14]

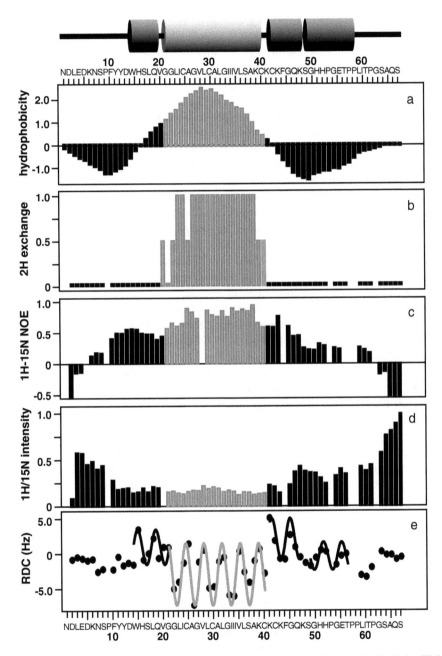

FIGURE 8.4 Summary of the solution NMR data for the FXYD protein Mat8 in SDS micelles. The secondary structure is shown at top, with the transmembrane helix in light gray, and polar helices in dark gray. (a) Hydrophobicity as a function of residue number. (b) $^1H/^2H$ exchange as a function of residue number. (c) Heteronuclear NOE as a function of residue number. (d) Resonance intensity as a function of residue number. (e) RDC as a function of residue number.

between structured helical residues and those in more mobile loop and terminal regions of the FXYD protein Mat8.

RDCs are extremely useful both for structure refinement and for the de novo determination of protein folds.[71–76] During refinement, these measurements supplement an already large number of chemical shifts, approximate distance measurements, and dihedral angle restraints. Among the principal advantages of anisotropic spectral parameters in solution NMR spectroscopy is that they can report on the global orientations of separate domains of a protein and of individual bonds relative to a reference frame, which reflects the preferred alignment of the molecule in the magnetic field. This does not preclude their utility in characterizing the local backbone structure of a protein molecule.

The RDCs and RCSAs measured in solution NMR experiments provide direct angular restraints with respect to a molecule-fixed reference frame.[37,40,41] They are analogous to the nonaveraged dipolar couplings and chemical shift anisotropies measured in solid-state NMR experiments.[77–79] These orientation restraints are the principal mechanism for overcoming the limitations resulting from having few reliable long-range NOEs available as distance restraints — often encountered with samples of membrane-bound proteins in micelles.

Dipolar waves are very effective at identifying the helical residues in membrane proteins and the relative orientations of the helical segments, and they also serve as indices of the helix regularity in proteins.[42] The magnitudes of the RDCs are plotted as a function of residue number and are fitted to a sine wave with a period of 3.6 residues.[36,42,43] The quality of fit is monitored by a scoring function in a four-residue sliding window and by the phase of the fit. As shown in Figure 8.4e for Mat8 in SDS micelles, the one-dimensional dipolar waves extracted from RDCs measured on weakly aligned samples with solution NMR experiments are very effective at mapping protein structure. Dipolar waves from solution NMR data give relative orientations of helices in a common molecular frame. However, dipolar waves from solid-state NMR data give absolute measurements of helix orientations because the polypeptides are immobile and the samples have a known alignment in the magnetic field.

8.5 NMR IN LIPID BILAYERS

Glass-supported oriented phospholipid bilayers containing membrane proteins accomplish the principal requirements of immobilizing and orienting the protein for solid-state NMR structure determination. The planar lipid bilayers are supported on glass slides and are oriented in the NMR probe so that the bilayer normal is parallel to the field of the magnet, as shown in Figure 8.5a. The choice of lipid can be used to control the lateral spacing between neighboring phospholipid molecules, as well as the vertical spacing between bilayers. The use of phospholipids with unsaturated chains leads to more expanded and fluid bilayers, and the addition of negatively charged lipids increases interbilayer repulsions, leading to larger interstitial water layers between bilayer leaflets.

Samples of membrane proteins in lipid bilayers oriented on glass slides can be prepared by deposition from organic solvents followed by evaporation and lipid

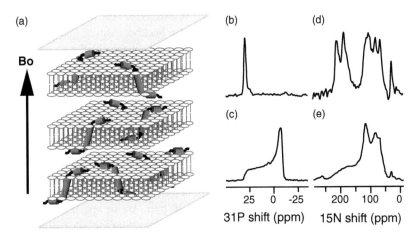

FIGURE 8.5 Effect of sample orientation on the solid-state NMR spectra of isotopically labeled proteins. (a) The glass-supported phospholipid bilayer samples are oriented in the NMR probe so that the bilayer normal is parallel to the direction of the magnetic field (Bo). (b) Oriented phospholipid bilayers give single-line, one-dimensional ^{31}P chemical shift NMR spectra, whereas (c) spherical lipid bilayer vesicles give powder patterns. (d) The one-dimensional ^{15}N chemical shift NMR spectrum of uniformly ^{15}N-labeled FXYD CHIF in oriented lipid bilayers displays multiple resonances, compared with (e) the powder pattern that is obtained for the same protein in unoriented lipid bilayer vesicles. The ^{15}N chemical shifts are referenced to 0 ppm for liquid ammonia.

hydration, or by fusion of reconstituted unilamellar lipid vesicles with the glass surface.[80] The choice of solvents in the first method and of detergents in the second is critical for obtaining highly oriented lipid bilayer preparations. In all cases, the thinnest available glass slides are used to obtain the best filling factor in the coil of the probe. With carefully prepared samples, it is possible to obtain ^{15}N resonance line widths of less than 3 ppm.[81] Notably, these line widths are less than those typically observed in single crystals of peptides, demonstrating that the proteins in the bilayers are very highly oriented, with mosaic spreads of less than about 2°.

The FXYD samples were prepared by mixing ^{15}N-labeled protein and phospholipids in organic solvents, spreading the solution on the surface of the glass slides, removing the solvents under vacuum, and hydrating the sample in a water-saturated atmosphere.[14] Each sample was wrapped in parafilm and then sealed in thin polyethylene film before insertion in the NMR probe. The degree of phospholipid bilayer alignment can be assessed with solid-state ^{31}P NMR spectroscopy of the lipid phosphate headgroup. The ^{31}P NMR spectra obtained for samples of the FXYD protein CHIF are characteristic of a liquid-crystalline bilayer arrangement in both oriented (Figure 8.5b) and unoriented (Figure 8.5c) samples. The spectrum from the oriented sample has a single peak near 30 ppm, as expected for highly oriented bilayers.

The one-dimensional ^{15}N chemical shift NMR spectra of ^{15}N-labeled membrane proteins in oriented bilayers display significant resolution throughout the frequency range of the ^{15}N amide chemical shift. For example, the spectrum in Figure 8.5d was obtained from a sample of uniformly ¹⁵N-labeled CHIF in oriented bilayers. The resonance intensity near 200 ppm arises from backbone amide sites in the CHIF transmembrane helix that have their NH bonds nearly perpendicular to the plane of the membrane, whereas the intensity near 80 ppm is from sites in the N- and C-termini of the protein, with NH bonds nearly parallel to the membrane surface. The peak near 35 ppm results from the amino groups of the lysine sidechains and the N-terminus. This spectrum is strikingly different from that of an unoriented sample, which provides no resolution among resonances (Figure 8.5e). Most of the backbone sites are structured and immobile on the timescale of the ¹⁵N chemical shift interaction (10 kHz), contributing to the characteristic amide powder pattern between about 220 and 60 ppm. Some of the CHIF backbone sites, probably near the N- and C-termini, are mobile, and give rise to the resonance band centered near 120 ppm. Therefore, although certain resonances near 120 ppm, in the spectrum of oriented CHIF, may reflect specific orientations of their corresponding sites, some others arise from mobile backbone sites.

The one-dimensional ¹⁵N chemical shift spectra of PLM, Mat8, and CHIF (Figure 8.6) provide a view of the FXYD protein architecture in membranes. A preliminary analysis of the solid-state NMR data is possible because both CD and NMR spectroscopy in micelles show that the overall secondary structure of these proteins is α-helical. Membrane proteins in lipid bilayers are largely immobile on NMR timescales; therefore, their resonances are not motionally averaged but have frequencies that reflect the orientation of their respective sites relative to the direction of the magnetic field. In our samples, the lipid bilayer plane is perpendicular to the magnetic field direction, and therefore, each resonance frequency reflects the orientation of its corresponding protein site in the membrane.[80]

In each of the three spectra, the resonance intensity near 200 ppm is from the transmembrane helices of PLM, Mat8, and CHIF, whereas the intensity near 80 ppm is from sites in amphipathic helices and loops at the N and C termini. For all three proteins, the narrow dispersion of ¹⁵N resonances centered at 200 ppm indicates that the transmembrane helix crosses the lipid bilayer membrane with a very small tilt angle.

Amide hydrogen exchange rates are useful for identifying residues that are involved in hydrogen bonding and that are exposed to water. The amide hydrogens in transmembrane helices can have very slow exchange rates resulting from strong hydrogen bonds in the low dielectric of the lipid bilayer environment, and their ¹⁵N chemical shift NMR signals persist for days after exposure to D_2O.[82] Faster exchange rates are observed for transmembrane helices that are not tightly hydrogen bonded and that are in contact with water because they participate in channel pore formation,[83] in addition to being observed for other water-exposed helical regions of proteins with weaker hydrogen-bonded networks. When the CHIF sample was exposed to D_2O, the amide hydrogens in the transmembrane helix did not exchange, whereas those in the rest of the protein did, and their resonances disappeared from the spectrum (Figure 8.6d). This indicates that the CHIF transmembrane helix is not water exposed, but rather, forms a tight hydrogen-bonding network that is resistant to hydrogen exchange.

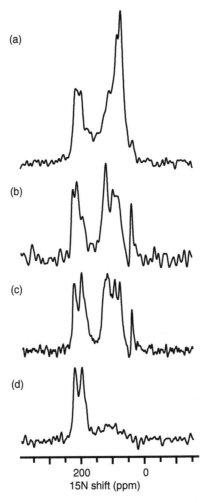

FIGURE 8.6 Solid-state NMR ^{15}N chemical shift spectra of uniformly ^{15}N-labeled FXYD proteins, (a) PLM, (b) Mat8, and (c,d) CHIF, in planar oriented lipid bilayers. Resonances near 200 ppm are from amino acid residues in the transmembrane helix of each protein. Amide hydrogens in the transmembrane helix are resistant to hydrogen exchange and are visible in the spectra obtained after hydration with H_2O (a,b,c), as well as in the spectrum of CHIF obtained after exposing the sample to D_2O, in which the other resonances from exchangeable hydrogens disappear (d). The ^{15}N chemical shifts are referenced to 0 ppm for liquid ammonia.

When membrane proteins are incorporated in planar lipid bilayers that are oriented in the field of the NMR magnet, the frequencies measured in their multi-dimensional solid-state NMR spectra contain orientation-dependent information that can be used for structure determination.[80] The PISEMA (polarization inversion with spin exchange at the magic-angle) experiment gives high-resolution, two-dimensional, [1]H-^{15}N dipolar coupling/^{15}N chemical shift correlation spectra of oriented membrane proteins in which the individual resonances contain orientation restraints for structure determination.[84] PISEMA spectra of membrane proteins in oriented lipid bilayers also provide sensitive indices of protein secondary structure and topology because they exhibit characteristic wheel-like patterns of resonances, called PISA wheels, that reflect helical wheel projections[85] of residues in both α-helices and β-sheets.[77-79] When a PISA wheel is observed, no assignments are needed to determine the tilt of a helix, and a single resonance assignment is sufficient to determine the helix rotation in the membrane. This information is extremely useful for determining the supramolecular architectures of membrane proteins and their assemblies.

The shape and position of the PISA wheel in the spectrum depends on the protein secondary structure and its orientation relative to the lipid bilayer surface, as well as the amide N-H bond length and the magnitudes and orientations of the principal elements of the amide ^{15}N chemical shift tensor. This direct relationship between spectrum and structure makes it possible to calculate solid-state NMR spectra for specific structural models of proteins and provides the basis for a method of backbone structure determination from a limited set of uniformly and selectively ^{15}N-labeled samples.[5,86]

The PISA wheels calculated for an α-helix and a β-strand oriented at varying degrees in a lipid bilayer are shown in Figure 8.7. When the helix or strand cross the membrane, with their long axes exactly parallel to the lipid bilayer normal and to the magnetic field direction (0°), all of the amide sites have an identical orientation relative to the direction of the applied magnetic field, and therefore all of the resonances overlap with the same dipolar coupling and chemical shift frequencies. Tilting the helix or strand away from the membrane normal introduces variations in the orientations of the amide NH bond vectors in the magnetic field and leads to dispersion of the ^1H-^{15}N dipolar coupling and ^{15}N chemical shift frequencies, manifest in the appearance of PISA wheel resonance patterns in the spectra. Because helices and strands yield clearly different resonance patterns, with circular wheels for helices and twisted wheels for strands, these spectra represent signatures of the secondary structure.[78] The spectra also demonstrate that it is possible to determine the tilt of an α-helix or β-strand in lipid bilayers without resonance assignments.

PISA wheels have been observed in the PISEMA spectra of many uniformly ^{15}N-labeled α-helical membrane proteins,[2,4-7,87] including the FXYD protein CHIF (Figure 8.9). Because the PISA wheels also reflect helical wheel representations of the protein, it is possible to obtain the helix or strand rotation in the membrane once a single resonance assignment is obtained. This is illustrated in Figure 8.8 for the spectra calculated for Bacteriorhodopsin, and for the integral membrane porin from *Rhodobacter capsulatus*.

The structure of Bacteriorhodopsin consists of seven α-helices that span the membrane.[88] The PISEMA spectrum calculated from the structure is shown in Figure 8.8a, and the resonances for α5, the fifth helix, trace a PISA wheel (Figure 8.8b) that mirrors the helical wheel projection of the helix (Figure 8.8c). The helical wheel is arranged so that its amide nitrogen atoms coincide with their corresponding resonances in the calculated spectrum, and this arrangement is as predicted by the assigned NMR resonances. Thus, the polarity of the PISA wheel provides a direct

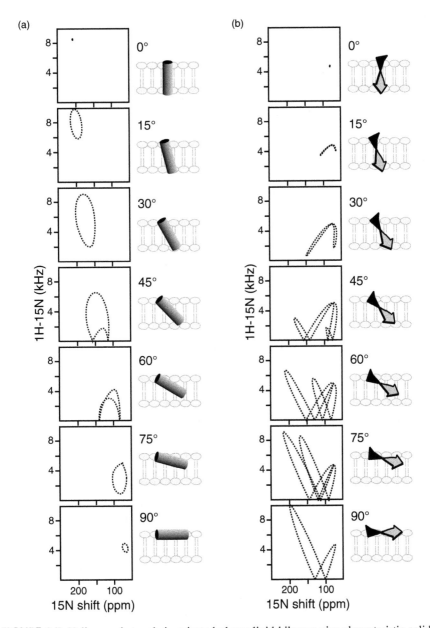

FIGURE 8.7 Helices and strands in oriented planar lipid bilayers give characteristic solid-state NMR spectra called PISA wheels. The ^{1}H-^{15}N dipolar coupling/^{15}N chemical shift PISEMA spectra were calculated for (a) an ideal α-helix with uniform dihedral angles (phi/psi = 65°/40°), and (b) an ideal β-strand with uniform dihedral angles (phi/psi = 135°/140°), at different tilts relative to the magnetic field direction and the membrane normal. The ^{15}N chemical shifts are referenced to 0 ppm for liquid ammonia. Spectra were calculated as described.[78]

measure of both the helix tilt and the rotation about its long axis within the membrane. Because a single resonance assignment is sufficient to index the PISA wheel, a single assignment is also sufficient to determine helix rotation in the membrane.

The structure of porin from *R. capsulatus* consists of 16 β-strands arranged to form a β-barrel pore through the membrane.[89] The PISEMA spectrum calculated from the PDB coordinates of the structure (Figure 8.8d) displays only modest resolution among its 269 resonances (301 residues excluding the N-terminus and the 38 prolines) and demonstrates the need for higher-dimensional spectroscopy. However, the majority of resonances fall on PISA wheel traces, characteristic of β-strands tilted by 30°–60°. For example, the 35° degree tilt of the fourteenth β-strand, β14, is immediately apparent from its PISEMA spectrum, which maps onto the twisted PISA wheel calculated for an ideal β-strand with a tilt of 35° (Figure 8.7e). Residues from opposite sides of the strand occupy different wings of the spectrum in a manner similar to that of its "beta" wheel representation in Figure 8.7f. Even-numbered residues face the polar interior of the protein and occupy the upfield wing of the twisted PISA wheel, whereas odd-numbered residues face the hydrophobic membrane interior and occupy the downfield wing of the spectrum. This segregation of resonances from pore-facing or lipid-facing residues is seen in the spectra from all strands.[78]

8.6 FXYD CHIF IN LIPID BILAYERS

The two-dimensional PISEMA spectrum of CHIF in lipid bilayers is shown in Figure 8.9. Each amide site in the protein contributes one correlation peak, characterized by ^{1}H-^{15}N dipolar coupling and ^{15}N chemical shift frequencies that reflect NH bond orientation relative to the membrane. For CHIF, the PISA wheel that is observed in the region from 6 to 10 kHz and 180 to 220 ppm of the PISEMA spectrum provides definitive evidence that the protein associates with the lipid bilayer as a transmembrane helix. To estimate the tilt of the CHIF transmembrane helix, we compared the experimental spectrum with those calculated for an ideal α-helix, with 3.6 residues per turn and identical backbone dihedral angles for all residues (phi, psi = −57°, 47°), tilted at 10°, 15°, and 20° relative to the lipid bilayer normal. This comparative analysis demonstrates that the CHIF helix is tilted by about 15° in the membrane (or 75° from the membrane surface).

The expression and purification of the recombinant FXYD family proteins PLM, CHIF, and Mat8 enable NMR structural studies to be performed in bilayer and micelle environments. The combined restraints from solution and solid-state NMR studies provide a view of the protein structures in membranes: the hydrophobic residues from Val19 to Lys40 form a helix that traverses the membrane at an angle of 15°, whereas the N and C termini form helices that rest on the membrane surface (Figure 8.10).

The FXYD membrane proteins regulate ion, osmolyte, and fluid homeostasis in a variety of tissues and are emerging as auxiliary tissue–specific and physiological state–specific subunits of the Na,K-ATPase. The ability to produce milligram quantities of pure FXYD proteins also opens the door for functional studies that together with structure determination, can provide important structure–activity correlations.

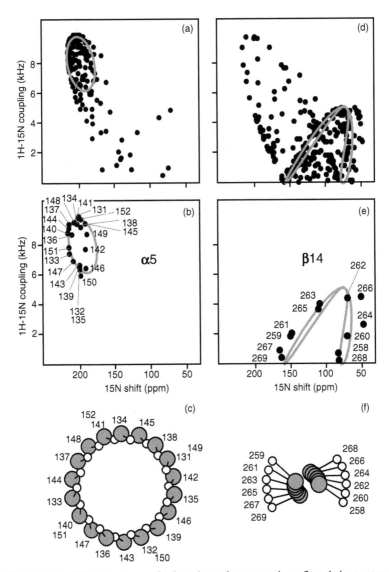

FIGURE 8.8 The PISEMA spectra of oriented membrane proteins reflect their structures and topologies. Spectra were calculated for (a–c) bacteriorhodopsin (PDB file 1C3W) and (d–f) integral membrane porin from *Rhodobacter capsulatus* (PDB file 2POR) in oriented lipid bilayers. (a) PISEMA spectrum calculated for uniformly ^{15}N-labeled bacteriorhodopsin. (b) The PISEMA spectrum extracted for helix α5, which has a tilt of 15° in the membrane, fits to a PISA wheel calculated for an ideal α-helix with a 15° tilt (gray trace). (c) The pattern of resonances in the PISA wheel reflects the helical wheel projection of helix α5. (d) PISEMA spectrum calculated for uniformly ^{15}N-labeled porin. (e) The spectrum extracted for β14 fits on the twisted PISA wheel that was calculated for an ideal β-strand with a tilt of 35° (gray trace). (f) The pattern of resonances in the PISA wheel reflects the wheel projection of strand β14. The arrows mark the direction of the magnetic field and the lipid bilayer normal. The ^{15}N chemical shifts are referenced to 0 ppm for liquid ammonia. Spectra were calculated as described.[78]

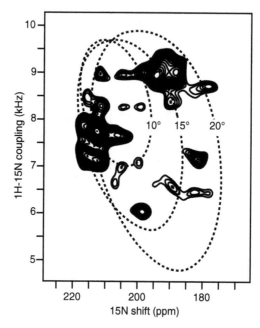

FIGURE 8.9 The experimental ^{1}H/^{15}N solid-state NMR PISEMA spectrum of CHIF is super-imposed on the PISA wheels calculated for ideal α-helices with tilts of 10°, 15°, and 20° in the lipid bilayer. The spectrum fits on the 15° PISA wheel, thus determining the tilt of the CHIF transmembrane helix. ^{15}N chemical shifts are referenced to 0 ppm for liquid ammonia.

For example, the reconstitution of Na,K-ATPase activity in the presence of FXYD proteins would be an important step in understanding the mechanism of pump regulation, whereas the incorporation of FXYD proteins in lipid bilayers would enable ion channel activities to be characterized by measuring specific ionic currents.

8.7 CONCLUSIONS

The combination of solution and solid-state NMR is a powerful approach for the structure determination of proteins in planar lipid bilayers. The method requires the production of milligram quantities of isotopically labeled protein, followed by pro-tein reconstitution in lipid micelles and oriented lipid bilayers. Substantial progress has been made in the areas of both recombinant protein expression and sample preparation, so that the spectra of several membrane proteins have now been recorded for structure determination, and the high-resolution structures of six membrane proteins have been determined in bilayers.[1–7]

Solid-state NMR PISA wheels and solution NMR dipolar waves provide pow-erful and visually accessible indices of membrane protein secondary structure and topology. PISA wheels are observed for many α-helical membrane proteins in lipid bilayers, and we predict that they are also likely to be observed for oriented β-stranded proteins, although this case remains to be examined experimentally. Although the definitive answer about the topology of a membrane protein can only

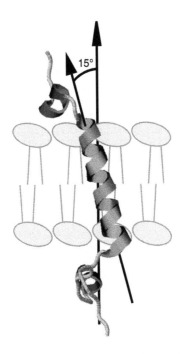

FIGURE 8.10 The structure of FXYD CHIF in lipid bilayers obtained from the combined solution and solid-state NMR data. The transmembrane helix has a tilt of 15°.

come from its three-dimensional structure, obtained from the resolution and sequential assignment of resonances from all residues, these patterns enable valuable structural information to emerge from the spectroscopic studies before complete three-dimensional structure determination and also provide the basis for structure determination methods.[4,5]

ACKNOWLEDGMENTS

Our research was supported by grants from the National Institutes of Health (R01CA082864) and the Department of the Army Breast Cancer Research Program (DAMD17-00-1-0506). The NMR studies used the Burnham Institute NMR Facility, supported by a grant from the National Institutes of Health (P30CA30199) and the Biomedical Technology Resources for Solid-State NMR of Proteins at the University of California San Diego, supported by a grant from the National Institutes of Health (P41EB002031).

REFERENCES

1. Ketchem, R.R., Hu, W. and Cross, T.A., High-resolution conformation of gramicidin A in a lipid bilayer by solid-state NMR, *Science*, 261, 1457, 1993.

2. Opella, S.J., Marassi, F.M., Gesell, J.J., Valente, A.P., Kim, Y., Oblatt-Montal, M. and Montal, M., Structures of the M2 channel-lining segments from nicotinic acetylcholine and NMDA receptors by NMR spectroscopy, *Nat. Struct. Biol.*, 6, 374, 1999.

3. Valentine, K.G., Liu, S.F., Marassi, F.M., Veglia, G., Opella, S.J., Ding, F.X., Wang, S.H., Arshava, B., Becker, J.M. and Naider, F., Structure and topology of a peptide segment of the 6th transmembrane domain of the *Saccharomyces cerevisiae* alpha-factor receptor in phospholipid bilayers, *Biopolymers*, 59, 243, 2001.

4. Wang, J., Kim, S., Kovacs, F. and Cross, T.A., Structure of the transmembrane region of the M2 protein H(+) channel, *Protein Sci.*, 10, 2241, 2001.

5. Marassi, F.M. and Opella, S.J., Simultaneous assignment and structure determination of a membrane protein from NMR orientational restraints, *Protein Sci.*, 12, 403, 2003.

6. Park, S.H., Mrse, A.A., Nevzorov, A.A., Mesleh, M.F., Oblatt-Montal, M., Montal, M., and Opella, S.J., Three-dimensional structure of the channel-forming trans-membrane domain of virus protein "u" (Vpu) from HIV-1, *J. Mol. Biol.*, 333, 409, 2003.

7. Zeri, A.C., Mesleh, M.F., Nevzorov, A.A. and Opella, S.J., Structure of the coat protein in fd filamentous bacteriophage particles determined by solid-state NMR spectroscopy, *Proc. Natl. Acad. Sci. USA*, 100, 6458, 2003.

8. Sweadner, K.J. and Rael, E. The FXYD gene family of small ion transport regulators or channels: cDNA sequence, protein signature sequence, and expression, *Genomics*, 68, 41, 2000.

9. Palmer, C.J., Scott, B.T. and Jones, L.R., Purification and complete sequence determination of the major plasma membrane substrate for cAMP-dependent protein kinase and protein kinase C in myocardium, *J. Biol. Chem.*, 266, 11126, 1991.

10. Attali, B., Latter, H., Rachamim, N. and Garty, H., A corticosteroid-induced gene expressing an "IsK-like" K+ channel activity in *Xenopus* oocytes, *Proc. Natl. Acad. Sci. USA*, 92, 6092, 1995.

11. Morrison, B.W., Moorman, J.R., Kowdley, G.C., Kobayashi, Y.M., Jones, L.R. and Leder, P., Mat-8, a novel phospholemman-like protein expressed in human breast tumors, induces a chloride conductance in *Xenopus* oocytes, *J. Biol. Chem.*, 270, 2176, 1995.

12. Vaarala, M.H., Porvari, K., Kyllonen, A. and Vihko, P., Differentially expressed genes in two LNCaP prostate cancer cell lines reflecting changes during prostate cancer progression, *Lab. Invest.*, 80, 1259, 2000.

13. Moorman, J.R., Palmer, C.J., John, J.E., 3rd, Durieux, M.E. and Jones, L.R., Phospholemman expression induces a hyperpolarization-activated chloride current in Xenopus oocytes, *J. Biol. Chem.*, 267, 14551, 1992.

14. Crowell, K.J., Franzin, C.M., Koltay, A., Lee, S., Lucchese, A.M., Snyder, B.C. and Marassi, F.M., Expression and characterization of the FXYD ion transport regulators for NMR structural studies in lipid micelles and lipid bilayers, *Biochim. Biophys. Acta*, 1645, 15, 2003.

15. Miroux, B. and Walker, J.E., Over-production of proteins in *Escherichia coli*: mutant hosts that allow synthesis of some membrane proteins and globular proteins at high levels, *J. Mol. Biol.*, 260, 289, 1996.

16. Di Guan, C., Li, P., Riggs, P.D. and Inouye, H., Vectors that facilitate the expression and purification of foreign peptides in Escherichia coli by fusion to maltose-binding protein, *Gene*, 67, 21, 1988.

17. Smith, D.B. and Johnson, K.S., Single-step purification of polypeptides expressed in *Escherichia coli* as fusions with glutathione S-transferase, *Gene*, 67, 1988.

18. Staley, J.P. and Kim, P.S., Formation of a native-like subdomain in a partially folded intermediate of bovine pancreatic trypsin inhibitor, *Protein Sci.*, 3, 1822, 1994.

19. Pautsch, A., Vogt, J., Model, K., Siebold, C. and Schulz, G.E., Strategy for membrane protein crystallization exemplified with OmpA and OmpX, *Proteins*, 34, 167, 1999.
20. Opella, S.J., Ma, C. and Marassi, F.M., Nuclear magnetic resonance of membrane-associated peptides and proteins, *Methods Enzymol.*, 339, 285, 2001.
21. Gross, E. and Witkop, B., Selective cleavage of the methionyl peptide bonds in ribonuclease with cyanogen bromide, *J. Am. Chem. Soc.*, 83, 1510, 1961.
22. Andrade, M.A., Chacon, P., Merelo, J.J. and Moran, F., Evaluation of secondary structure of proteins from UV circular dichroism spectra using an unsupervised learning neural network, *Protein Eng.*, 6, 383, 1993.
23. Henry, G.D. and Sykes, B.D., Methods to study membrane protein structure in solution, *Methods Enzymol.*, 239, 515, 1994.
24. Williams, K.A., Farrow, N.A., Deber, C.M. and Kay, L.E., Structure and dynamics of bacteriophage IKe major coat protein in MPG micelles by solution NMR, *Biochemistry*, 35, 5145, 1996.
25. Almeida, F.C. and Opella, S.J., fd coat protein structure in membrane environments: structural dynamics of the loop between the hydrophobic trans-membrane helix and the amphipathic in-plane helix, *J. Mol. Biol.*, 270, 481, 1997.
26. Gesell, J., Zasloff, M. and Opella, S.J., Two-dimensional 1H NMR experiments show that the 23-residue magainin antibiotic peptide is an alpha-helix in dodecylphosphocholine micelles, sodium dodecylsulfate micelles, and trifluoroethanol/water solution, *J. Biomol. NMR*, 9, 127, 1997.
27. MacKenzie, K.R., Prestegard, J.H. and Engelman, D.M., A transmembrane helix dimer: structure and implications, *Science*, 276, 131, 1997.
28. Arora, A., Abildgaard, F., Bushweller, J.H. and Tamm, L.K., Structure of outer membrane protein A transmembrane domain by NMR spectroscopy, *Nat. Struct. Biol.*, 8, 334, 2001.
29. Fernandez, C., Adeishvili, K. and Wuthrich, K., Transverse relaxation-optimized NMR spectroscopy with the outer membrane protein OmpX in dihexanoyl phosphatidylcholine micelles, *Proc. Natl. Acad. Sci. USA*, 98, 2358, 2001.
30. Hwang, P.M., Choy, W.Y., Lo, E.I., Chen, L., Forman-Kay, J.D., Raetz, C.R., Prive, G.G., Bishop, R.E. and Kay, L.E., Solution structure and dynamics of the outer membrane enzyme PagP by NMR, *Proc. Natl. Acad. Sci. USA*, 99, 13560, 2002.
31. Ma, C., Marassi, F.M., Jones, D.H., Straus, S.K., Bour, S., Strebel, K., Schubert, U., Oblatt-Montal, M., Montal, M. and Opella, S.J., Expression, purification, and activities of full-length and truncated versions of the integral membrane protein Vpu from HIV-1, *Protein Sci.*, 11, 546, 2002.
32. Mascioni, A., Karim, C., Barany, G., Thomas, D.D. and Veglia, G., Structure and orientation of sarcolipin in lipid environments, *Biochemistry*, 41, 475, 2002.
33. Oxenoid, K., Sonnichsen, F.D. and Sanders, C.R., Topology and secondary structure of the N-terminal domain of diacylglycerol kinase, *Biochemistry*, 41, 12876, 2002.
34. Sorgen, P.L., Cahill, S.M., Krueger-Koplin, R.D., Krueger-Koplin, S.T., Schenck, C.C. and Girvin, M.E., Structure of the Rhodobacter sphaeroides light-harvesting 1 beta subunit in detergent micelles, *Biochemistry*, 41, 31, 2002.
35. Krueger-Koplin, R.D., Sorgen, P.L., Krueger-Koplin, S.T., Rivera-Torres, I.O., Cahill, S.M., Hicks, D.B., Grinius, L., Krulwich, T.A. and Girvin, M.E., An evaluation of detergents for NMR structural studies of membrane proteins, *J. Biomol. NMR*, 28, 43, 2004.
36. Mesleh, M.F., Lee, S., Veglia, G., Thiriot, D.S., Marassi, F.M. and Opella, S.J., Dipolar waves map the structure and topology of helices in membrane proteins, *J. Am. Chem. Soc.*, 125, 8928, 2003.

37. Lee, S., Mesleh, M.F. and Opella, S.J., Structure and dynamics of a membrane protein in micelles from three solution NMR experiments, *J. Biomol. NMR*, 26, 327, 2003.

38. Marassi, F.M. and Opella, S.J., Simultaneous assignment and structure determination of a membrane protein from NMR orientational restraints, *Protein Sci.*, 12, 403, 203.

39. Oxenoid, K., Kim, H.J., Jacob, J., Sonnichsen, F.D. and Sanders, C.R., NMR assignments for a helical 40 kDa membrane protein, *J. Am. Chem. Soc.*, 126, 5048, 2004.

40. Bax, A., Kontaxis, G. and Tjandra, N., Dipolar couplings in macromolecular structure determination, *Meth. Enzymol.*, 339, 127, 2001.

41. Prestegard, J.H. and Kishore, A.I., Partial alignment of biomolecules: an aid to NMR characterization, *Curr. Opin. Chem. Biol.*, 5, 584, 2001.

42. Mesleh, M.F., Veglia, G., DeSilva, T.M., Marassi, F.M. and Opella, S.J., Dipolar waves as NMR maps of protein structure, *J. Am. Chem. Soc.*, 124, 4206, 2002.

43. Mesleh, M.F. and Opella, S.J., Dipolar waves as NMR maps of helices in proteins, J. Magn. Reson., 163, 288, 2003.

44. Meier, S., Haussinger, D. and Grzesiek, S., Charged acrylamide copolymer gels as media for weak alignment, *J. Biomol. NMR*, 24, 351, 2002.

45. Chou, J.J., Gaemers, S., Howder, B., Louis, J.M. and Bax, A., A simple apparatus for generating stretched polyacrylamide gels, yielding uniform alignment of proteins and detergent micelles, *J. Biomol. NMR*, 21, 377, 2001.

46. Tycko, R., Blanco, F.J. and Ishii, Y., Alignment of biopolymers in strained gels: a new way to create detectable dipole-dipole couplings in high-resolution biomolecular NMR, *J. Am. Chem. Soc.*, 122, 9340, 2000.

47. Sass, H.J., Musco, G., Stahl, S.J., Wingfield, P.T. and Grzesiek, S., Solution NMR of proteins within polyacrylamide gels: diffusional properties and residual alignment by mechanical stress or embedding of oriented purple membranes, *J. Biomol. NMR*, 18, 303, 2000.

48. Shortle, D. and Ackerman, M.S., Persistence of native-like topology in a denatured protein in 8 M urea, *Science*, 293, 487, 2001.

49. Chou, J.J., Kaufman, J.D., Stahl, S.J., Wingfield, P.T. and Bax, A., Micelle-induced curvature in a water-insoluble HIV-1 Env peptide revealed by NMR dipolar coupling measurement in stretched polyacrylamide gel, *J. Am. Chem. Soc.*, 124, 2450, 2002.

50. Sanders, C.R., Sonnichsen, F.D. and Oxenoid, K., Tackling complex membrane proteins using solution NMR, paper presented at the 20th International Conference of Magnetic Resonance in Biological Systems, Toronto, Canada, 2002.

51. Meier, S., Haussinger, D., Jensen, P., Rogowski, M. and Grzesiek, S., High-accuracy residual 1HN-13C and 1HN-1HN dipolar couplings in perdeuterated proteins, *J. Am. Chem. Soc.*, 125, 44, 2003.

52. Battiste, J.L. and Wagner, G., Utilization of site-directed spin labeling and high-resolution heteronuclear nuclear magnetic resonance for global fold determination of large proteins with limited nuclear overhauser effect data. *Biochemistry*, 39, 5355, 2000.

53. Papavoine, C.H., Konings, R.N., Hilbers, C.W. and van de Ven, F.J., Location of M13 coat protein in sodium dodecyl sulfate micelles as determined by NMR, *Biochemistry*, 33, 12990, 1994.

54. Van Den Hooven, H.W., Doeland, C.C., Van De Kamp, M., Konings, R.N., Hilbers, C.W. and Van De Ven, F.J., Three-dimensional structure of the antibiotic nisin in the presence of membrane-mimetic micelles of dodecylphosphocholine and of sodium dodecylsulphate, *Eur. J. Biochem.*, 235, 382, 1996.

55. Jarvet, J., Zdunek, J., Damberg, P. and Graslund, A., Three-dimensional structure and position of porcine motilin in sodium dodecyl sulfate micelles determined by 1H NMR, *Biochemistry*, 36, 8153, 1997.

56. Damberg, P., Jarvet, J. and Graslund, A., Micellar systems as solvents in peptide and protein structure determination, *Methods Enzymol.*, 339, 271, 2001.

57. Kutateladze, T.G., Capelluto, D.G., Ferguson, C.G., Cheever, M.L., Kutateladze, A.G., Prestwich, G.D. and Overduin, M., Multivalent mechanism of membrane insertion by the FYVE domain, *J. Biol. Chem.*, 279, 3050, 2004.

58. Mori, S., Abeygunawardana, C., Johnson, M.O. and Vanzijl, P.C.M., Improved sensitivity of HSQC spectra of exchanging protons at short interscan delays using a new fast HSQC (FHSQC) detection scheme that avoids water saturation, *J. Magn. Reson. B*, 108, 94, 1995.

59. Yamazaki, T., Lee, W., Arrowsmith, C.H., Muhandiram, D.R. and Kay, L.E., A suite of triple resonance NMR experiments for the backbone assignment of 15N, 13C, 2H labeled proteins with high sensitivity, *J. Am. Chem. Soc.*, 116, 11655, 1994.

60. Wittekind, M. and Mueller, L., HNCACB, a high-sensitivity 3D NMR experiment to correlate amide protons and nitrogen resonances with the alpha- and beta-carbon resonances in proteins, *J. Magn. Reson. B*, 101, 201, 1993.

61. Grzesiek, S. and Bax, A., Correlating backbone amide and side chain resonances in larger proteins by multiple relayed triple resonance NMR, *J. Am. Chem. Soc.*, 114, 6291, 1992.

62. Clore, G.M. Gronenborn, A.M., Determination of three-dimensional structures of proteins and nucleic acids in solution by nuclear magnetic resonance spectroscopy, *Crit. Rev. Biochem. Mol. Biol.*, 24, 479, 1989.

63. Wuthrich, K., Determination of three-dimensional protein structures in solution by nuclear magnetic resonance: an overview, *Methods Enzymol.*, 177, 125, 1989.

64. Ferentz, A.E. and Wagner, G., NMR spectroscopy: a multifaceted approach to macromolecular structure, *Q. Rev. Biophys.*, 33, 29, 2000.

65. Czerski, L., Vinogradova, O. and Sanders, C.R., NMR-Based amide hydrogen-deuterium exchange measurements for complex membrane proteins: development and critical evaluation, *J. Magn. Reson.*, 142, 111, 2000.

66. Veglia, G., Zeri, A.C., Ma, C. and Opella, S.J. Deuterium/hydrogen exchange factors measured by solution nuclear magnetic resonance spectroscopy as indicators of the structure and topology of membrane proteins, *Biophys. J.* 82, 2176, 2002.

67. Bogusky, M.J., Leo, G.C. and Opella, S.J., Comparison of the dynamics of the membrane-bound form of fd coat protein in micelles and in bilayers by solution and solid-state nitrogen-15 nuclear magnetic resonance spectroscopy, *Proteins*, 4, 123, 1988.

68. Boguski, M.J., Schiksnis, R.A., Leo, G.C. and Opella, S.J., Protein backbone dynamics by solid-state and solution 15N NMR spectroscopy, *J. Magn. Reson.*, 72, 186, 1987.

69. Gust, D., Moon, R.B. and Roberts, J.D., Applications of natural-abundance nitrogen-15 nuclear magnetic resonance to large biochemically important molecules, *Proc. Natl. Acad. Sci. USA*, 72, 4696, 1975.

70. Farrow, N.A., Zhang, O., Forman-Kay, J.D. and Kay, L.E., A heteronuclear correlation experiment for simultaneous determination of 15N longitudinal decay and chemical exchange rates of systems in slow equilibrium, *J. Biomol. NMR*, 4, 727, 1994.

71. Tolman, J.R., Flanagan, J.M., Kennedy, M.A. and Prestegard, J.H., Nuclear magnetic dipole interactions in field-oriented proteins: information for structure determination in solution, *Proc. Natl. Acad. Sci. USA,* 92, 9279, 1995.

72. Clore, G.M. and Gronenborn, A.M., New methods of structure refinement for macromolecular structure determination by NMR, *Proc. Natl. Acad. Sci. USA*, 95, 5892–5898, 1998.

73. Hus, J.C., Marion, D. and Blackledge, M.J., *De novo* determination of protein structure by NMR using orientational and long-range order restraints, *J. Mol. Biol.*, 298, 927, 2000.

74. Mueller, G.A., Choy, W.Y., Yang, D., Forman-Kay, J.D., Venters, R.A. and Kay, L.E., Global folds of proteins with low densities of NOEs using residual dipolar couplings: application to the 370-residue maltodextrin-binding protein, *J. Mol. Biol.*, 300, 197, 2000.

75. Fowler, C.A., Tian, F., Al-Hashimi, H.M. and Prestegard, J.H., Rapid determination of protein folds using residual dipolar couplings, *J. Mol. Biol.*, 304, 447, 2000.

76. Delaglio, F., Kontaxis, G. and Bax, A., Protein structure determination using molecular fragment replacement and NMR dipolar couplings, *J. Am. Chem. Soc.*, 122, 2142, 2000.

77. Wang, J., Denny, J., Tian, C., Kim, S., Mo, Y., Kovacs, F., Song, Z., Nishimura, K., Gan, Z., Fu, R., et al., Imaging membrane protein helical wheels, *J. Magn. Reson.*, 144, 162, 2000.

78. Marassi, F.M., A simple approach to membrane protein secondary structure and topology based on NMR spectroscopy, *Biophys. J.*, 80, 994, 2001.

79. Marassi, F.M. and Opella, S.J., A solid-state NMR index of helical membrane protein structure and topology, *J. Magn. Reson.*, 144, 150, 2000.

80. Marassi, F.M., NMR of peptides and proteins in membranes, *Concepts Magn. Reson.*, 14, 212, 2002.

81. Marassi, F.M., Ramamoorthy, A. and Opella, S.J., Complete resolution of the solid-state NMR spectrum of a uniformly 15N-labeled membrane protein in phospholipid bilayers, *Proc. Natl. Acad. Sci. USA*, 94, 8551, 1997.

82. Franzin, C.M., Choi, J., Zhai, D., Reed, J.C. and Marassi, F.M., Structural studies of apoptosis and ion transport regulatory proteins in membranes, *Magn. Reson. Chem.*, 42, 172, 2004.

83. Tian, C., Gao, P.F., Pinto, L.H., Lamb, R.A. and Cross, T.A., Initial structural and dynamic characterization of the M2 protein transmembrane and amphipathic helices in lipid bilayers, *Protein Sci.*, 12, 2597, 2003.

84. Wu, C.H., Ramamoorthy, A. and Opella, S.J., High-resolution heteronuclear dipolar solid-state NMR spectroscopy, *J. Magn. Reson. A*, 109, 270, 1994.

85. Schiffer, M. and Edmunson, A.B., Use of helical wheels to represent the structures of proteins and to identify segments with helical potential, *Biophys. J.*, 7, 121, 1967.

86. Marassi, F.M. and Opella, S.J., Using pisa pies to resolve ambiguities in angular constraints from PISEMA spectra of aligned proteins, *J. Biomol. NMR*, 23, 239, 2002.

87. Marassi, F.M., Ma, C., Gesell, J.J. and Opella, S.J., Three-dimensional solid-state NMR spectroscopy is essential for resolution of resonances from in-plane residues in uniformly (15)N-labeled helical membrane proteins in oriented lipid bilayers, *J. Magn. Reson.*, 144, 156, 2000.

88. Luecke, H., Schobert, B., Richter, H.T., Cartailler, J.P. and Lanyi, J.K., Structure of bacteriorhodopsin at 1.55 A resolution, *J. Mol. Biol.*, 291, 899, 1999.

89. Weiss, M.S., Kreusch, A., Schiltz, E., Nestel, U., Welte, W., Weckesser, J. and Schulz, G.E., The structure of porin from Rhodobacter capsulatus at 1.8 A resolution, *FEBS Lett.*, 280, 379, 1991.

9 Solid-State ^{19}F-Nuclear Magnetic Resonance Analysis of Membrane-Active Peptides

Anne S. Ulrich, Parvesh Wadhwani,
Ulrich H.N. Dürr, Sergii Afonin, Ralf W. Glaser,
Erik Strandberg, Pierre Tremouilhac,
Carsten Sachse, Marina Berditchevskaia, and
Stephan Grage

CONTENTS

ABBREVIATIONS: $\chi 1$: side chain torsion angle around Cα-Cβ; θ: angle between the spin interaction tensor and the membrane normal; φ: second polar angle describing the alignment of an asymmetric spin interaction tensor with respect to the membrane normal; τ: angle describing the tilt of a peptide in the membrane; ρ: azimuthal rotation angle of a peptide in the membrane; 3F-Ala: 3-fluoro-alanine; 4F-Phg: 4-fluoro-phenylglycine; Aib: aminoisobutyric acid; CSA: chemical shift anisotropy; CF3-Ala: 2-trifluoromethyl-alanine; CF_3-Phg: 4-trifluoromethyl-phenylglycine; DIC: diisopropylcatbodiimide; DLPC: dilauryl-phosphatidylchonline; DMPC: dimyristoyl-phosphatidylcholine; DPPC: dipalmitoyl-phosphatidylcholine; F_3-Ala: 3,3,3-trifluoro-alanine; HOBt: O-(1H-benzoxytriazole); MAS: magic-angle spinning; N: membrane normal; r: distance between two ^{19}F-labels; POPC: palmitoyl-oleoyl-phosphatidylcholine; REDOR: rotational echo double resonance; TFA: trifluoroacetic acid; TFE: trifluoroethanol

KEY WORDS: biomembranes, ^{19}F-NMR, fluorine labeling, antimicrobial peptides fusogenic peptides

9.1 BACKGROUND

This chapter is focused on solid-state ^{19}F-nuclear magnetic resonance (NMR) studies of membrane-active peptides, including an outline of synthetic labeling strategies and an overview of the results on several biologically relevant systems. The methodological considerations and fundamental ^{19}F-NMR parameters have been recently discussed in a complementary review;[1] hence, no NMR details or technical explanations will be given here. Instead, the emphasis lies on the general approach of resolving the structures and monitoring structural changes of peptides in model membranes in a highly sensitive manner.

9.1.1 MEMBRANE-ACTIVE PEPTIDES

Membrane proteins are characterized by typical structural motifs such as transmembrane α-helical bundles or β barrels, which keep the protein fully immersed and stable in the lipid bilayer. However, there are also many biological examples in which comparatively short amphiphilic peptides (10–50 amino acids) interact with membranes transiently or in a dynamic fashion. These sequences may be functional entities *per se*, such as antimicrobial and cytotoxic peptides, or they may be part of larger proteins, such as fusogenic peptides and membrane recruitment segments. Several models are illustrated in Figure 9.1, showing how various types of amphiphilic sequences can interact with lipid bilayers. The peptide structures are often found to be α-helical, but they may also have a β-sheet conformation or any other complex fold. In every case, the global architecture exposes a specific pattern of hydrophilic and hydrophobic, charged and uncharged H-bonding shielded residues on the peptide surface to face the appropriate aqueous and lipid environment.

Remarkably, many peptides have been reported to undergo structural changes in response to membrane binding, to the local environment, to their own concentration, or to an external trigger such as a pH change. Therefore, in trying to understand the function of a membrane-active peptide, it is not sufficient to draw a static picture of its three-dimensional structure, but also the dynamic aspects of its membrane

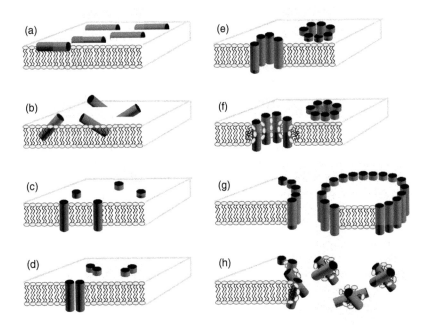

FIGURE 9.1 Amphiphilic peptides can be engaged in many different types of interactions with lipid membranes, depending on the distribution of polar and hydrophobic amino acids: (a) planar surface-bound state, (b) immersion with an oblique tilt angle, (c) transmembrane alignment, (d) dimer of transmembrane segments, (e) transmembrane pore or barrel stave, (f) toroidal wormhole pore, (g) formation of bicelles, (h) formation of micelles.

interactions and multiple membrane-bound states have to be characterized in detail. This includes conformational interconversions (e.g., from disordered to ordered, or from α-helix to β-sheet), structural realignments (e.g., from flat on the surface to membrane immersed), self-assembly processes (e.g., from monomeric to oligomeric), and so on. Although crystallography has been very successful in structural biology — even for membrane proteins — the dynamic behavior of peptides is much harder to deal with by diffraction methods. Other spectroscopic techniques such as fluorescence, circular dichroism, infrared, or ESR remain rather qualitative with regard to the three-dimensional structural resolution achieved. Solid-state NMR may thus be considered the most suitable approach for obtaining local structural details as well as overall global information about peptides in a quasi-native lipid environment.[2–3] Isotope labeling is usually required for these kinds of NMR studies. Site-specific as well as uniform labeling schemes are traditionally employed with ^{2}H, ^{13}C, and ^{15}N in peptides and proteins, whereas ^{31}P is naturally abundant in phospholipids. More recently, significant improvements in terms of sensitivity and detectability have been made by studying selectively ^{19}F-labeled biomolecules,[1] which will be reviewed here.

9.1.2 SOLID-STATE ^{19}F-NMR

The choice of ^{19}F as an NMR label is motivated mainly by the considerable increase in sensitivity by one to two orders of magnitude compared with conventional labels.[4-5] Instead of using several milligrams of peptide for a single experiment, ^{19}F-NMR spectra can be acquired with 40 μg (≈20 nmol) in an overnight measurement, allowing for example to access peptide/lipid ratios of 1:3000.[6] Additional advantages of ^{19}F lie in the extended distance range accessible in dipolar measurements, and in the fact that this nucleus has no natural abundance background.

The peptide–lipid systems characterized here will be based on two types of NMR parameters, as illustrated in Figure 9.2, namely, orientational constraints in terms of an angle θ between the spin interaction tensor and the membrane normal (plus another angle φ for asymmetric tensors), and intra- or intermolecular distance constraints r between two spins. A small number of orientational constraints suffices to define the alignment and mobility of a well-folded peptide in a membrane. This is done in terms of the peptide tilt angle τ, its azimuthal rotation angle ρ, and an order parameter S_{mol} describing the extent of motional averaging (see Figure 9.2). Distance constraints are useful to characterize secondary structures, and they are essential to obtain direct information on intermolecular interactions.

As the basics of ^{19}F-NMR structure analysis have been recently reviewed,[1,3] here we only note that orientational constraints are usually measured in static samples that are often macroscopically aligned, such that the membrane normal N is parallel

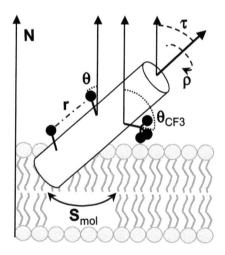

FIGURE 9.2 Illustration of the relevant ^{19}F-NMR labels and structural parameters that are used to describe the alignment and mobility of a well-folded peptide in a lipid bilayer. The labels have to be rigidly attached to the peptide backbone and are shown as black spheres, either as single substituents or CF_3-groups. Local orientational constraints are measured in terms of an angle θ (or θ_{CF3}) between the substituent axis and the membrane normal N (dotted lines), and a distance r can be measured between two such labels (dash-dot). The alignment of the entire peptide is described by its tilt angle τ and azimuthal rotation ρ (dashed lines) and by a molecular order parameter S_{mol}.

to the static magnetic field direction. The value of θ is then obtained by analyzing the CSA (chemical shift anisotropy) of a single ^{19}F-label,[4,7] from the dipolar coupling between two such labels,[8] or from the dipolar coupling within a CF_3-group.[9,10] Distance measurements always rely on the dipolar coupling strength between two labels. They can be carried out either under magic-angle spinning (MAS) by heteronuclear REDOR (rotational echo double resonance) or homonuclear dipolar recoupling techniques,[11–13] or in a static sample by a homonuclear CPMG experiment.[14,15]

9.2 CHEMICAL LABELING STRATEGIES FOR ^{19}F-NMR

Labeling of peptides with ^{19}F is not trivial, neither in view of peptide synthesis nor with regard to biological integrity. When replacing a proton with a highly electronegative ^{19}F substituent, it is essential to check whether this may have introduced any structural distortions or disturbed the function. Labeling of aliphatic and aromatic amino acids is usually tolerable and may only contribute to a slightly enhanced hydrophobicity, whereas ^{19}F near any functional group should be avoided. For NMR structure analysis, however, it is not just sufficient to attach a harmless ^{19}F-label onto a suitable side chain. In fact, the reporter group should be placed at a strategic position where it represents the structure and mobility of the whole peptide. A suitable ^{19}F-label therefore should be rigidly attached to the peptide backbone, should be small enough not to disturb the overall fold, and should maintain the same hydrophobicity as the native substituent. According to these criteria, several candidates have been identified among the multitude of fluorinated amino acids[16] and used for NMR structure analysis of peptides. These are 4F-L-phenylglycine (4F-Phg), 4-CF_3-L-phenylglycine (CF_3-Phg), 3F-L-alanine (F-Ala), 3,3,3-F_3-L-alanine (F_3-Ala), and 2-CF_3-alanine (CF_3-Ala). We will mostly use the short names and omit the stereochemical assignment wherever it is obvious.

In the following sections, we discuss the specific chemical challenges of incorporating each of these amino acids into synthetic peptides, and point out their particular advantages and complications for structure analysis. One general remark concerns all samples prepared for ^{19}F-NMR: it is essential to avoid fluorinated solvents such as TFE (trifluoroethanol) or TFA (trifluoroacetic acid). Even though TFA is an indispensable cleavage agent in Fmoc solid-phase peptide chemistry and is commonly used as an additive in HPLC, the final purification step needs to be performed on a TFA-free column. Instead of using 0.1% TFA, adequate resolution can usually be obtained with 5 mM HCl in the gradient.[17] Experience indicates that even laboratory pipettes and solvent bottles with broken seals may get contaminated by TFA; hence, it should be carefully stored and kept away from any ^{19}F-NMR sample.

9.2.1 4F-PHENYLGLYCINE

Phenylglycine (Phg) is a non-proteinogenic amino acid that occurs in a number of natural products, drugs, and artificial sweeteners. In contrast to phenylalanine, the phenyl ring is directly attached to the α-carbon, and the ^{19}F-label is most suitably attached in *para*-position. During peptide synthesis, the electronegative

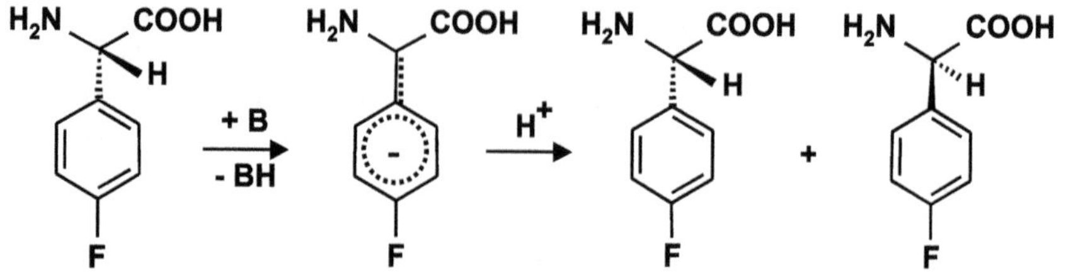

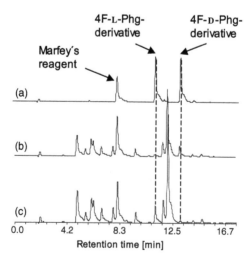

FIGURE 9.3 Base-catalyzed racemization of 4F-L-Phg during peptide synthesis.

[19]F-substituent of 4F-Phg renders the α-hydrogen rather acidic, so it is readily abstracted under the basic conditions employed in standard Fmoc protocols. The deprotonated form of 4F-Phg is resonance-stabilized, as shown in Figure 9.3, which leads to extensive racemization of the original L-amino acid. Therefore we prefer to employ a racemic mixture of 4F-DL-Phg from the start, as it has usually been possible to resolve the resulting epimeric peptides by HPLC in (semi-)preparative quantities.[17] The distinctive UV/Vis spectrum of 4F-Phg, with a maximum at 263 nm, helps to monitor the HPLC run and to determine peptide concentration (ε_{263} = 473 ± 13 M[1] cm[1]).

Following peptide synthesis and purification, the D- and L-derivatives of 4F-Phg can be assigned using Marfey's reagent on the acid-hydrolyzed peptides, as the diastereomeric amino acid derivatives are readily distinguished by analytical HPLC (see Figure 9.4). Even though racemization and HPLC purification is a nuisance, the resulting pair of epimeric peptides has other advantages. For example, the different labels in D- and L-configuration may provide two independent orientational constraints.[9,10] Furthermore, the sterically obstructive 4F-D-Phg may serve as a useful

FIGURE 9.4 Assignment of epimeric B18 peptides containing racemized 4F-Phg by means of HPLC: (A) control run of the free amino acid after derivatization with Marfey's reagent, (B) acid-hydrolyzed peptide bearing 4F-D-Phg, (C) acid-hydrolyzed peptide bearing 4F-L-Phg.

probe to examine peptide folding and self-assembly, as it cannot be accommodated within an oligomeric β-sheet structure, unlike the sterically compatible L-form.[17]

From the NMR point of view, 4F-L-Phg is a useful label for measuring distances, as, for example, demonstrated with gramicidin S by CPMG experiments.[8,14,15] Orientational constraints can also be extracted from the ^{19}F CSA tensor, which is asymmetric and therefore described by two angles θ and φ. As the phenyl ring does not tend to be rotationally averaged, the side chain torsion angle χ_1 of the Cα-Cβ bond needs to be known for this kind of analysis.[8] It turns out that the preferred value of χ_1 differs in α-helical and β-sheet conformations, and hence a comparatively large error is associated with the CSA analysis.[1,6] Nonetheless, several peptide structures have been comprehensively described, using either a surplus number of 4F-Phg labels[7] or by supplementing the ^{19}F-NMR data with specific measurements on other labels such as ^{15}N.[18]

9.2.2 4-CF$_3$-PHENYLGLYCINE

Similar to 4F-Phg, CF$_3$-Phg also has a tendency to racemize during Fmoc peptide synthesis, and HPLC separation of peptide epimers has been possible. As the deprotonated intermediate is stabilized only by inductive, but not by mesomeric effects, there is hope to avoid racemization using base-free coupling agents such as DIC/HOBt. Alternatively, it may be preferable to couple preformed di- or tripeptide units containing CF$_3$-L-Phg, as they may be more efficiently separated by HPLC than the complete polypeptide sequences.

With regard to its use in ^{19}F-NMR structure analysis, CF$_3$-Phg is a very convenient label for measuring orientational constraints. In contrast to 4F-Phg, there is no ambiguity associated with the side chain torsion angle χ_1, as the CF$_3$-group is motionally averaged, which renders the tensors axially symmetric, as illustrated in Figure 9.5. The anisotropic dipolar coupling within the CF$_3$-group, together with the

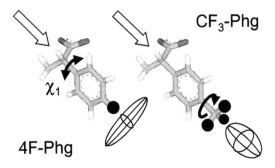

FIGURE 9.5 Comparison of the ^{19}F-NMR labels 4F-Phg (left) and CF$_3$-Phg (right). The CSA tensor of the single ^{19}F-substituent is asymmetric in 4F-Phg, and the side chain torsion angle χ_1 may be ambiguous. Fast rotation of the CF$_3$-group renders both the effective CSA tensor and the intragroup dipolar coupling tensor axially symmetric.

sign from the CSA, provides a unique measure of the effective angle θ_{CF3} between the CF$_3$-axis and the membrane normal (see Figure 9.2).[9,10] This orientational constraint is readily obtained from a simple one-pulse ^{19}F-NMR spectrum, without any need for chemical shift referencing[6,19] nor any ambiguity about the value of χ_1.

9.2.3 3F-ALANINE

With 3F-Alanine, the main difficulty in peptide synthesis arises from its tendency to eliminate hydrogen fluoride (HF) under basic conditions, yielding dehydro-alanine, as shown in Figure 9.6. Elimination could be adequately suppressed by using deuterated 2-^{2}H-3F-Ala and performing the coupling at 10°C, as the kinetic isotope effect reduces the spontaneous loss of HF (P. Wadhwani, unpublished data).

^{19}F-NMR structure analysis has benefited much from the virtually nonperturbing size of 3F-Ala and the proximity of the ^{19}F-label to the peptide backbone. Given that the CH$_2$F-group is motionally averaged,[20,21] it is very suitable for REDOR distance measurements on peptides.[11-13] Qualitative orientational changes have also been detected via 3F-Ala, but quantitative angular constraints cannot be reliably obtained as long as the CSA tensor alignment in the molecular frame has not been fully characterized.[1,12]

9.2.4 3,3,3-F$_3$-ALANINE

The risk of HF elimination during peptide synthesis is even more pronounced for F$_3$-Ala (see Figure 9.7) than for 3F-Ala. Nevertheless, we managed to incorporate this amino acid into synthetic peptides using base-free conditions, and have acquired the first promising ^{19}F-NMR spectra in reconstituted membranes (P. Tremouilhac, unpublished data). The obvious advantages of F$_3$-Ala are its small size, which should

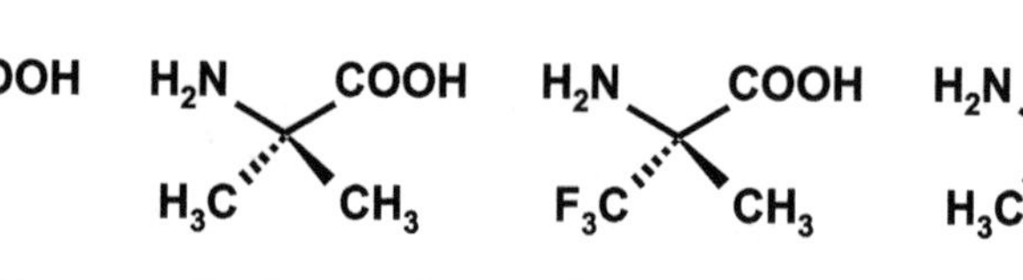

FIGURE 9.6 Base-catalyzed HF elimination from 3F-Ala during peptide synthesis.

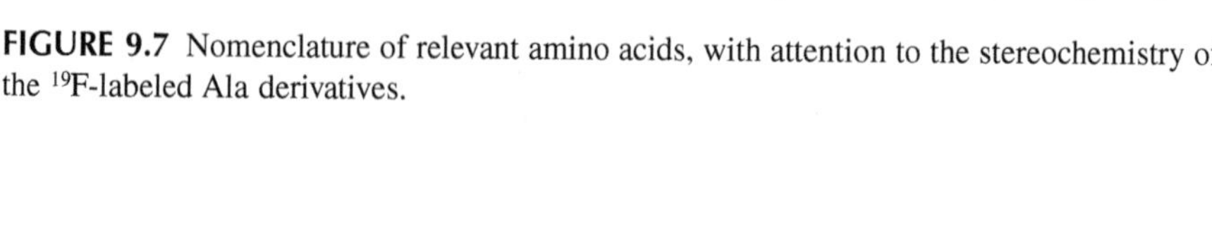

FIGURE 9.7 Nomenclature of relevant amino acids, with attention to the stereochemistry of the ^{19}F-labeled Ala derivatives.

be less perturbing than any rigid Phg-substituent, combined with the convenient ^{19}F-NMR properties of a rotating CF$_3$-group.

9.2.5 2-CF$_3$-ALANINE

CF$_3$-Ala may be considered a ^{19}F-labeled derivative of aminoisobutyric acid (Aib), which is a natural amino acid in non-ribosomally produced peptides (see Figure 9.7). Unlike 3F-Ala, the α-carbon in CF$_3$-Ala carries no acidic proton, and hence elimination of HF cannot occur. The main difficulty in incorporating CF$_3$-Ala into synthetic peptides is the low reactivity of this sterically obstructed amino acid, giving low yields with standard protocols.[22] To avoid termination of peptide synthesis, we found it advantageous to presynthesize appropriate di- or tripeptide units, which can then be coupled by standard protocols.[23]

Structure analysis of a peptaibol labeled with CF$_3$-Ala is in progress, offering the same methodological advantages as CF$_3$-Phg and F$_3$-Ala. One of the fundamental questions that needed clarification concerns the assignment of the R-and S-stereoisomers (see Figure 9.7). To this aim, small building blocks containing CF$_3$-Ala were first separated by HPLC before incorporating them into the full sequence by fragment condensation. The structures of the resulting epimeric peptides were subsequently compared using liquid-state ^{1}H-NMR in 70% TFE. Specific NOE cross-peaks of the methyl-group of CF$_3$-Ala were identified in both epimers, and hence their relative intensities could be used to discriminate the R- and S-forms (R.W. Glaser, unpublished data).

9.3 STRUCTURE ANALYSIS OF MEMBRANE-ACTIVE PEPTIDES

The alignment of a rigid molecular entity (such as an α-helix or a β-strand) in the membrane is fully characterized by three parameters, namely, two Euler angles (τ, ρ) and an order parameter (S_{mol}) (see Figure 9.2). It is convenient to define the tilt angle τ as the inclination between the molecular pseudo-axis (e.g., helix axis) and the membrane normal. The azimuthal angle ρ describes the rotation necessary to position the amphiphilic surface relative to the membrane. In the following sections, several different peptide structures will be analyzed by calculating a τ-ρ map, whose meaning is schematically illustrated in Figure 9.8.[1,3,7–10,18,24] A grid search has to be performed for all combinations of τ and ρ, and the deviation of the simulated resonance frequencies from the set of experimental ^{19}F-NMR data is displayed as a surface profile. Regions below a certain threshold (as a result of experimental errors and ambiguities of the assumed peptide conformation) are then taken to represent possible solutions of the true alignment. For this kind of analysis, a specific molecular conformation needs to be assumed. If there exists a sufficient number of NMR constraints, a unique solution should be found. The secondary structure, too, can be deduced via such τ-ρ analysis, as incorrect conformations should give no acceptable minimum. The τ-ρ map also has to be evaluated systematically as a function of the molecular order parameter S_{mol}, for which the best-fit value should be stated together with the final solution of the peptide alignment.

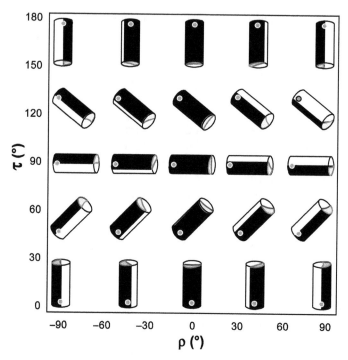

FIGURE 9.8 Schematic illustration of the meaning of the tilt angle τ and azimuthal rotation ρ of an amphiphilic peptide, depicted in a τ-ρ map. The membrane normal points up (parallel to $\tau = 0°$), and $\rho = 0°$ is defined in the chosen peptide coordinate frame. The chirality/directionality of the peptide is signified by a grey dot at one terminus. Because the two bilayer leaflets cannot be discriminated, it is $(\tau, \rho) = (180° - \tau, 180° + \rho)$.

Altogether, the alignment and motional behavior of a peptide with a known secondary structure can be comprehensively characterized from a limited number of orientational constraints, and its folded conformation may be confirmed by distance measurements. The angular analysis can be based on a single asymmetric ^{19}F CSA tensor provided that the peptide is immobilized in the lipid bilayer and $S_{mol} = 1$.[4] Alternatively, the CSA data of several labeled positions need to be collected.[7,18] Orientational constraints from axially symmetric dipolar interactions need to be acquired from at least three labeled positions, and typically four to five data points give enough information to discard multiple solutions.[9,10]

Several different peptide structures have been resolved to date by solid-state ^{19}F-NMR, and are supported by other spectroscopic and functional studies. In the following examples, we highlight some representative biological systems in which ^{19}F-NMR has provided an answer to a structural question and contributed to a better understanding of function.

9.3.1 Fusogenic Peptide B18

Membrane fusion events in intracellular vesicle transport, viral infection, or fertilization are triggered by complex protein machineries, which bring two lipid bilayers together and mediate their merging. In many of these proteins a short peptide sequence is responsible and sufficient for triggering fusion; hence, these fusogenic peptides are valuable tools to study fundamental aspects of fusion. The 18–amino acid peptide "B18" (LGLLLRHLRHHSNLLANI) was identified as the minimal fusogenic sequence from the sea urchin fertilization protein "bindin," which is contained in the acrosomal vesicle of the sperm cell and gets exposed on the membrane surface during fusion with an oocyte (Ulrich et al. 1998). The peptide *per se* is able to induce fusion of uncharged liposomes *in vitro*, and its interactions with lipid membranes have been thoroughly characterized.[7,17,26–31]

One of the characteristic features of B18 and related viral fusion peptides is their pronounced conformational plasticity, i.e., the ability to adopt different structures depending on the lipid environment, pH, ionic conditions, and peptide concentration.[26,32,33] The fusogenic activities of B18 from fluorescence fusion assays and electron microscopy have been correlated with structural data from circular dichroism, infrared, and NMR spectroscopy. These results indicate that the peptide is functionally active in a conformation made up of two α-helical segments connected by a flexible loop.[29] Inactive B18, however, is self-assembled into amyloid-like fibrils with a characteristic β-sheet conformation, as seen by x-ray diffraction and electron microscopy.[26,30] Solid-state ¹³C-MAS investigations on B18 have demonstrated a concentration-dependent conversion from an α-helical into a β-sheet conformation in POPC above a peptide/lipid ratio of 1:25.[31] Likewise, ¹⁹F-NMR of 4F-Phg labeled peptides in oriented dimyristoyl-phosphatidylcholine/dimyristoyl-phosphatidl-glyc-erol (DMPC)/(DMPG) membranes showed that B18 assumes a well-defined helical structure at peptide/lipid ratios lower than 1:100, but aggregation takes place at higher concentration and with increasing age of the sample.[7] In the aggregated fibrillar state the uniform alignment of B18 is lost, even though the lipids maintain their lamellar morphology and a high quality of orientation. Solid-state ²H-NMR showed that the amyloid-like fibrils penetrate into the hydrophobic region of the bilayer rather than interact electrostatically with the lipid head groups.[30]

Using ¹⁹F-NMR, the alignment of B18 in its monomeric helical state was determined at a peptide/lipid ratio of 1:150 in DMPC/DMPG, based on nine different peptide analogues with 4F-Phg labels. By analyzing this large number of orientational constraints, the expected α-helical conformation of the two segments could be confirmed, and it was demonstrated that both termini are well structured rather than frayed out.[7] The τ-ρ maps in Figure 9.9 show that the N-terminal α-helix is obliquely immersed into the bilayer ($\tau \approx 53°$, $\rho \approx 67°$, $S_{mol} \approx 0.8$), whereas the C-terminal segment remains flat on the membrane surface (with $\tau \approx 91°$, $\rho \approx 19°$, $S_{mol} \approx 0.5$), which is fully consistent with the hydrophobicity profile of the folded peptide.

The "boomerang"-like membrane immersion of the B18 peptide, illustrated in Figure 9.10, is similar to the picture of a membrane-embedded viral fusion peptide described by ESR.[34] It may explain how fusion might occur according to the three commonly accepted prerequisites: first, the peripherally bound C-terminal helix causes dehydration of the lipid headgroups, as confirmed by FT-IR,[28] which is a necessary condition for two opposing membranes to approach each other closely. Second, the oblique penetration of the N-terminal helix into the hydrophobic bilayer leads to a considerable perturbation of the lipid acyl chains, thereby helping them

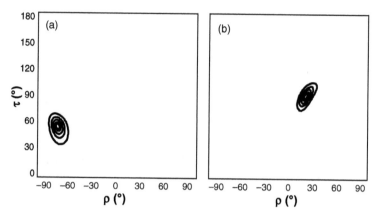

FIGURE 9.9 Experimental τ-ρ maps for the N-terminal (left) and the C-terminal (right) α-helices of the fusogenic peptide B18, as determined from five and four, respectively, orientational constraints (4F-Phg labels) by solid-state ¹⁹F-NMR.

FIGURE 9.10 Interaction of the fusogenic B18 peptide with membranes, as observed by ¹⁹F-NMR: the unstructured peptide in aqueous solution folds into two α-helical segments on membrane binding, one of which inserts at an oblique angle into the lipid bilayer (putative pre-fusion state).[7] At high peptide concentration it aggregates into extended amyloid-like β-sheet fibrils, which remain closely bound to the membrane (putative postfusion state).[30]

to overcome the activation energy barrier of the molecular rearrangements during fusion. Finally, the tilted segment may induce significant local membrane curvature and stabilize the highly curved and possibly inverted intermediate states of fusion.

The concentration-dependent conversion from a monomeric α-helical to an aggregated β-sheet conformation raises the question of which of these structures is functionally relevant.[32,35,36] The occurrence of fibrils is reminiscent of pathological amyloid deposits associated with various disease states, in which an excess of proteinaceous "waste" products accumulate in an aggregated form. Hence they appear less likely to be relevant for triggering membrane fusion. In contrast, oligomeric intermediates have been postulated for viral fusion events, as several trimeric proteins need to engage in a concerted action at the same time.[37,38] Our functional tests on the B18 peptides labeled 4F comparison with either an L- or D-epimer of Phg[37,38] indicate that the rate-limiting step of fusion does not need to involve a structured state at all, and that an oligomeric β-sheet can be excluded.[17] That is because B18 peptides containing 4F-D-Phg exhibit virtually the same fusion activity as their epimeric analogues with 4F-L-Phg. As the stiff D-enantiomeric side chain is sterically too obstructive to be accommodated in an oligomeric β-sheet, we suggest that membrane fusion involves a flexible intermediate state. It may be speculated that a stable monomeric α-helical state trapped in an NMR sample corresponds to prefusion conditions, whereas an aggregated β-sheet conformation reflects the post-fusion state (see Figure 9.10).

9.3.2 Antimicrobial Peptide Gramicidin S

The action of cationic peptide antibiotics is attributed to a permeabilization of bacterial membranes, given that these molecules possess an overall amphiphilic structure (see Figure 9.1).[3] Gramicidin S is a cyclic β-sheet peptide ([VOL^{D}F-PVOL^{D}FP]$_{cyclo}$) produced by *Bacillus brevis*, offering some pharmaceutical potential provided that its selectivity can be further improved.[39,40] To find out how the peptide is aligned in membranes and how it behaves in different lipid environments, 4F-Phg labels were substituted at the two equivalent positions of either Leu (see Figure 9.11) or Val.

^{19}F-NMR analysis in oriented membranes showed that (L-/L-substituted) gram-icidin S undergoes a concentration-dependent realignment in DMPC. At a low peptide/lipid ratio (≤1:80), it is bound flat to the membrane surface, as summarized in Figure 9.12, with its symmetry axis parallel to the bilayer normal (τ = 0°, ρ being

4F-D-Phg

FIGURE 9.11 Architecture of the cyclic peptide gramicidin S, in which the two equivalent Leu residues have been substituted by 4F-L/ L-Phg (left), and by 4F-D/ L-Phg (right).[7]

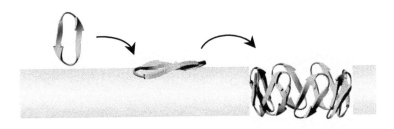

FIGURE 9.12 Interaction of the antimicrobial peptide gramicidin S with membranes, as observed by ^{19}F-NMR: the cationic peptide is electrostatically attracted to negatively charged membranes, and it binds flat to the lipid bilayer surface (irrespective of charge) via its hydrophobic side chains.[8,17] At high peptide concentration it can realign and self-assemble into an oligomeric β-barrel pore.[18]

unrestricted). In the liquid crystalline state, the peptide is highly mobile ($S_{mol} \approx 0.3$) and is engaged in long-axial rotation about the membrane normal. On the other hand, in the gel phase it is immobilized ($S_{mol} \approx 1.0$) but well ordered.[8] Interestingly, at high concentration ($\geq 1{:}40$) gramicidin S can flip upright in the membrane to form a putative β-barrel pore ($\tau \approx 80°$, $\rho \approx 45°$, $S_{mol} \approx 1.0$).[42] Oligomerization was postulated not only because of the concentration dependence, but also in view of the distinctive molecular mobility (the entire assembly does not wobble but undergoes fast long-axial rotation). Oligomerization offers the possibility of favorable intermolecular H-bonding, as depicted in Figure 9.12. A complementary ^{15}N-NMR measurement was used to validate the azimuthal rotation angle ρ (corresponding to the slant of the β-strands), given the ambiguity in the side-chain torsion angle χ_1 of 4F-Phg. Note that the number of monomers in our model is not accessible by NMR, but a hexameric complex stabilized by H-bonds has been recently observed by crystallographic analysis of a modified gramicidin S analogue.[43]

Because the two distinct orientational states of gramicidin S can be readily discriminated in the ^{19}F-NMR spectrum, it was possible to monitor the peptide's realignment under many different conditions. Oligomerization is favored at temperatures close to the lipid phase transition, where bilayer defects are most abundant.[42] The tendency to realign also varies considerably with the lipid acyl-chain length, as seen in Figure 9.13. In the short-chain lipid DLPC (with 12 carbons), the temperature range encompasses about 30°C, whereas in DMPC (with 14 carbons), the window is only 10°C, and in DPPC (with 16 carbons), no realignment is observed at all and the peptide remains flat on the membrane surface throughout. It was also shown by ^{19}F-NMR that gramicidin S no longer binds to membranes in a well-defined alignment in the presence of more than 20% cholesterol. Because eukaryotic membranes consist of up to 50% cholesterol, whereas bacteria do not contain any, the

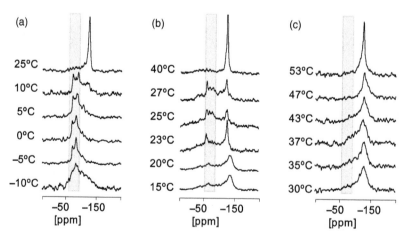

FIGURE 9.13 Representative ^{19}F-NMR spectra of gramicidin S labeled with 4F-Phg in lipid bilayers of (a) DLPC with short lipid chains, (b) DMPC with intermediate chain length, and (c) DPPC with long acyl chains. Realignment of the peptide (signals corresponding to the shaded region) is favored in short-chain lipids at temperatures near the phase transition.[42]

antimicrobial function and hemolytic side effects of gramicidin S seem to be controlled by the lipid composition of the respective membranes.[39–41]

By solid state ¹⁹F-NMR, it has thus been shown that gramicidin S is able to bind to model membranes in two different states: a monomeric peripheral alignment and an oligomeric membrane-immersed β-barrel. Which of these states is responsible for the antimicrobial action, however, is not apparent from the NMR data alone. A comparison of the peptide analogues substituted with the L- and D-epimers of 4F-Phg may provide some insight into these functional aspects, as the sterically confined D-side chain is unable to participate in oligomerization and pore formation via H-bonding.

9.3.3 ANTIMICROBIAL PEPTIDE PGLa

The antimicrobial peptide PGLa (GMASKAGAIAGKIAKVALKAL-NH₂) from the skin of *Xenopus laevis* possesses an amphiphilic structure when folded as an α-helix (see Figure 9.14). To confirm its fold and determine its alignment in lipid membranes, a series of peptide analogues was synthesized with either 4F-Phg or CF₃-Phg substituted for a single Ile or Ala.[6,9,10,17] As noted above, the CF₃-group offered significant advantages over the single F-substituent (see Figure 9.5) because the axially symmetric dipolar coupling (including its sign) of a CF₃-group can be readily obtained from a simple one-pulse spectrum, as illustrated in Figure 9.14. Analysis of the CSA tensor of 4F-Phg, in contrast, is challenged by the need for accurate chemical shift referencing and by some ambiguity in the exact value of the side chain torsion angle χ_1.[6,19] In either case, care has to be taken in assessing the structural and functional integrity of the labeled peptide analogues. For example, when substituted for Ala8 on the hydrophilic face of PGLa, only 4F-Phg, and not CF₃-Phg, was tolerated, as the latter peptide analogue was structurally and functionally disturbed according to circular dichroism spectroscopy and antimicrobial assays.[10,17]

A comprehensive structure analysis of PGLa in DMPC bilayers at a peptide/lipid ratio of 1:200 was carried out using the non-perturbing CF₃-Phg labels in four different positions.[10] As expected from previous ¹⁵N-NMR analysis,[44] the helix was found to be aligned flat in the membrane ($\tau \approx 95°$), and the azimuthal rotation angle ($\rho \approx 110°$) agrees well with the amphiphilic surface of PGLa. The peptide is monomeric and mobile in the fluid lipid bilayer ($S_{mol} \approx 0.6$), undergoing long-axial rotation at 35°C. In the lipid gel phase at 15°C, it is slightly less mobile ($S_{mol} \approx 0.7$), and long-axial rotation has ceased. Interestingly, with increasing peptide concentration, PGLa changes its orientation, as demonstrated by a concentration-dependent series of ¹⁹F-NMR measurements from 1:3000 to 1:8.[6] At a peptide-lipid ratio of 1:100, the flat helix alignment prevails at temperatures both well above and well below the lipid phase transition. When approaching about 30°C in DMPC, however, the peptide switches reversibly into a distinctly different orientation.[42] Unlike gramicidin S, where two distinct populations could be detected simultaneously, an interconversion of PGLa occurs on the NMR timescale because a broadened signal is observed at intermediate temperatures. At high peptide concentrations such as 1:50, the new structure persists at all temperatures. In this state, the peptide was found to be aligned at an oblique tilt angle of $\tau \approx 125°$ with respect to the membrane

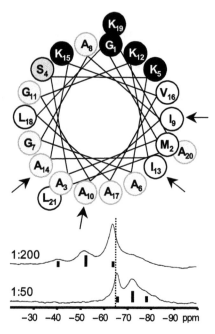

FIGURE 9.14 Helical-wheel representation of the antimicrobial peptide PGLa, which folds into an amphiphilic α-helix on membrane binding. The positions labeled individually with CF₃-Phg are indicated by arrows. The ^{19}F-NMR spectrum of the label in position 13 shows a change in the value and sign of the dipolar splitting when the lipid/peptide ratio is varied, indicating a realignment of the peptide.[9,10] The dipolar triplets are indicated as stick diagrams, and the isotropic chemical shift is marked by a dotted line.

normal, and with ρ ≈ 90° still being consistent with the polarity of the helix (see Figure 9.15).[9] Even at high concentration, the tilted peptide undergoes long-axial rotation and its order parameter is similar to that at low-concentration, hence it cannot be extensively aggregated.

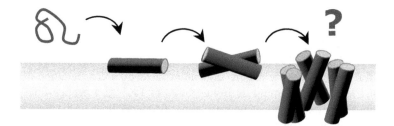

FIGURE 9.15 Interaction of the antimicrobial peptide PGLa with lipid membranes as observed by ^{19}F-NMR: it binds flat to the bilayer surface as an amphiphilic α-helix but it changes its tilt angle with increasing peptide concentration as a result of the formation of putative dimers.[9,10] It may be speculated that the PGLa dimers can assemble further into a toroidal wormhole, which has been implicated for the closely related peptide K3.[11,12]

The observed oblique alignment of PGLa was a rather unexpected result, and an explanation had to be found for how the peptide is stabilized at this unusual tilted angle. Therefore, it was suggested that the peptide may self-assemble into dimers in the membrane, which had been demonstrated for two related peptides magainin and K3 (see following) by transferred-NOE and REDOR NMR, respectively.[11,12,45] For PGLa, a symmetric antiparallel (head-to-tail) arrangement, as illustrated in Figure 9.15, would be consistent with the occurrence of single ^{19}F-NMR signals in the oriented samples.[9] It is tempting to speculate that self-association may take place via a Gly-rich patch exposed on one side of the tilted helix, as a similar motif is known to aid the dimerization of other, transmembrane helices. The functional relevance of the tilted and presumably dimeric PGLa structure is not yet clear. Either the oblique immersion may cause considerable acyl-chain perturbation and lead directly to membrane disruption or it might represent an intermediate state in the formation of a toroidal wormhole in analogy to the case of K3 discussed below.[11,12]

The concentration-dependent switch from the flat to the tilted PGLa alignment was independently confirmed by ^{2}H- and ^{15}N-NMR measurements of deuterated Ala labels or ^{15}N-Gly, being more favorable in terms of accuracy but much less sensitive than ^{19}F-NMR.[46,47] When compared with the unperturbed ^{2}H-NMR structure, the ^{19}F-NMR results proved to be almost as accurate at low peptide concentration, but less so at high concentration. We thus conclude that the most useful NMR strategy for determining reliable structures over a wide range of conditions consists of a combination of labeling schemes. First, a systematic search is performed by highly sensitive ^{19}F-NMR to scan many different sample compositions and temperatures, which is then to be followed by a precise ^{2}H- or ^{15}N-NMR analysis under well-specified conditions where the relevant structures occur.

The behavior of PGLa was monitored in different types of lipid bilayers, to draw conclusions on its mode of action and explain its selectivity toward bacterial membranes. Unlike gramicidin S, the realignment of PGLa was found to be independent of the acyl-chain length of the saturated lipids used.[48] Only a slight difference was observed in the presence of 50% negatively charged DMPG.[17] Neither hydrophobic bilayer cohesion nor electrostatic effects therefore seem to influence the tendency and concentration range of peptide realignment in the membrane. In contrast, by solution state ^{19}F-NMR, it was demonstrated that the initial binding of the cationic PGLa to lipid vesicles is exclusively controlled by its electrostatic attraction to negatively charged lipids.[17] These simple NMR experiments agree well with the general concept of why antimicrobial peptides are selective against negatively charged bacterial membranes as opposed to zwitterionic eukaryotic cell membranes.

9.3.4 ANTIMICROBIAL PEPTIDE K3

Based on the amphiphilic properties of PGLa, an α-helical model peptide K3 [(KIAGKIA)$_3$–NH$_2$] had been designed that exhibits improved antimicrobial activity and reduced hemolytic side effects.[49,50] This peptide was labeled with 3F-Ala in one of the native Ala side chains, which causes minimal perturbation, as well as with ^{13}C-Ala for REDOR analysis by Schaefer et al.[11,12] The simple ^{19}F-NMR spectra of K3 in oriented membranes (see Figure 9.16) show that this peptide, too, undergoes

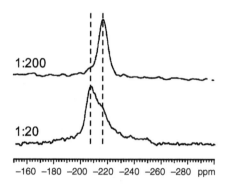

FIGURE 9.16 Static single-pulse ^{19}F-NMR spectra of the antimicrobial model peptide K3 with a 3F-Ala label in position Ala10, showing a change in membrane alignment for different peptide/lipid ratios.

a change in its alignment or mobility at a peptide–lipid ratio of between 1:200 and 1:20. An extensive set of REDOR experiments provided several ^{13}C-^{19}F distance constraints, which could be translated into a detailed molecular picture. First, a set of short-range distances demonstrated that K3 forms parallel (head-to-head) dimers at a high peptide–lipid ratio. Furthermore, the simultaneous observation of long-range distances indicated that these dimers are in close proximity to other, mono-meric K3 peptides.[12] Finally, a combination of ^{13}C-^{19}F, ^{13}C-^{31}P, and ^{15}N-^{31}P distances between selectively labeled lipids and peptides showed that K3 is in close contact with both the lipid headgroups and acyl chain tails.[11] This collection of structural data suggests the formation of a toroidal wormhole at high peptide concentration, where the pore contains both monomeric as well as dimeric peptides. On the basis of the refined picture of K3 (but without knowing the absolute tilt angles) and on the observed realignment of PGLa (but without having measured intermolecular distances), a generalized course of events on membrane binding is schematically suggested in Figure 9.15 for this class of α-helical amphiphilic peptides.

9.4 PERSPECTIVES

Over recent years, sufficient experience with ^{19}F-labeling and NMR analysis of membrane-associated peptides has been gained, allowing a critical comparison with the more established solid-state NMR approaches. We feel that some of the draw-backs with ^{19}F (i.e., non-trivial chemical synthesis and unusual spectrometer hardware) are more than compensated for by the excellent sensitivity and the robust-ness of the simple ^{19}F-NMR measurements. Of course, the risk of conformational and functional perturbance by the various ^{19}F-labels discussed above has to be borne in mind and tested for every new system to be investigated. However, we have not yet come across any case in which this would have prevented a thorough character-ization of the peptide conformation, alignment, and mobility in a lipid bilayer. A rather promising label with minimal structural perturbation and optimal NMR prop-erties appears to be F$_3$-Ala for acquiring orientational constraints and 3F-Ala for REDOR distance constraints.

Several biologically relevant peptide–lipid systems have been characterized to date in a highly sensitive manner by ¹⁹F-NMR, varying systematically many of the experimental parameters that can be addressed in model membranes, such as peptide concentration, lipid composition, temperature, pH, and so forth. It turns out that peptide concentration is one of the most critical parameters that governs the structural changes observed in all of the systems outlined above. In the case of the fusogenic peptide B18, increasing concentration induces a conformational switch from a monomeric kinked α-helix to a fibrillar β-sheet; for the antimicrobial gramicidin S, it causes self-assembly and realignment of the peptide into a β-barrel pore; and for PGLa and K3, it triggers dimerization and possibly pore formation as another step. Such comprehensive NMR characterization of several representative membrane-active peptides has not yet been performed to our knowledge under a comparably wide range of conditions. This is most likely attributed to the lower sensitivity or to the presence of background signals that limit the application of other isotope-labeling schemes. We expect that the structural characterization of ¹⁹F-labeled peptides will further benefit from the possibility of observing them in native membranes, and possibly even *in vivo* in suitably prepared systems. Such NMR experiments could provide valuable insights into the behavior of externally added peptides, drugs, or toxins in a representative cellular environment, which could then be compared with and interpreted in terms of the structural data obtained in model membranes.

ACKNOWLEDGMENTS

We are grateful to all past and present members of the group who have contributed to this work directly or indirectly by their efforts in research and infrastructure, specifically Olaf Zwernemann, Dahlia Fischer, Daniel Maisch, Dorit Grasnick, Silvia Gehrlein, Dieter Dennerlein, and Ralf Eisenhuth for the peptide synthesis; Raiker Witter, Ulrich Sternberg, Jürgen Bialy, Johannes Peters, Markus Schmitt, Peter Herrmann, Gerd Klinger, Hartmut Heinzmann, Wolfgang Möck, and Jesus Salgado for the NMR activities; and Birgid Langer, Christian Lange, Sonja Müller, Soraya Benamira, Steffi Vollmer, Christian Weber, Marco Ieronimo, and Tom Eisele for the biochemistry. We thank our collaborators for their support and advice, especially Jake Schaefer, Beate Koksch, Ronald McElhaney, Charles Glabe, Karl-Heinz Gührs, Ute Möllmann, and Reinhard Ulrich. We also acknowledge financial support by the DFG, FCI, TMWFK, the University of Jena, and the CFN Karlsruhe.

REFERENCES

1. Ulrich, A.S., Solid state 19F-NMR methods for studying biomembranes, *Prog. NMR Spectr.*, 46, 1–21, 2005.
2. Opella, S.J. and Marassi, F.M., Structure determination of membrane proteins by NMR spectroscopy, *Chem. Rev.*, 104, 3587–3606, 2004.
3. Strandberg, E. and Ulrich, A.S., NMR methods for studying membrane-active antimicrobial peptides, *Concepts Magn. Reson.*, 23A, 89–120, 2004.

4. Grage, S.L., Wang, J., Cross, T.A. and Ulrich, A.S., Structure analysis of fluorine-labelled tryptophan side-chains in gramicidin A by solid state 19F-NMR, *Biophys. J.*, 83, 3336–335, 2002.

5. Ulrich, A.S., High resolution 1H and 19F solid state NMR, in Encyclopedia of Spectroscopy and Spectrometry, J. Lindon, G. Tranter and J. Holmes, Editors. 2000, Academic Press: London. p. 813-825.

6. Glaser, R.W. and Ulrich, A.S., Susceptibility corrections in solid-state NMR experiments with oriented membrane samples. Part I: Applications, *J. Magn. Reson.*, 164, 104–14, 2003.

7. Afonin, S., Dürr, U.H.N., Glaser, R.W. and Ulrich, A.S., "Boomerang"-like insertion of a fusogenic peptide in a lipid membrane revealed by solid-state 19F NMR, *Magn. Reson. Chem.*, 42, 195–203, 2004.

8. Salgado, J., Grage, S.L., Kondejewski, L.H., Hodges, R.S., McElhaney, R.N. and Ulrich, A.S., Membrane-bound structure and alignment of the antimicrobial beta-sheet peptide gramicidin S derived from angular and distance constraints by solid state 19F-NMR, *J. Biomol. NMR*, 21, 191–208, 2001.

9. Glaser, R.W., Sachse, C., Dürr, U.H.N., Wadhwani, P., Afonin, S., Strandberg, E. and Ulrich, A.S., Concentration-dependent re-alignment of the antimicrobial peptide PGLa in lipid membranes observed by solid state 19F-NMR, *Biophys. J.*, 88, 3392–3397, 2004.

10. Glaser, R.W., Sachse, C., Dürr, U.H.N., Wadhwani, P. and Ulrich, A.S., Orientation of the antimicrobial peptide PGLa in lipid membranes determined from 19F-NMR dipolar couplings of 4-CF3-phenylglycine labels, *J. Magn. Reson.*, 168, 153–63, 2004.

11. Toke, O., Maloy, W.L., Kim, S.J., Blazyk, J. and Schaefer, J., Secondary structure and lipid contact of a peptide antibiotic in phospholipid bilayers by REDOR, *Biophys. J.*, 87, 662–74, 2004.

12. Toke, O., O'Connor, R.D., Weldeghiorghis, T.K., Maloy, W.L., Glaser, R.W., Ulrich, A.S. and Schaefer, J., Structure of (KIAGKIA)3 aggregates in phospholipid bilayers by solid-state NMR, *Biophys. J.*, 87, 675–687, 2004.

13. Grage, S.L., Watts, J.A. and Watts, A., 2H[19F] REDOR for distance measurements in biological solids using a double resonance spectrometer, *J. Magn. Reson.*, 166, 1–10, 2004.

14. Grage, S.L. and Ulrich, A.S., Structural parameters from 19F homonuclear dipolar couplings, obtained by multipulse solid-state NMR on static and oriented systems, *J. Magn. Reson.*, 138, 98–106, 1999.

15. Grage, S.L. and Ulrich, A.S., Orientation-dependent 19F dipolar couplings within a trifluoromethyl group are revealed by static multipulse NMR in the solid state, *J. Magn. Reson.*, 146, 81–8, 2000.

16. Kukhar, V.P. and Soloshonok, V.A., Fluorine-Containing Amino-Acids. 1995, Chichester: Wiley.

17. Afonin, S., Glaser, R.W., Berditchevskaia, M., Wadhwani, P., Gührs, K.H., Möllmann, U., Perner, A. and Ulrich, A.S., 4-Fluorophenylglycine as a label for 19F NMR structure analysis of membrane-associated peptides, *Chembiochem.*, 4, 1151–63, 2003.

18. Afonin, S., Ph.D. Thesis, University of Karlsruhe, 2003.

19. Ulrich, R., Glaser, R.W. and Ulrich, A.S., Susceptibility corrections in solid state NMR experiments with oriented membrane samples. Part II: Theory, *J. Magn. Reson.*, 164, 115–27, 2003.

20. Holl, S.M., Marshall, G.R., Beusen, D.D., Kociolec, K., Redlinski, A.S., Leplawy, M.T., McKay, R.A. and Schaefer, J., Determination of an 8A interatomic distance in a helical peptide by solid-state NMR spectroscopy, *J. Am. Chem. Soc.*, 114, 4830–4833, 1992.

21. Goetz, J.M., Wu, J.H., Lee, A.F. and Schaefer, J., Two-dimensional transferred-echo double resonance study of molecular motion in a fluorinated polycarbonate, Solid State NMR, 12, 87–95, 1998.

22. Thust, S. and Koksch, B., Protease-catalyzed peptide synthesis for the site-specific incorporation of alpha-fluoroalkyl amino acids into peptides, *J. Org. Chem.*, 68, 2290–6, 2003.

23. Maisch, D., Diploma Thesis, University of Karlsruhe, 2004.

24. Dürr, U.H.N., Ph.D. Thesis, University of Karlsruhe, 2004.

25. Ulrich, A.S., Otter, M., Glabe, C. and Hoekstra, D., Membrane fusion is triggered by a distinct peptide sequence of the sea urchin fertilization protein bindin, *J. Biol. Chem.*, 273, 16748–16755, 1998.

26. Ulrich, A.S., Tichelaar, W., Förster, G., Zschörnig, O., Weinkauf, S. and Meyer, H.W., Ultrastructural characterization of peptide-induced membrane fusion and peptide self-assembly in the bilayer, *Biophys. J.*, 77, 829–841, 1999.

27. Binder, H., Arnold, K., Ulrich, A.S. and Zschörnig, O., Interaction of Zn2+ with phospholipid membranes, *Biophys. Chem.*, 90, 57–74, 2001.

28. Binder, H., Arnold, K., Ulrich, A.S. and Zschörnig, O., The effect of Zn2+ on the secondary structure of a histidine-rich fusogenic peptide and its interaction with lipid membranes, *Biochim. Biophys. Acta*, 1468, 345–358, 2000.

29. Glaser, R.W., Grüne, M., Wandelt, C. and Ulrich, A.S., Structure analysis of a fusogenic peptide sequence from the sea urchin fertilization protein bindin, *Biochemistry*, 38, 2560–9, 1999.

30. Grage, S.L., Afonin, S., Grüne, M. and Ulrich, A.S., Interaction of the fusogenic peptide B18 in its amyloid-state with lipid membranes studied by solid state NMR, *Chem. Phys. Lipids*, 132, 65–77, 2004.

31. Barre, P., Zschörnig, O., Arnold, K. and Huster, D., Structural and dynamical changes of the bindin B18 peptide upon binding to lipid membranes. A solid-state NMR study, *Biochemistry*, 42, 8377–86, 2003.

32. Bodner, M.L., Gabrys, C.M., Parkanzky, P.D., Yang, J., Duskin, C.A. and Weliky, D.P., Temperature dependence and resonance assignment of 13C NMR spectra of selectively and uniformly labeled fusion peptides associated with membranes, *Magn. Reson. Chem*, 42, 187–94, 2004.

33. Pecheur, E.I., Sainte-Marie, J., Bienvenue, A. and Hoekstra, D., Peptides and membrane fusion: towards an understanding of the molecular mechanism of protein-induced fusion, *J. Membr. Biol.*, 167, 1–17, 1999.

34. Han, X., Bushweller, J.H., Cafiso, D.S. and Tamm, L.K., Membrane structure and fusion-triggering conformational change of the fusion domain from influenza hemagglutinin, *Nat. Struct. Biol.*, 8, 715–20, 2001.

35. Yang, J. and Weliky, D.P., Solid-state nuclear magnetic resonance evidence for parallel and antiparallel strand arrangements in the membrane-associated HIV-1 fusion peptide, *Biochemistry*, 42, 11879–90, 2003.

36. Gabrys, C.M., Yang, J. and Weliky, D.P., Analysis of local conformation of membrane-bound and polycrystalline peptides by two-dimensional slow-spinning rotor-synchronized MAS exchange spectroscopy, *J. Biomol. NMR*, 26, 49–68, 2003.

37. Lau, W.L., Ege, D.S., Lear, J.D., Hammer, D.A. and DeGrado, W.F., Oligomerization of fusogenic peptides promotes membrane fusion by enhancing membrane destabilization, Biophys. J., 86, 272–84, 2004.

38. Yang, R., Yang, J. and Weliky, D.P., Synthesis, enhanced fusogenicity, and solid state NMR measurements of cross-linked HIV-1 fusion peptides, *Biochemistry*, 42, 3527–35, 2003.

39. Bradshaw, J., Cationic antimicrobial peptides: issues for potential clinical use, *BioDrugs*, 17, 233–40, 2003.

40. McInnes, C., Kondejewski, L.H., Hodges, R.S. and Sykes, B.D., Development of the structural basis for antimicrobial and hemolytic activities of peptides based on gramicidin S and design of novel analogs using NMR spectroscopy, *J. Biol. Chem.*, 275, 14287–94, 2000.

41. Lee, D.L., Powers, J.P., Pflegerl, K., Vasil, M.L., Hancock, R.E. and Hodges, R.S., Effects of single D-amino acid substitutions on disruption of beta-sheet structure and hydrophobicity in cyclic 14-residue antimicrobial peptide analogs related to gramicidin S, *J. Pept. Res.*, 63, 69–84, 2004.

42. Afonin, S., Glaser, R.W., Sachse, C., Wadhwani, P. and Ulrich, A.S., Re-alignment of representative antimicrobial peptides in different lipid environments, submitted.

43. Grotenbreg, G.M., Timmer, M.S., Llamas-Saiz, A.L., Verdoes, M., van der Marel, G.A., van Raaij, M.J., Overkleeft, H.S. and Overhand, M., An unusual reverse turn structure adopted by a furanoid sugar amino acid incorporated in gramicidin S, *J. Am. Chem. Soc.*, 126, 3444–6, 2004.

44. Bechinger, B., Zasloff, M. and Opella, S.J., Structure and dynamics of the antibiotic peptide PGLa in membranes by solution and solid-state nuclear magnetic resonance spectroscopy, *Biophys. J.*, 74, 981–7, 1998.

45. Wakamatsu, K., Takeda, A., Tachi, T. and Matsuzaki, K., Dimer structure of magainin 2 bound to phospholipid vesicles, *Biopolymers*, 64, 314–27, 2002.

46. Strandberg, E., Wadhwani, P., Tremouilhac, P. and Ulrich, A.S., 2H-NMR analysis of the antimicrobial PGLa peptide tilted at an oblique angle in the lipid membrane, submitted.

47. Tremouilhac, P., Diploma Thesis, University of Karlsruhe, 2003.

48. Sachse, C., Diploma Thesis, University of Jena, 2003.

49. Maloy, W.L. and Kari, U.P., Structure-activity studies on magainins and other host defense peptides, *Biopolymers*, 37, 105–22, 1995.

50. Hirsh, D.J., Hammer, J., Maloy, W.L., Blazyk, J. and Schaefer, J., Secondary structure and location of a magainin analogue in synthetic phospholipid bilayers, *Biochemistry*, 35, 12733–41, 1996.

10 Solid-State Nuclear Magnetic Resonance Spectroscopic Studies of Magnetically Aligned Phospholipid Bilayers

Gary A. Lorigan

CONTENTS

ABSTRACT Solid-state NMR spectroscopic studies of integral membrane proteins are a rapidly growing area of research. Two high-resolution solid-state NMR methods that are used to increase both sensitivity and resolution are: (1) magic-angle spinning (MAS) and (2) aligning the sample with respect to the direction of the static magnetic field. Generally, the most common membrane protein alignment technique is carried out utilizing glass plates in a special solid-state NMR probe. This solid-state NMR review chapter is focused on the utilization of magnetically aligned phospholipid bilayers (bicelles) that spontaneously align in the magnetic field.

KEY WORDS: *dynamics, membrane, membrane protein, solid-state NMR spectroscopy*

10.1 MEMBRANE PROTEINS

Membrane proteins (which make up approximately one-third of the total number of known proteins) are responsible for many important properties and functions of biological systems: they transport ions across the membrane, they act as receptors, and they have pertinent roles in the assembly, fusion, and structure of cells and viruses. Despite the abundance and clear importance of integral membrane proteins, very little information about these systems exists. Structural studies of these membrane proteins represent one of the final frontiers in structural biology. X-ray crystallography is the premiere technique that is used to elucidate structural information of biologically significant protein systems. However, this technique has not been very successful in providing structural information about membrane protein systems. The hydrophobic surfaces associated with membrane-bound protein systems make the crystallization process extremely difficult. Although researchers are making progress with x-ray techniques, still only a handful of membrane protein structures have been obtained via x-ray crystallography.[1-6] Important scientific breakthroughs in the development and information on membrane protein structure recently led to a Nobel Prize (Agre and MacKinnon) in Chemistry in 2003.[2-12] Alternatively, solid-state nuclear magnetic resonance (NMR) spectroscopy is a powerful technique that can be used to provide structural, orientational, and dynamic information about membrane protein systems in lipid bilayers.[13-15] Proteins with slow molecular motions (on the NMR timescale) can be studied with solid-state NMR spectroscopy. In solution NMR spectroscopy, this information is lost and cannot be used to define the structural characteristics of the protein. To abstract the chemical shift anisotropy and dipolar coupling structural information, specialized solid-state NMR spectroscopic techniques have been developed that significantly reduce the NMR spectral line width. Two high-resolution solid-state NMR methods that are employed to increase both sensitivity and resolution are magic-angle spinning (MAS), in which the sample is rapidly spun at an angle of 54.7° with respect to the magnetic field, and orienting the sample with respect to the direction of the static magnetic field.[16] For these two techniques, the signal sensitivity and resolution can be dramatically increased via cross-polarization, which transfers the ^1H magnetization via the dipolar coupling from the abundant protons to the observed nucleus (e.g., ^{13}C or ^{15}N) with a lower gyromagnetic ratio and high-power ^1H-decoupling.[17]

The alignment NMR approach coupled with isotopic labeling (^2H, ^{15}N, or ^{19}F) allows structural, dynamic, and helical tilt information on integral membrane proteins to be ascertained.[18-24] Solid-state NMR spectroscopy yields orientational, angle, and distance parameters that can be transformed into a structural model of the protein or peptide system being investigated. The nuclear spin interaction associated with a peptide amide nitrogen (^{15}N) is highly anisotropic. In randomly oriented samples, this anisotropy results in a spectrum commonly referred to as a "powder pattern" because of the multiple angles and orientations that the nuclei have taken with respect to the magnetic field. Therefore, it is difficult to ascertain any type of structural information from the spectra. However, if the sample can be strategically oriented with respect to the magnetic field, the inherent anisotropies (^{15}N or ^{13}C) or quadrupolar splittings (^2H)

can be used to obtain high-resolution spectra with definitive characteristics that can be used to provide orientational and structural information.[14]

Traditionally, the alignment of membrane proteins in solid-state NMR spectroscopic studies has been obtained by mechanically orienting the lipid bilayers on glass plates.[25-27] This review focuses on the application of an alternate alignment approach in which the phospholipid bilayers spontaneously align in a static magnetic field.

10.2 MAGNETICALLY ALIGNED PHOSPHOLIPID BILAYERS (BICELLES)

Magnetically aligned phospholipid bilayers (bicelles) have been demonstrated to be useful models for studying the structural and dynamic properties of membrane systems and integral membrane proteins using solid-state NMR spectroscopic techniques.[28-62] In addition, bicelles are advantageous for spin-label electron paramagnetic resonance spectroscopy (EPR) studies of membrane protein systems.[63-67] The magnetic alignment of bicelles is the result of the anisotropy of the overall magnetic susceptibility of the system. The negative sign of the diamagnetic susceptibility anisotropy tensor ($\Delta\chi < 0$) for phospholipid bilayers dictates that the bicelles align with their bilayer normal oriented perpendicular to the direction of the static magnetic field.[36] Figure 10.1a shows a ^{2}H NMR spectrum of magnetically aligned 1,2-dimyristoyl-*sn*-glycero-3-phosphocholine/1,2-dihexanoyl-*sn*-glycero-3-phosphocholine (DMPC/DMPCd$_{54}$/DHPC) phospholipid bilayers. The sharp line widths and quadrupolar splittings indicate that the lipid bilayers are aligned in the magnetic field such that the membrane normal is perpendicular with respect to B$_0$.[36] The well-resolved ^{2}H peaks result from the different motional properties of the CD$_3$ and CD$_2$ acyl groups of DMPCd$_{54}$ aligned in the magnetic field. The addition of paramagnetic lanthanide ions with a large positive $\Delta\chi$ (Eu^{3+}, Er^{3+}, Tm^{3+}, and Yb^{3+}) can cause the bicelles to flip 90°, such that the average bilayer normal is colinear with the direction of the static magnetic field.[28] Figure 10.1b shows the ^{2}H NMR spectrum of magnetically aligned Yb^{3+}-doped (DMPC/DMPCd$_{54}$/DHPC) phospholipid bilayers. In this case, the resultant quadrupolar splitting doubles for the CD$_3$ and CD$_2$ groups on the acyl chains of the DMPCd$_{54}$ phospholipids, indicating that the bilayer normal is colinear with the static magnetic field. One advantage of flipping the phospholipid bicelles is that the spectral resolution is dramatically increased because the spectral width is spread over a wider frequency range. In addition, in a uniaxially aligned membrane protein system, the highly anisotropic spectral data can yield the orientation and structure of different segments of the protein with respect to the magnetic field and the lipid bilayer.[13,14] The structural characteristics of magnetically aligned phospholipid bilayers has already been discussed in the literature and varies depending on the sample temperature, hydration level, sample composition, and addition of lanthanide ions.[36,37,68-71]

In addition, the development of dilute magnetically aligned aqueous liquid-crystalline media has dramatically increased the refinement of structural studies of globular proteins with high-resolution NMR spectroscopy.[72-80] In general, in an isotropic solution, internuclear dipolar couplings average to zero as a result of

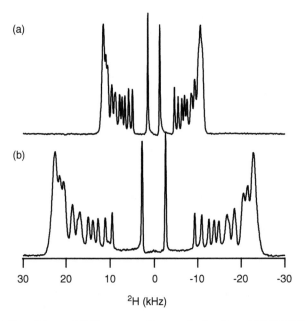

FIGURE 10.1 ^{2}H solid-state NMR spectra of magnetically aligned DMPC/DHPC phospholipid bilayers investigated with and without the paramagnetic lanthanide ion Yb^{3+}. (a) Bicelle spectrum contains 0% Yb^{3+} and is consistent with the bilayer normal oriented perpendicular to the static magnetic field. (b) Bicelle spectrum contains 5% molar Yb^{3+} (when compared with DMPC) and is consistent with the bilayer normal oriented parallel with the respect to the direction of the static magnetic field.

rotational molecular motion. By dissolving proteins in a dilute magnetically oriented aqueous liquid-crystalline medium (bicelle), a tunable degree of solute alignment with respect to the magnetic field can be created while retaining both the resolution and sensitivity of the regular isotropic NMR spectrum. In this system, the dipolar couplings no longer average to zero and can be accurately measured. This approach has been shown to significantly improve the accuracy of structures determined by solution NMR spectroscopy and to extend the size limit for protein structure determination.[72,73]

To better understand the magnetic alignment process of bicelles, we need to examine the various temperature-dependent phase behavior of the phospholipids. The phases of lanthanide-doped magnetically aligned phospholipid bilayers can be easily investigated as a function of temperature using ^{2}H solid-state NMR spectroscopy.[28–30,51,81] Previously, Mironov and coworkers stated that for some liquid crystalline systems, the nematic phase was much easier to magnetically align than the smectic phase.[82] This argument was used to explain the results of the magnetic alignment behavior for their lanthanide-containing metallomesogen.[82] Mironov and coworkers further stated that many liquid-crystalline substances magnetically align when the samples are cooled while undergoing phase transitions from the isotropic phase to the nematic phase to the smectic phase.[82] This magnetic alignment process in many cases does not occur if the sample is heated to the smectic phase because

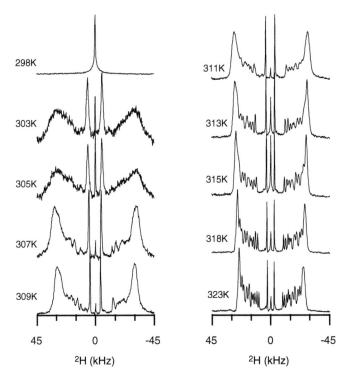

FIGURE 10.2 Normalized solid-state ^{2}H NMR spectra of a Tm^{3+}-doped bicelle sample show-ing the various temperature-dependent phases of the Tm^{3+}-doped bicelle system. ^{2}H NMR spectra were taken at various static temperatures in a resonant frequency of 46.07 MHz. The temperature at which each spectrum was taken is noted on the left of the spectrum.

of hysteresis effects.[82] Firestone and coworkers showed that their liquid-crystalline system magnetically aligned at a low temperature, corresponding to a hexagonal phase, and that it was necessary to raise the temperature slowly from the hexagonal phase to a highly viscous lamellar phase that "locked in" the macroscopic align-ment.[83,84] A ^{2}H NMR study and a neutron diffraction study observed a lanthanide-induced nematic-to-smectic phase transition for parallel-aligned Tm^{3+}-doped bicelles.[29,85]

Figure 10.2 shows several solid-state ^{2}H NMR spectra of a Tm^{3+}-doped DMPC/DHPC bicelle sample containing a small amount of DMPC$_{d54}$ taken at various static temperatures from 298 to 323 K. The sample was allowed to ther-mally equilibrate at these various static temperatures for 10 min before a ^{2}H NMR spectrum was acquired. The solid-state ^{2}H NMR spectra taken between 298 and 302 K exhibit an isotropic phase (only the spectra at 298 K are shown in Figure 10.2), which has been described as either rapidly tumbling bicelles accom-panied by higher fluidity of the sample or a mixed micelle phase.[86] Recent studies on the morphology of DMPC/DHPC bicelle phases have been conducted using pulsed field gradient NMR spectroscopy (PFG NMR) and small angular neutron scattering (SANS) techniques support the rapidly tumbling bicelle model.[68,69]

Solid-state ^{2}H NMR spectra taken at temperatures between 303 and 305 K revealed broad unresolved peaks indicative of randomly dispersed bicelles. This powder-like spectrum at a temperature of 303 K marks the gel-to-liquid crystalline phase transition for our bicelle system but was not stable in the L_α phase at these temperatures. This main phase transition temperature (T_m) is greater than that observed for pure DMPC without added Tm^{3+}, which may be explained by the Tm^{3+} ions invoking greater order and packing of the phospholipid acyl chains.[29,87] ^{2}H NMR spectra taken at temperatures between 307 and 311 K exhibited well-resolved peaks and are in good agreement with magnetically aligned parallel-oriented DMPC phospholipid bilayers in the nematic L_α phase. ^{2}H NMR spectra taken at a temperature of 313 K or above showed increased resolution and order and were ascribed by Prosser and coworkers and Katsaras and coworkers as the formation of a lanthanide-induced lamellar L_α smectic phase.[29,85]

From the solid-state ^{2}H NMR spectra in Figure 10.2, the bicelles are in an isotropic phase at 298 K. The nature of this isotropic phase may be either a phase-separated mixture of DHPC-DMPC mixed micelles or DMPC bilayers in the gel phase, or small, rapidly tumbling bicelles in the gel phase.[68,88,89] In the former case, the spherical geometry of mixed micelles lack magnetic anisotropy and therefore will not align in the presence of a magnetic field. Lipid dispersions of pure DMPC (i.e., multilamellar DMPC bilayers) also will not align in the gel phase.[63] In the latter case, small, rapidly tumbling DMPC/DHPC bicelles accompanied by a low viscosity similar to that of pure water have a high degree of thermal motion and cannot maintain any degree of alignment. The low viscosity also lowers the degree of cooperativity between the different bicelles possibly because of significant amounts of bulk water filling the interbicelle space.

In the gel phase, the phospholipid acyl chains and long molecular axis are tilted 30° with respect to the normal of the phospholipid bilayers.[90] Tilting may also be present in bicelles, and the angle may be even greater than 30° because of the greater hydration levels used for our bicelle samples as compared to those used in the literature.[90] Tilting the phospholipid acyl chains would significantly decrease the magnitude of the magnetic susceptibility anisotropy of the bicelles relative to its maximum limit (i.e., when the long molecular axis is parallel with the bicelle normal).[82]

The typical bicelle consists of long-chain phospholipids such DMPC and a detergent such as DHPC. The size and structure of the bicelle changes as the molar ratio between the long-chain (n_1) and short-chain phospholipid (n_2), $q = n_1/n_2$ increases. Phospholipids that are present in a bilayer undergo various temperature-dependent phase transitions.[66] Above T_m, the phospholipids are in a relatively fluid liquid crystalline state. The lamellar liquid-crystalline phase is also termed the L_α phase. The transition from the gel phase to the L_α phase above T_m is a requirement for magnetically aligned phospholipid bilayers and can be explained in terms of the strong forces in the polar head sheets and the weak forces between the acyl chains. When a phospholipid bilayer system is heated above its T_m, the van der Waals forces between the hydrocarbons become weak compared to the thermal motions.[88] Thus, the chains are transformed into a state of disorder with a high degree of gauche-conformation.[88]

Because a bilayer arrangement of phospholipids is characterized by a distinct central hydrophobic region bounded by two polar interfacial regions, we expect that the thickness of the hydrophobic region will have influence on the structure and function of transmembrane proteins. Despite the success of using DMPC/DHPC bicelle systems to study the structural and dynamic properties of membrane proteins with solid-state NMR spectroscopy, there are size limitations on the length and size of proteins to be studied.[91] Natural biological lipids are dominated by chains 16–18 carbons in length. In addition, α-helical transmembrane sections of proteins generally contain 25 residues, which may be too long to assemble into the standard DMPC/DHPC bicelle matrix. Thus, it is necessary to develop a method to magnetically align longer-chain phospholipids in a static magnetic field.

Because of the longer acyl chains, the lipid bicelles of such a system will extend the thickness of the lipid bilayer and have the potential to provide a better model for a biological membrane and possibly enhance protein activity for solid-state NMR spectroscopic investigations. Figure 10.3a shows a ^{2}H NMR spectrum of a $q = 2.0$ 1-palmitoyl-2-stearoyl-sn-glycero-3-phosphatidylcholine (PSPC)/DHPC bicelle sample at 50°C prepared in the absence of lanthanide ions. To change the sign of the net magnetic susceptibility anisotropy tensor of the bicelles so that their normal is parallel to the applied magnetic field, 5% molar Yb^{3+} with respect to PSPC was added to the

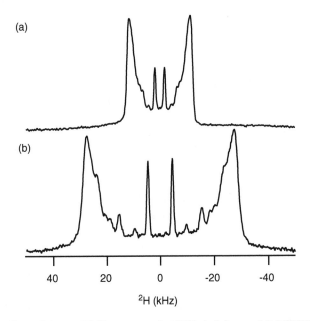

FIGURE 10.3 ^{2}H solid-state NMR spectra of a 25% (w/w) $q = 2.0$ PSPC/DHPC bicelle sample investigated with and without the paramagnetic lanthanide ion Yb^{3+}. (a) Bicelle spectrum contains 0% Yb^{3+} and is consistent with the bilayer normal being perpendicular to the magnetic field. (b) Bicelle spectrum contains 5% molar Yb^{3+} (when compared with PSPC) and is consistent with the bilayer normal being parallel with the direction of the static magnetic field.

PSPC/DHPC bicelle sample, and the results are shown in Figure 10.3b. The quadrupolar splittings observed in the ^{2}H NMR spectrum are spread out and are much better resolved. The spectrum indicates that the PSPC/DHPC bicelles have flipped such that their bilayer normal is now parallel with the magnetic field. The ^{2}H line shape and breadth of the magnetically aligned lanthanide-doped PSPC/DHPC bicelle sample observed in Figure 10.2b is similar to previous DMPC/DHPC/lanthanide spectra in the literature.[28,33] The breadth is slightly larger for the PSPC bicelle spectra in both the parallel and perpendicular orientations when compared with the ^{2}H spectra of standard DMPC/DHPC bicelle samples. The PSPC/DHPC bicelle spectra have a wider quadrupolar splitting than the corresponding DMPC/DHPC bicelle spectra because the L_α phase PSPC phospholipids are closer to T_M when the aligned phase occurs.

Temperature plays a very important role in the magnetic alignment of PSPC/DHPC phospholipid bilayers. ^{2}H NMR studies were carried out for a 25% w/w PSPC/DHPC phospholipid bicelle sample with $q = 1.8$ over a temperature range from 40 to 60°C.[59] At 40 and 45°C, the spectra reveal only an isotropic component. At 50°C, the ^{2}H NMR spectrum indicates that the PSPC/DHPC phospholipid bicelles are magnetically aligned such that the normal of the lipid bilayer is perpendicular to the static magnetic field. Above 50°C, the spectra reveal a powder-like spectrum with a large isotropic component centered at 0 kHz. Characteristics of PSPC/DHPC bicelle alignment are observed only at 50°C, whereas the DMPC/DHPC phospholipid bilayer arrays align between 34° and 43°C. DMPC/DHPC bicelles are known to magnetically align above the gel to a liquid-crystalline T_m of pure DMPC ($T_m = 23$°C).[36,37,88] Similarly, our results indicate that the PSPC/DHPC bicelles align above the T_m of pure PSPC (approximately 48°C).[26] In addition, the range of temperatures at which the PSPC/DHPC bicelles align is very narrow when compared with the DMPC/DHPC system, which could be a result of the instability of the system above or below the T_m value of 48°C.[36,37,88] These results indicate that longer chain phospholipids such as PSPC, when mixed with detergents such as DHPC, can be magnetically aligned, and they can serve as potential model systems for studying longer integral membrane proteins.

10.2.1 INCORPORATING CHOLESTEROL INTO MAGNETICALLY ALIGNED PHOSPHOLIPID BILAYERS

Cholesterol is a major constituent of eukaryotic cell membranes. The distribution of cholesterol varies among the membranes, depending on different locations and stages of cell development, indicating that cholesterol plays an integral role in cell biology and metabolism.[92] Cholesterol is implicated in a lot of diseases, such as heart disease, stroke, and Alzheimer's.[82,93] The effects of cholesterol on model membranes have been studied extensively by a variety of techniques including molecular dynamics simulations, solid-state NMR spectroscopy, EPR spectroscopy, x-ray diffraction, neutron diffraction, differential scanning calorimetric spectroscopy, and Fourier transform infrared spectroscopy.[94–101] It is believed that cholesterol acts as a regulator by modulating the fluidity of cellular membranes. Cholesterol can enhance the mechanical strength of the membrane and alter the gel-to-liquid crystalline (L_α) T_m of the phospholipids.[102] Molecular dynamics simulations of phospholipid bilayers as a function of cholesterol concentration indicate a significant increase in ordering the phospholipid chains and a reduced fraction of *gauche* conformations, as well as a reduction in lateral diffusion of the phospholipids.[103,104] Deuterium NMR relaxation studies on phospholipid bilayers containing cholesterol indicate that axial rotations of the phospholipid molecules occur at a higher rate than in pure phospholipid bilayers. In addition, the rigid cholesterol molecule appears to undergo slower axial rotation than the corresponding phospholipid molecules.[105] Thus, it is important to probe the structural and dynamic effects of cholesterol on magnetically aligned phospholipid bilayers from the perspective of both the phospholipids and cholesterol molecules.[58,62]

Previously, most cholesterol-membrane studies were based on unoriented or mechanically aligned membrane samples.[99,106] The use of multilamellar vesicles in solid-state NMR spectroscopic studies gives rise to broad powder type spectra. To obtain pertinent structural and dynamic information requires deconvoluting the spectra through the so-called dePakeing process.[107] Alternatively, uniaxially aligned bilayer membranes in solid-state NMR spectroscopic studies reveal high-resolution spectra and pertinent structural and dynamic information. Traditionally, macroscopic uniaxial orientation of membrane bilayers can be accomplished mechanically by stacking phospholipids bilayers between glass plates. The ^{2}H NMR quadrupolar splittings can yield the dynamic information on deuterium-labeled segments of the phospholipid acyl chains.

To understand the interaction between the cholesterol and the phospholipids, molecular order parameters (S_{mol}) are used to define the structural and dynamical properties of the corresponding molecules in the membrane. An ensemble of molecules gives rise to $S_{mol} = 0$ for unrestricted motions of every individual molecule. A S_{mol} value of 1 indicates that all the molecules are perfectly aligned and motionally restricted in one direction.[29,58,108,109] One of the advantages of using the bicelle technique is that the samples are easy to prepare and avoid using dePakeing algorithms, making it easier to calculate the molecular order parameters directly from the quadrupolar splittings of each deuteron from NMR spectra.

Deuterium-labeled cholesterol-d$_6$ NMR studies have also been carried out on DMPC/DHPC/cholesterol bicelle samples. Cholesterol has been shown to orient in phospholipids bilayers with its polar hydroxyl group in close vicinity to the head groups of phospholipid molecules, and with its alkyl side chain extending toward the bilayer center.[110] The amount of cholesterol-d$_6$ in the bicelle samples is titrated from 5 mol% to 20 mol% with respect to DMPC. The ^{2}H NMR spectra of 10 mol% cholesterol-d$_6$ at 318 K is displayed in Figure 10.4a. The resolved peaks are simulated using the DMFIT simulation program and are displayed as dotted lines in Figure 10.4b.[111] The peak assignments are based on previous work.[106] The inner doublets with the smallest quadrupolar splitting (4.2 kHz) represent the 6-^{2}H deuteron. The second smallest quadrupolar splittings (48.1 and 50.9 kHz) represent the 4-^{2}H$_{eq}$ and 2-^{2}H$_{eq}$ deuterons on the fused ring of cholesterol, and the last set of doublets (73.4, 76.1, and 79.2 kHz) correspond to the quadrupolar splitting of the 4,2-^{2}H$_{ax}$ deuterons and the deuteron at the C$_3$-^{2}H position, respectively. The molecular order parameters are calculated as described previously.[60] The six quadrupolar splittings from the different C-^{2}H bonds give almost

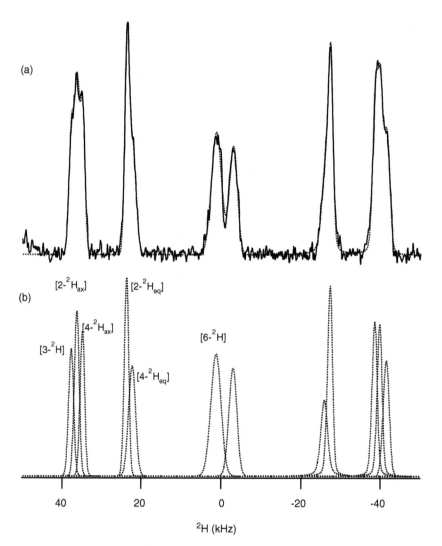

FIGURE 10.4 ^{2}H solid-state NMR spectra of Tm^{3+}-doped magnetically aligned DMPC/DHPC/cholesterol-d$_6$ phospholipids bilayers at 318 K. Cholesterol-d$_6$ is incorporated into the magnetically aligned bilayers as the deuterium label. The concentration of cholesterol-d$_6$ is 10 mol% with respect to DMPC. The simulations were carried out using the DMFIT simulation program. (a) The solid-line spectrum represents the ^{2}H NMR spectrum. The dotted-line spectrum is the summed simulated NMR spectrum. (b) The dotted line in the spectrum represents individual peaks that are simulated.

the same molecular order parameter (S$_{mol}$) values within a small standard deviation. The final result is the average of the six values of S$_{mol}$. The quadrupolar splittings and S$_{mol}$ values confirm that the cholesterol ring is rather rigid in the phospholipid bilayer under these conditions.

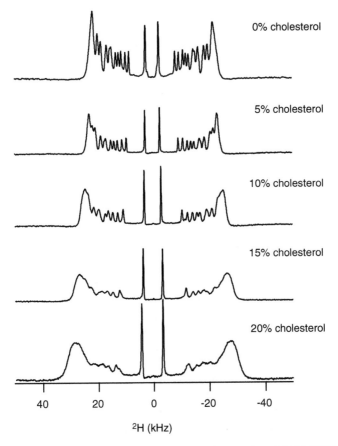

FIGURE 10.5 ^{2}H-NMR spectra of Tm^{3+}-doped magnetically aligned DMPC/DHPC/DMPC-d$_{54}$/cholesterol phospholipids bilayers at 318 K. DMPC-d$_{54}$ is incorporated into the magnetically aligned bicelles as the deuterium label. The cholesterol concentration is 0, 5, 10, 15, and 20 mol% with respect to DMPC.

For ^{2}H-labeled cholesterol, the molecular order parameters will only give information about the flexibility of the lipid chains indirectly. However, by incorporating DMPC-d$_{54}$ into the bicelle samples, the lipid chain dynamics can be probed with ^{2}H solid-state NMR studies. The bicelle samples are analyzed in both the parallel and perpendicular orientations. The molecular order parameters calculated from both bicelle samples are similar. However, only the parallel-aligned phospholipid bilayer data will be presented because of the increased resolution. Figure 10.5 shows a series of ^{2}H NMR spectra of parallel-aligned DMPC/DHPC phospholipid bilayer samples doped with DMPC-d$_{54}$, investigated as a function of cholesterol concentration at 318 K. The ^{2}H NMR spectrum of the bicelle sample with 0 mol% cholesterol with respect to DMPC is well resolved. The good resolution is clearly indicative of well-aligned phospholipid bilayers. As the cholesterol concentration increases, the spectra become broader and start to lose resolution. The bicelle samples prepared in the absence of cholesterol yield well-resolved spectra at 308 K and at a slightly lower temperature.

The bicelle samples prepared with 10 mol% cholesterol with respect to DMPC yield well-resolved spectra at a temperature of 318 K or higher. However, for the bicelle samples with 20 mol% cholesterol, well-resolved spectra were obtained only at temperatures above 318 K. Thus, the minimum temperature required for alignment of DMPC/DHPC phospholipid bilayers increases as the cholesterol concentration increases from 0 to 20 mol%. This result is consistent with EPR spectroscopic results using the cholestane spin label.[60]

Figure 10.6 displays the molecular order parameter S_{mol} profile versus the carbon atom position along the acyl chains of DMPC-d_{54} placed into magnetically aligned phospholipid bilayers. The S_{mol} for the individual C-^{2}H bonds of the methylene groups and the terminal methyl groups of the acyl chains is directly evaluated from the corresponding quadrupolar splittings of each group. The quadrupolar splittings for the deuterons in the plateau region are estimated by integration of the last broad peak according to the literature.[112] Figure 10.6a shows the molecular order parameter S_{mol} profile of varying cholesterol concentrations as a function of the carbon number. The molecular order parameters decrease gradually toward the end of the hydrocarbon chain. Cholesterol enhances the ordering of the entire acyl chain within the phospholipid bilayers as the amount of cholesterol in the bicelle sample increases from 0 to 20 mol% with respect to DMPC. The increase in ordering of the methylene groups near the top of the hydrocarbon chain is almost the same as the increase in ordering of the methyl groups at the end of the hydrocarbon chain as the cholesterol concentration increases. These results indicate that the ordering effect of cholesterol to the phospholipid molecules is uniform along the entire acyl chains of the phospholipid molecules. Figure 10.6b shows the temperature effect on the molecular order parameter of DMPC-d_{54} at 15 mol% cholesterol with respect to DMPC. The molecular order parameters decrease as the temperature increases. The decrease in ordering of the methylene groups and methyl groups at different positions along the acyl chain of the phospholipid molecules are almost identical.

The normalized order parameter over the 13 quadrupolar splittings, calculated using $<S_{mol}> = (1/13) \Sigma_{i=2}^{14} S^i_{mol}$, can be used to characterize the ordering of the entire myristoyl chain.[29] Figure 10.7a displays the cholesterol concentration dependence of the normalized order parameter profile as a function of temperature. In Figure 10.7b, the normalized order parameter is plotted versus cholesterol concentration at various temperatures. The variable $<S_{mol}>$ decreases with increasing temperature but increases with increasing cholesterol concentration. Both graphs indicate that the ordering effect of cholesterol on the membranes is more significant at low temperatures. In contrast, the disordering effect on the membrane caused by increasing the temperature is more significant at high cholesterol concentrations. These results agree well with the EPR spectroscopic results obtained on the cholestane spin label.[60]

The deuterium-labeled cholesterol-d_6 NMR spectrum yields six different quadrupolar splittings corresponding to C-^{2}H bonds at six different positions of the sterol ring. However, the molecular order parameters calculated from the six quadrupolar splittings are similar to each other in magnitude. The result confirms that cholesterol molecules are fairly rigid inside the DMPC/DHPC phospholipid bilayers. The magnitude of change of ordering is uniform along the entire acyl chains from carbon 2

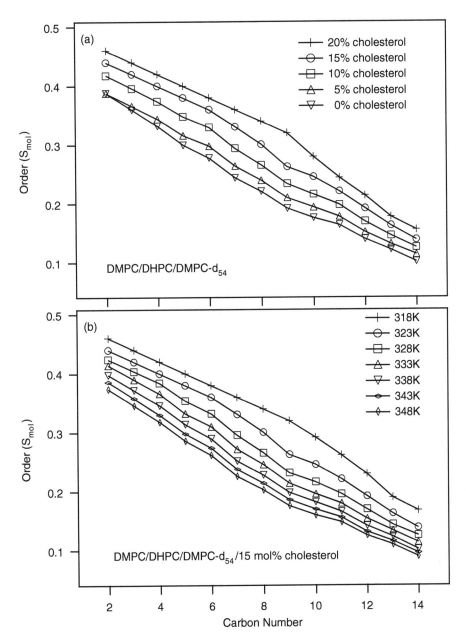

FIGURE 10.6 Molecular order parameter profiles with respect to the position along the acyl chain of deuterium-labeled DMPC-d_{54} incorporated into Tm^{3+}-doped DMPC/DHPC/cholesterol bicelle samples. (a) The cholesterol concentrations are 0, 5, 10, 15, and 20 mol% with respect to DMPC. The sample temperature is 323 K. (b) The cholesterol concentration is 15 mol% with respect to DMPC. The temperature range is from 318 to 348 K. The DMPC/DHPC/cholesterol bicelles with 15 mol% cholesterol at 308 and 313 K are not aligned.

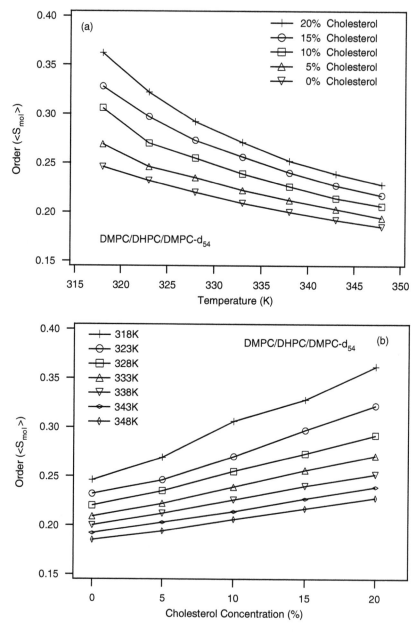

FIGURE 10.7 (a) The average molecular order parameter profiles over the entire length of the acyl chain of DMPC-d$_{54}$ as a function of temperature in magnetically aligned DMPC/DHPC/cholesterol bicelles at cholesterol concentrations of 0, 5, 10, 15, and 20 mol% with respect to DMPC. (b) The average molecular order parameter profiles over the entire length of the acyl chain of DMPC-d$_{54}$ as a function of cholesterol concentration in magnetically aligned DMPC/DHPC/cholesterol bicelles at different temperatures. Both graphs are from bicelle samples that are aligned with their normal being parallel to the static magnetic field.

to 14. The phospholipid acyl chain carbons at positions 2–10 have been estimated to lie close to the sterol ring structure of cholesterol, and the carbons at positions 11–14 are close to the side-chain tail of cholesterol.[110] Analysis of the data indicates that the cholesterol side chain located at the end of the sterol ring is also rigid and has a similar ordering effect on the phospholipid molecules. The effective length of cholesterol has been estimated to correspond to a 17-carbon all-trans hydrocarbon chain,[110] whereas DMPC has 14 carbons on one of its acyl chains. Therefore, the side chain of cholesterol is probably highly packed and tangled in the middle of the phospholipid bilayers, which induces ordering of the membrane in that region.

The degree of ordering increases as the amount of cholesterol in the DMPC/DHPC bilayer samples increases. ^{2}H NMR spectra are sensitive to motions with correlation times between 10^{-3} and 10^{-9} s.[113] Cholesterol molecules have a higher degree of ordering and slower motion than the corresponding phospholipids molecules in the same bicelle sample. The dynamics of the phospholipid membrane can be characterized by three correlation times corresponding to the rotation about the principal diffusion axis of the molecule (chain rotation), rotation about this axis (chain fluctuation or wobbling), and a trans-gauche isomerization of the acyl chain. The flat cholesterol molecule with cylindrical symmetry probably has relatively higher activation energy for molecular rotation about its molecular axis than the corresponding phospholipid molecules. The molecular rotation of phospholipid molecules will slow down because of the close contact of cholesterol molecules with slower motions. At the same time, the rigidity of the cholesterol molecules can also restrict the trans-gauche isomerization of the acyl chains of the phospholipid molecules next to it. As the temperature increases, the molecules gain more energy; thus, both the intermolecular motion (chain rotation and fluctuation) and intramolecular motion will increase. Therefore, the molecular ordering decreases when the temperature increases. The data in this review clearly indicate that cholesterol can be easily incorporated into magnetically aligned phospholipid bilayers and investigated with solid-state NMR spectroscopy.

10.2.2 Membrane Proteins Inserted into Magnetically Aligned Phospholipid Bilayers

Integral membrane proteins have been successfully incorporated into magnetically aligned phospholipid bilayers and investigated with solid-state NMR spectroscopy.[30,33,45,46] We have used a transmembrane peptide containing the transmembrane (TM-A) domain (SSYYIVHDAIIAYIFYFLADKYI) of the integral membrane diverged acetylenase microsomal Δ12-desaturase (CREP-1) to illustrate the feasibility of using magnetically aligned phospholipid bilayers for solid-state NMR studies on integral membrane proteins.[114] The procedure for synthesizing and purifying the peptide has been established in the literature. Circular dichorism (CD) studies on the integral membrane protein TM-A have indicated that the peptide is dominantly α-helical.[114] Figure 10.8a shows the solid-state ^{2}H NMR spectrum of ^{2}H-labeled CD$_3$ Ala-56 CREP-1 TM-A peptide inserted into magnetically aligned phospholipid bilayers. The bilayer normal is perpendicular to the static magnetic field, and the resultant quadrupolar splitting is 14.8 kHz. The well-resolved

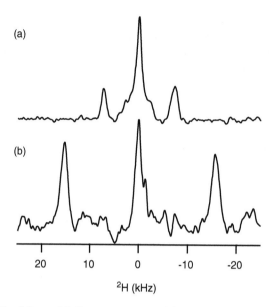

FIGURE 10.8 ^{2}H solid-state NMR spectra of magnetically aligned DMPC/DHPC phospho-lipid bilayers containing 2 mg TM-A deuterated at Ala-56. The spectra were acquired at 38°C without (a) and with ytterbium(III) ions (b). The quadrupolar splitting approximately doubles when comparing the sample with the bilayer normal perpendicular to the magnetic field normal (a) with the sample with the bilayer normal parallel to the magnetic field normal (b).

quadrupolar splittings and line shape indicate that the hydrophobic peptide is well oriented in the static magnetic field when compared with randomly dispersed samples (data not shown). The addition of Yb^{3+} to the bicelle flips the bicelle 90°, such that the bilayer normal is now parallel with B$_0$. At this orientation, the quadrupolar splitting doubles to 30.0 kHz (Figure 10.8b). The peak centered at 0 kHz is a result of the residual D$_2$O from the ^{2}H-depleted water. The absence of any powder pattern components indicates that the TM-A peptide is well oriented in the phospholipid bilayer nanotube arrays. ^{2}H NMR analysis of the aligned quadrupolar splittings indicates that the helical tilt of the TM-A CREP-1 peptide is 6° ± 6° with respect to the bilayer normal.[45,46] Thus, magnetically aligned phospholipid bilayers are an excellent model membrane system for conducting oriented solid-state NMR experiments of integral membrane proteins or peptides.

The sample was prepared using DMPC and DHPC phospholipids at a q ratio of 3.5. The q ratio represents the mole ratio of the long-chain phospholipid (DMPC) to the short-chain phospholipid (DHPC). DMPC (72 mg) was placed into a pear-shaped flask, and DHPC (15 mg) was added to a second pear-shaped flask. ^{2}H'-labeled Ala-56 TM-A (2 mg) was dissolved in trifluoroethano (TFE) (200 μL) and added to the second flask. Both flasks were concentrated to dryness using a rotary evaporator and placed under vacuum overnight. N-(2-hydroxyethyl)-piperazine-N'-2-ethanesulfonic acid (HEPES) buffer, pH 7.0 (270 μL, 100 mM), made from deuterium-depleted water, was added to the second flask. The flask was vortexed, sonicated, frozen, and thawed as necessary to homogenize the sample. The solubilized sample was then transferred to the first flask, and the same steps were repeated until the sample was translucent. The sample was then transferred to a flat-bottom 21-mm NMR tube (5 mm outside diameter) using a Pasteur pipette. Ytterbium(III) chloride, prepared with deuterium-depleted water, was added to 5 mol% with respect to DMPC for experiments in which the bilayer normal of the phospholipids was aligned parallel to the static magnetic field.

ACKNOWLEDGMENTS

G.A.L. acknowledges his research group for all of their help in preparing this review article. This work was supported by a National Science Foundation CAREER Award (CDE-0133433), a National Institutes of Health Grant (GM60259–01), and an American Heart Association Scientist Development Grant (0130396). The 500-MHz wide-bore NMR spectrometer was obtained from National Science Foundation Grant 10116333.

REFERENCES

1. Chang, G., Spencer, R.H., Lee, A.T., Barclay, M.T. and Rees, D.C., Structure of the MscL homolog from *Mycobacterium tuberculosis*: a gated mechanosensitive ion channel, *Science*, 282(5397), 2220–2226, 1998.
2. Doyle, D.A., Cabral, J.M., Pfuetzner, R.A., Kuo, A., Gulbis, J.M., Cohen, S.L., Chait, B.T. and MacKinnon, R., The structure of the potassium channel: molecular basis of K+ conduction and selectivity, *Science*, 280(5360), 69–77, 1998.
3. Jiang, Y.X., Lee, A., Chen, J.Y., Cadene, M., Chait, B.T. and MacKinnon, R., Crystal structure and mechanism of a calcium-gated potassium channel, *Nature*, 417(6888), 515–522, 2002.
4. Jiang, Y.X., Lee, A., Chen, J.Y., Cadene, M., Chait, B.T. and MacKinnon, R., The open pore conformation of potassium channels, *Nature*, 417(6888), 523–526, 2002.
5. Jiang, Y.X., Lee, A., Chen, J.Y., et al., X-ray structure of a voltage-dependent K+ channel, *Nature*, 423(6935), 33–41, 2003.
6. Jiang, Y.X., Ruta, V., Chen, J.Y., Lee, A. and MacKinnon, R., The principle of gating charge movement in a voltage-dependent K+ channel, *Nature*, 423(6935), 42–48, 2003.
7. MacKinnon, R., Cohen, S.L., Kuo, A.L., Lee, A. and Chait, B.T., Structural conservation in prokaryotic and eukaryotic potassium channels, *Science*, 280(5360), 106–109, 1998.
8. MacKinnon, R., Nothing automatic about ion-channel structures, *Nature*, 416(6878), 261–262, 2002.
9. Dutzler, R., Campbell, E.B., Cadene, M., Chait, B.T. and MacKinnon, R., X-ray structure of a ClC chloride channel at 3.0 angstrom reveals the molecular basis of anion selectivity *Nature*, 415(6869), 287–294, 2002.
10. Dutzler, R., Campbell, E.B. and MacKinnon, R., Gating the selectivity filter in ClC chloride channels, *Science*, 300(5616), 108–112, 2003.
11. Agre, P., Johnson, P.F. and McKnight, S.L., Cognate DNA-binding specificity retained after leucine zipper exchange between Gcn4 and C/Ebp, *Science*, 246(4932), 922–926, 1989.

12. Agre, P., Lee, M.D., Devidas, S. and Guggino, W.B., Aquaporins and ion conductance, *Science*, 275(5305), 1490–1490, 1997.

13. Opella, S.J. and McDonnell, P.A., NMR structural studies of membrane proteins, *NMR Proteins*, 159–184, 1993.

14. Opella, S.J., NMR and membrane proteins, *Nat. Struct. Biol.*, 4(Suppl.), 845–848, 1997.

15. Cross, T.A. and Opella, S.J., Solid-state NMR structural studies of peptides and proteins in membranes, *Curr. Opin. Struct. Biol.*, 4(4), 574–581, 1994.

16. Smith, S.O. and Peersen, O.B., Solid-state NMR approaches for studying membrane-protein structure, *Annu. Rev. Biophys. Biomol. Struct.*, 21, 25–47, 1992.

17. Pines, A., Gibby, M. and Waugh, J.S., Proton-enhanced NMR of spins in solids, *J. Chem. Phys.*, 59, 569–590, 1973.

18. Marassi, F.M., Ramamoorthy, A. and Opella, S.J., Complete resolution of the solid-state NMR spectrum of a uniformly 15N-labeled membrane protein in phospholipid bilayers, *Proc. Natl. Acad. Sci. USA*, 94(16), 8551–8556, 1997.

19. Marassi, F.M. and Opella, S.J., NMR structural studies of membrane proteins, *Curr. Opin. Struc. Biol.*, 8(5), 640–648, 1998.

20. Marassi, F.M. and Opella, S.J., A solid-state NMR index of helical membrane protein and topology, *J. Magn. Res.*, 144, 150–155, 2000.

21. Marassi, F.M. and Opella, S.J., Using Pisa pies to resolve ambiguities in angular constraints from PISEMA spectra of aligned proteins, *J. Biomol. NMR*, 23(3), 239–242, 2002.

22. Marassi, F.M., NMR of peptides and proteins in oriented membranes, *Concepts Magn. Reson.*, 14(3), 212–224, 2002.

23. Marassi, F.M. and Crowell, K.J., Hydration-optimized oriented phospholipid bilayer samples for solid-state NMR structural studies of membrane proteins, *J. Magn. Reson.*, 161(1), 64–69, 2003.

24. Marassi, F.M. and Opella, S.J., Simultaneous assignment and structure determination of a membrane protein from NMR orientational restraints, *Protein Sci.*, 12(3), 403–411, 2003.

25. Bechinger, B., Kim, Y., Chirlian, L.E., Gesell, J., Neumann, J.M., Montal, M., Tomich, J., Zasloff, M. and Opella, S.J., Orientations of amphipathic helical peptides in membrane bilayers determined by solid-state NMR spectroscopy, *J. Biomol. NMR*, 1(2), 167–173, 1991.

26. Bechinger, B. and Opella, S.J. Flat-coil probe for NMR spectroscopy of oriented membrane samples, *J. Magn. Reson.*, 95, 585–588, 1991.

27. Bechinger, B., Zasloff, M. and Opella, S.J., Structure and interactions of magainin antibiotic peptides in lipid bilayers: a solid-state nuclear magnetic resonance investigation, *Biophys. J.*, 62(1), 12–14, 1992.

28. Prosser, R.S., Hunt, S.A., DiNatale, J.A. and Vold, R.R., Magnetically aligned membrane model systems with positive order parameter: switching the sign of S_{zz} with paramagnetic ions, *J. Am. Chem. Soc.*, 118, 269–270, 1996.

29. Prosser, R.S., Hwang, J.S. and Vold, R.R., Magnetically aligned phospholipid bilayers with positive ordering: a new model membrane system, *Biophys. J.*, 74, 2405–2418, 1998.

30. Prosser, R.S., Bryant, H., Bryant, R.G. and Vold, R.R., Lanthanide chelates as bilayer alignment tools in NMR studies of membrane-associated peptides, *J. Magn. Reson.*, 141, 256–260, 1999.

31. Losonczi, J.A. and Prestegard, J.H., Improved dilute bicelle solutions for high-resolution NMR of biological macromolecules, *J. Biomol. NMR*, 12(3), 447–451, 1998.

32. Losonczi, J.A. and Prestegard, J.H., Nuclear magnetic resonance characterization of the myristoylated, N-terminal fragment of ADP-ribosylation factor 1 in a magnetically oriented membrane array, *Biochemistry*, 37(2), 706–716, 1998.

33. Howard, K.P. and Opella, S.J., High-resolution solid-state NMR spectra of integral membrane proteins reconstituted into magnetically oriented phospholipid bilayers, *J. Magn. Reson. B*, 112, 91–94, 1996.

34. King, V., Paeker, M. and Howard, K.P. Pegylation of magnetically oriented lipid bilayers, *J. Magn. Reson.*, 142, 177–182, 2000.

35. Parker, M.A., King, V. and Howard, K.P., Nuclear magnetic resonance study of doxorubicin binding to cardiolipin containing magnetically oriented phospholipid bilayers, *Biochim. Biophys. Acta-Biomembr.*, 1514(2), 206–216, 2001.

36. Sanders, C.R., Hare, B.J., Howard, K.P. and Prestegard, J.H., Magnetically-oriented phospholipid micelles as a tool for the study of membrane-associated molecules, *Prog. NMR Spectroscop.*, 26, 421–444, 1994.

37. Sanders, C.R. and Prosser, R.S., Bicelles: a model membrane system for all seasons, *Structure*, 15, 1227–1234, 1998.

38. Prestegard, J.H., New techniques in structural NMR–anisotropic interactions, *Nat. Struct. Biol.*, 5, 517–522, 1998.

39. Struppe, J. and Vold, R.R., Dilute bicellar solutions for structural NMR work, *J. Magn. Reson.*, 135, 541–546, 1998.

40. Nevzorov, A.A., Mesleh, M.F. and Opella, S.J. Structure determination of aligned samples of membrane proteins by NMR spectroscopy, *Magn. Reson. Chem.*, 42(2), 162–171, 2004.

41. Sizun, C., Aussenac, F., Grelard, A. and Dufourc, E.J. NMR methods for studying the structure and dynamics of oncogenic and antihistaminic peptides in biomembranes, *Magn. Reson. Chem.*, 42(2), 180–186, 2004.

42. Aussenac, F., Laguerre, M., Schmitter, J.M. and Dufourc, E.J. Detailed structure and dynamics of bicelle phospholipids using selectively deuterated and perdeuterated labels. H-2 NMR and molecular mechanics study, *Langmuir*, 19(25), 10468–10479, 2003.

43. Sasaki, H., Fukuzawa, S., Kikuchi, J., Yokoyama, S., Hirota, H. and Tachibana, K. Cholesterol doping induced enhanced stability of bicelles, *Langmuir*, 19(23), 9841–9844, 2003.

44. Andersson, A. and Maler, L. Motilin-bicelle interactions: membrane position and translational diffusion, *FEBS Lett.*, 545(2–3), 139–143, 2003.

45. Whiles, J.A., Deems, R., Vold, R.R. and Dennis, E.A. Bicelles in structure-function studies of membrane-associated proteins, *Bioorgan. Chem.*, 30(6), 431–442, 2002.

46. Whiles, J.A., Glover, K.J., Vold, R.R. and Komives, E.A., Methods for studying transmembrane peptides in bicelles: consequences of hydrophobic mismatch and peptide sequence, *J. Magn. Reson.*, 158(1–2), 149–156, 2002.

47. Tan, C.B., Fung, B.M. and Cho, G.J. Phospholipid bicelles that align with their normals parallel to the magnetic field, *J. Am. Chem. Soc.*, 124(39), 11827–11832, 2002.

48. Glover, K.J., Whiles, J.A., Vold, R.R. and Melacini, G., Position of residues in transmembrane peptides with respect to the lipid bilayer: a combined lipid NOEs and water chemical exchange approach in phospholipid bicelles, *J. Biomol. NMR*, 22(1), 57–64, 2002.

49. Crowell, K.J. and Macdonald, P.M., Europium III binding and the reorientation of magnetically aligned bicelles: insights from deuterium NMR spectroscopy, *Biophys. J.*, 81(1), 255–265, 2001.

50. Sternin, E., Nizza, D. and Gawrisch, K. Temperature dependence of DMPC/DHPC mixing in a bicellar solution and its structural implications, *Langmuir*, 17(9), 2610–2616, 2001.
51. Prosser, R.S. and Shiyanovskaya, I.V. Lanthanide ion assisted magnetic alignment of model membranes and macromolecules, *Concepts Magn. Reson.*, 13(1), 19–31, 2001.
52. Sanders, C.R. and Oxenoid, K. Customizing model membranes and samples for NMR spectroscopic studies of complex membrane proteins, *Biochim. Biophys. Acta-Biomembr.*, 1508(1–2), 129–145, 2000.
53. Raffard, G., Steinbruckner, S., Arnold, A., Davis, J.H. and Dufourc, E.J. Temperature-composition diagram of dimyristoylphosphatidylcholine-dicaproylphosphatidylcholine "bicelles" self-orienting in the magnetic field. A solid state H-2 and P-31 NMR study, *Langmuir*, 16(20), 7655–7662, 2000.
54. Tian, F., Losonczi, J.A., Fischer, M.W.F. and Prestegard, J.H. Sign determination of dipolar couplings in field-oriented bicelles by variable angle sample spinning (VASS), *J. Biomol. NMR*, 15(2), 145–150, 1999.
55. Struppe, J., Komives, E.A., Taylor, S.S. and Vold, R.R. H-2 NMR studies of a myristoylated peptide in neutral and acidic phospholipid bicelles, *Biochemistry*, 37(44), 15523–15527, 1998.
56. Vold, R.R., Prosser, R.S. and Deese, A.J. Isotropic solutions of phospholipid bicelles: a new membrane mimetic for high-resolution NMR studies of polypeptides, *J. Biomol. NMR*, 9(3), 329–335, 1997.
57. Vold, R.R. and Prosser, R.S. Magnetically oriented phospholipid bilayered micelles for structural studies of polypeptides. Does the ideal bicelle exist? *J. Magn. Reson. Ser. B*, 113(3), 267–271, 1996.
58. Dave, P.C., Tiburu, E.K., Nusair, N.A. and Lorigan, G.A. Calculating order parameter profiles utilizing magnetically aligned phospholipid bilayers for ^{2}H solid-state NMR studies, *Solid State Nucl. Magn. Reson.*, 24, 137–149, 2003.
59. Tiburu, E.K., Moton, D.M. and Lorigan, G.A. Development of magnetically aligned phospholipid bilayers in mixtures of palmitoylsteroylphosphatidylcholine and dihexanoylphosphatidylcholine by solid-state NMR spectroscopy, *Biochim. Biophys. Acta*, 1512, 206–214, 2001.
60. Lu, J.X., Caporini, M.A. and Lorigan, G.A., The effects of cholesterol on magnetically aligned phospholipid bilayers: a solid-state NMR and EPR spectroscopy study, *J. Magn. Reson.*, 168, 18–30, 2004.
61. Nusair, N.A., Tiburu, E.K., Dave, P.C. and Lorigan, G.A., Investigating fatty acids inserted into magnetically aligned phospholipid bilayers using EPR and solid-state NMR spectroscopy, *J. Magn. Reson.*, 68(2), 228–237, 2004.
62. Tiburu, E.K., Dave, P.C. and Lorigan, G.A., Solid-state H-2 NMR studies of the effects of cholesterol on the acyl chain dynamics of magnetically aligned phospholipid bilayers, *Magn. Reson. Chem.*, 42(2), 132–138, 2004.
63. Mangels, M.L., Cardon, T.B., Harper, A.C., Howard, K.P. and Lorigan, G.A., Spectroscopic characterization of spin-labeled magnetically oriented phospholipid bilayers by EPR spectroscopy, *J. Am. Chem. Soc.*, 122, 7052–7058, 2000.
64. Mangels, M.L., Harper, A.C., Smirnov, A.I., Howard, K.P. and Lorigan, G.A. Investigating magnetically aligned phospholipid bilayers with EPR spectroscopy at 94 GHz, *J. Magn. Reson.*, 151, 253–259, 2001.
65. Cardon, T.B., Tiburu, E.K., Padmanabhan, A., Howard, K.P. and Lorigan, G.A., Magnetically aligned phospholipid bilayers at the parallel and perpendicular orientations for X-band spin-label EPR studies, *J. Am. Chem. Soc.*, 123, 2913–2914, 2001.

66. Cardon, T.B., Tiburu, E.K. and Lorigan, G.A. Magnetically aligned phospholipid bilayers in weak magnetic fields: optimization, mechanism, and advantages for X-band EPR studies, *J. Magn. Reson.*, 161, 77–90, 2003.

67. Caporini, M.A., Padmanabhan, A., Cardon, T.B. and Lorigan, G.A. Investigating magnetically aligned phospholipid bilayers with various lanthanide ions for X-band spin-label EPR studies, *Biochim. Biophys. Acta*, 1612, 52–58, 2003.

68. Nieh, M.P., Glinka, C.J., Krueger, S., Prosser, R.S. and Katsaras, J., SANS study of the structural phases of magnetically alignable lanthanide-doped phospholipid mixtures, *Langmuir*, 17(9), 2629–2638, 2001.

69. Gaemers, S. and Bax, A., Morphology of three lyotropic liquid crystalline biological NMR media studied by translational diffusion anisotropy, *J. Am. Chem. Soc.*, 123(49), 12343–12352, 2001.

70. Nieh, M.P., Glinka, C.J., Krueger, S., Prosser, R.S. and Katsaras, J., SANS study on the effect of lantanide ions and charged lipids on the morphology of phospholipid mixtures, *Biophys. J.*, 82, 2487–2498, 2002.

71. Wang, H., Nieh, M.P., Hobbie, E.K., Glinka, C.J. and Katsaras, J. Kinetic pathway of the bilayered-micelle t perforated-lamellae transition, *Phys. Rev. E*, 67, 1–4, 2003.

72. Bax, A. and Tjandra, N., High-resolution heteronuclear NMR of human ubiquitin in an aqueous liquid crystalline medium, *J. Biomol. NMR*, 10(3), 289–292, 1997.

73. Cornilescu, G., Marquardt, J.L., Ottiger, M. and Bax, A., Validation of protein structure from anisotropic carbonyl chemical shifts in a dilute liquid crystalline phase, *J. Am. Chem. Soc.*, 120(27), 6836–6837, 1998.

74. Ottiger, M. and Bax, A., Determination of relative N-H-N N-C, C-alpha-C, and C(alpha)-H-alpha effective bond lengths in a protein by NMR in a dilute liquid crystalline phase, *J. Am. Chem. Soc.*, 120(47), 12334–12341, 1998.

75. Ottiger, M. and Bax, A., Characterization of magnetically oriented phospholipid micelles for measurement of dipolar couplings in macromolecules, *J. Biomol. NMR*, 12(3), 361–372, 1998.

76. Ottiger, M., Delaglio, F., Marquardt, J.L., Tjandra, N. and Bax, A., Measurement of dipolar couplings for methylene and methyl sites in weakly oriented macromolecules and their use in structure determination, *J. Magn. Reson.*, 134(2), 365–369, 1998.

77. Ramirez, B.E. and Bax, A., Modulation of the alignment tensor of macromolecules dissolved in a dilute liquid crystalline medium, *J. Am. Chem. Soc.*, 120(35), 9106–9107, 1998.

78. Tjandra, N. and Bax, A., Direct measurement of distances and angles in biomolecules by NMR in a dilute liquid crystalline medium, *Science*, 278, 1697, 1997.

79. Tjandra, N., Omichinski, J.G., Gronenborn, A.M., Clore, G.M. and Bax, A., Use of dipolar H-1-N-15 and H-1-C-13 couplings in the structure determination of magnetically oriented macromolecules in solution, *Nat. Struct. Biol.*, 4, 732–738, 1997.

80. Wang, Y.X., Marquardt, J.L., Wingfield, P., Stahl, S.J., Lee-Huang, S., Torchia, D. and Bax, A., Simultaneous measurement of H-1-N-15, H-1-C-13, and N-15-C-13 dipolar couplings in a perdeuterated 30 kDa protein dissolved in a dilute liquid crystailine phase, *J. Am. Chem. Soc.*, 120(29), 7385–7386, 1998.

81. Prosser, R.S., Losonczi, J.A. and Shiyanovskaya, I.V., Use of a novel aqueous liquid crystalline medium for high- resolution NMR of macromolecules in solution, *J. Am. Chem. Soc.*, 120(42), 11010–11011, 1998.

82. Mironov, V.S., Galyametdinov, Y.G., Ceulemans, A. and Binnemans, K., On the magnetic anisotropy of lanthanide-containing metallomesogens, *J. Chem. Phys.*, 113(22), 10293–10303, 2000.

83. Firestone, M.A., Thiyagarajan, P. and Tiede, D.M., Structure and optical properties of a thermoresponsive polymer-grafted, liped-based complex fluid, *Langmuir*, 14, 4688–4698, 1998.

84. Firestone, M.A., Tiede, D.M. and Seifert, S., Magnetic field-induced ordering of a polymer-grafted biomembrane-mimetic hydrogel, *J. Phys. Chem. B*, 104(11), 2433–2438, 2000.

85. Katsaras, J., Donaberger, R.L., Swainson, I.P., Tennant, D.C., Tun, Z., Vold, R.R. and Prosser, R.S., Rarely observed phase transitions in a novel lyotropic liquid crystal system, *Phys. Rev. Lett.*, 78(5), 899–902, 1997.

86. Raffard, G., Steinbruckner, S., Arnold, A., Davis, J.H. and Dufourc, E.J., Temperature-composition diagram of dimyristoylphosphatidylcholine-dicaproylphosphatidylcholine "bicelles" self-orienting in the magnetic field. A solid state ^{2}H and ^{31}P NMR study, *Langmuir*, 16, 7655–7662, 2000.

87. Conti, J., Halladay, H.N. and Petersheim, M., An ionotropic phase transition in phosphatidylcholine: cation and anion cooperativity, *Biochim. Biophys. Acta*, 902, 53–64, 1987.

88. Sanders, C.R. and Schwonek, J.P., Characterization of magnetically orientable bilayers in mixtures of dihexanoylphosphatidylcholine and dimyristoylphosphatidylcholine by solid-state NMR, *Biochemistry*, 31(37), 8898–8905, 1992.

89. Sternin, E., Nizza, D. and Gawrisch, K., Temperature dependence of DMPC/DHPC mixing in a bicellar solution and its structural implications, *Langmuir*, 17, 2610–2616, 2001.

90. Faure, C., Bonakdar, L. and Dufourc, E.J., Determination of DMPC hydration in the Lα and Lβ phases by ^{2}H solid state NMR of D_2O, *FEBS Lett.*, 405, 263–266, 1997.

91. Czerski, L. and Sanders, C.R., Functionality of a membrane protein in bicelles, *Anal. Biol.*, 284, 327–333, 2000.

92. Yeagle, P.L., Cholesterol and the cell membrane, *Biochim. Biophys. Acta*, 822, 267–287, 1985.

93. Borroni, B., Pettenati, C., Bordonali, T., Akkawi, N., Luca, M.D. and Padovani, A., Serum cholesterol levels modulate long-term efficacy of cholinesterase inhibitors in Alzheimer disease, *Neurosci. Lett.*, 343, 213–215, 2003.

94. Chiu, S.W., Jakobsson, E., Mashl, J. and Scott, H.L., Cholesterol-induced modifications in lipid bilayers: a simulation study, *Biophys. J.*, 83, 1842–1853, 2002.

95. Kessel, A., Ben-Tal, N. and May, S., Interactions of cholesterol with lipid bilayers: the preferred configuration and fluctuations, *Biophys. J.*, 81, 643–658, 2001.

96. Jedlovszky, P. and Mezei, M., Effect of cholestrol on the properties of phospholipid membranes. 1. Structural features, *J. Phys. Chem. B*, 107, 5311–5321, 2003.

97. Epand, R.M., Bain, A.D., Sayer, B.G., Bach, D. and Wachtel, E., Properties of mixtures of cholesterol with phosphatidylcholine or with phosphatidylserine studied by ^{13}C magic-angle spinning nuclear magnetic resonance, *Biophys. J.*, 83, 2053–2063, 2002.

98. Taylor, M.G. and Smith, I.C.P., Reliability of nitroxide spin probes in reporting membrane properties: a comparison of nitroxide- and deuterium-labeled steroids, *Biochemistry*, 20, 5252–5255, 1981.

99. Dufourc, E.J., Parish, E.J., Chitrakorn, S. and Smith, I.C.P., Structural and dynamical details of cholesterol-lipid interaction as revealed by deuterium NMR, *Biochemistry*, 23, 6062–6071, 1984.

100. Lemmich, J., Mortensen, K., Ipsen, J.H., Honger, T., Bauer, R. and Mouritsen, O.G., The effect of cholestrol in small amounts on lipid-bilayer softness in the region of the main phase transition, *Eur. Biophys. J.*, 25, 293–304, 1997.

101. Rappolt, M., Vidal, M.F., Kriechbaum, M., Steinhart, M., Amenitsch, H., Bernstorff, S. and Laggner, P., Structural, dynamic and mechanical properties of POPC at low cholesterol concentration studied in pressure/temperature space, *Euro. Biophys. J. Biophys. Lett.*, 31(8), 575–585, 2003.

102. McMullan, R.K. and McElhaney, R.N., Physical studies of cholesterol-phospholipid interactions, *Curr. Opin. Colloid Interface Sci.*, 1, 83–90, 1996.

103. Hofsäβ, C., Lindahl, E. and Edholm, O., Molecular dynamics simulations of phospholipid bilayer with cholesterol. *Biophys. J.*, 84, 2192–2206, 2003.

104. Smondyrev, A.M. and Berkowits, M.L., Structure of dipalmitoylphosphatidylcholine/cholesterol bilayer at low and high cholesterol concentrations: molecular dynamics simulation, *Biophys. J.*, 77, 2075–2089, 1999.

105. Trouard, T.P., Nevzorov, A.A., Alam, T.M., Job, C., Zajicek, J. and Brown, M.F., Influence of cholesterol on dynamics of dimyristoylphosphatidylcholine bilayers as studied by deuterium NMR relaxation, *J. Chem. Phys.*, 110(17), 8802–8818, 1999.

106. Marsan, M.P., Muller, I., Ramos, C., Rodriquez, F., Dufourc, E.J., Czaplicki, J. and Milon, A., Cholesterol orientation and dynamics in dimyristoylphosphatidylcholine bilayers: a solid state deuterium NMR analysis, *Biophys. J.*, 76, 351–359, 1999.

107. Schäfer, H., Mädler, B. and Sternin, E., Determination of orientational order parameters from ^{2}H NMR spectra of magnetically partially oriented lipid bilayers, *Biophys. J.*, 74, 1007–1014, 1998.

108. Smith, I.C.P. and Butler, K.W., Oriented lipid systems as model membranes, in *Spin Labeling, Theory and Application*, Berliner, J.B., Ed., Academic Press, New York, 1976, p. 592.

109. Aussenac, F., Tavares, M. and Dufourc, E.J., Cholesterol dynamics in membranes of raft composition: a molecular point of view from ^{2}H and ^{31}P solid-state NMR, *Biochemistry*, 42, 1383–1390, 2003.

110. Ohvo-Rekilä, H., Ramstedt, B., Leppimäki, P. and Slotte, J.P., Cholesterol interactions with phospholipids in membranes, *Prog. Lipid Res.*, 41, 66–97, 2002.

111. Massiot, D., Fayon, F., Capron, M., King, I., LeCalve, S., Alonso, B., Durand, J.-O., Bujoli, B., Gran, Z. and Hoatson, G., Modelling one- and two-dimensional solid state NMR spectra, *Magn. Reson. Chem.*, 40, 70–76, 2002.

112. Huster, D., Arnold, K. and Garisch, K., Influence of docosahexaenoic acid and cholesterol on lateral lipid organization in phospholipid mixtures, *Biochemistry*, 37, 17299–17308, 1998.

113. Meier, P., Ohmes, E., Kothe, G., Blume, A., Weldner, J. and Elbl, H., Molecular order and dynamics of phospholipid membranes. a deuteron magnetic resonance study employing a comprehensive line-shape model, *J. Phys. Chem.*, 87, 4904–4912, 1983.

114. Minto, R.E., Gibbons, W.J., Cardon, T.B. and Lorigan, G.A., Synthesis and conformational studies of a transmembrane domain from a diverged microsomal delta(12)-desaturase, *Anal. Biochem.*, 308(1), 134–140, 2002.

11 Computational Aspects of Biological Solid-State Nuclear Magnetic Resonance Spectroscopy

Niels Chr. Nielsen

CONTENTS

ABSTRACT Numerical simulations play an increasingly important role in biological solid-state nuclear magnetic resonance (NMR) spectroscopy. This is ascribed to several factors. First, the pulse sequences used for efficient tailoring of the Hamiltonians to provide the desired coherence/polarization transfers or structural information become increasingly complicated in our wish to meet increasingly rigid specifications. Second, with an increasing tendency to use uniformly or extensively isotope-labeled samples to get all the information out of one or few samples, the spin systems that should be handled and interpreted inevitably grow larger than the more typical two-spin cases on which much of our technology is based. Third, with the two former complications there will be an interest in using software to design pulse sequences with optimum performance. Fourth, the extraction of parameters from one or several nuclear spin interactions from solid-state NMR spectra will often require computer simulations to provide accurate values for the parameters subsequently transferred to structural calculation software. Adding these elements up, it is foreseen that the development and availability of efficient software platforms for calculation of solid-state NMR experiments is — and, indeed, will be more so — a key element in the continued fast progress of biological solid-state NMR.

11.1 INTRODUCTION

During the last decades, solid-state nuclear magnetic resonance (NMR) spectroscopy has undergone a remarkable transition from being a physics/chemistry-oriented method for identification and characterization of small molecules primarily of chemical or materials science interest[1-4] to being a promising new tool in structural biology.[5-9] The last few years have witnessed a tremendous progress in the size of the systems amenable to analysis, going from isolated spins or spin-pairs in selectively ^{13}C or ^{15}N isotope-labeled molecules to uniformly ^{13}C, ^{15}N-labeled proteins in, at present, up to about 80–100 residues.[10-25] This progress, which still may be considered to represent an immature level not yet reflecting the expected upper limit of the size and complexity of the molecules amenable to studies, is ascribed to the steady development of increasingly powerful NMR instrumentation (higher fields, faster sample spinning, stronger radiofrequency fields, etc.), as well as advanced solid-state NMR pulse sequences. The latter includes a large variety of specialized pulse sequence building blocks which to very high precision tailor the otherwise very complex nuclear spin Hamiltonian into a manageable form. With a focus on extensively ^{13}C, ^{15}N isotope–enriched proteins, these pulse sequence elements provide the spin engineer with a toolbox for the assembly of advanced multidimensional solid-state NMR experiments with sufficient resolution power and information content. In this manner, it is possible to construct pulse sequences that in many respects resemble the multidimensional experiments that have rendered liquid-state NMR a prime technique in structural biology.[26-28]

The gradual maturation of solid-state NMR toward structural characterization of proteins is highly welcome and extremely interesting in the context of structural biology and proteomics research. The genome projects[29] have provided a wealth of information and raised a lot of new questions, including how we, in detail, characterize about half the proteins encoded for by the various genomes, as they are insoluble in water or impossible or difficult to crystallize. To the most important categories of such proteins belong membrane proteins, fibrils, protein assemblies, and extracellular matrix proteins, which in most cases are impossible to characterize in full functional form, using traditional methods such as liquid-state NMR and x-ray crystallography. For such proteins, solid-state NMR may prove to be a very important probe for structural information, although the challenge is big, and there is a need for substantial development of the technology and its interface to molecular biology and biophysics.

With solid-state NMR addressing increasingly complex molecules, and as a result of increasingly large isotope-labelled spin systems, the precise evaluation of the spin dynamics associated with even the simplest pulse sequence elements grows more difficult. This complexity obviously increases even more for advanced multiple-pulse elements and for the assembly of these elements into multidimensional NMR experiments containing numerous elements serving the fundamental transfers of coherence or polarization from spin to spin in the molecular skeleton of atoms. Although many of the elements have been developed on small spin systems — often spin-pairs, using analytical tools such as low-[30] and high-order[31] or even exact[32,33] effective Hamiltonian theory — the operation of these elements in multiple-spin systems and the delicate interplay between pulse sequence building blocks typically call for numerical evaluation. This need is addressed in this chapter, which serves to illustrate and demonstrate the requests for efficient computer programs that in a flexible manner, enable simulation of solid-state NMR experiments for essentially all types of spin systems present in biological systems. Such methods are relevant for experiment evaluation (i.e., which experiment is suited for a given application), experiment design (either as support to design by analytical means or as a method for design by itself), and experimental data interpretation.

11.2   THE OBJECTS AND THE NUMERICAL TOOLS

In our description of the tools for numerical simulations in biological solid-state NMR, it is instructive to address the system to be investigated and the properties of this system in relation to the experiments to be performed. With molecular structure being a major aim, our request is to characterize the nuclear and electronic surroundings of the nuclear spins under the consideration that the number of "active" spins may be altered by isotope labeling, mutation, and so forth. This will be denoted our *molecular object*. We will in the following text, to a large extent, restrict ourselves to the description of peptides (potentially in combination with ions and smaller molecules) as the molecular object — knowing that other targets, such as nucleic acids, RNA, DNA, and so on may be treated in a similar systematic fashion. The other element to address here is the assembly of an appropriate NMR pulse sequence to characterize this system, with the experiment assembled by an appropriate selection of known pulse sequence elements or from specific tailor-made elements having new or improved performance as part of the integral pulse sequence. This we denote our *NMR object*. The best combination of these objects provides the best experiment for structural characterization. Considering NMR on biological macromolecules, we see that both objects may be considered as a build-up of modules that highly simplifies the establishment of universal tools for their description and further development. The basic elements involved in our description are outlined in Figure 11.1, illustrating the relations between structure and NMR spectra as well as some of the tools discussed in the following text.

11.2.1   THE MOLECULAR OBJECT AND SIMMOL

The objective of biological solid-state NMR experiments is to provide information about structure or dynamics of the molecule subject to investigation. In a short historical perspective, the complexity of the object has shifted from very simple spin systems (although they may be present in large molecules) toward more complex systems as the NMR methodology has gradually improved. Following the

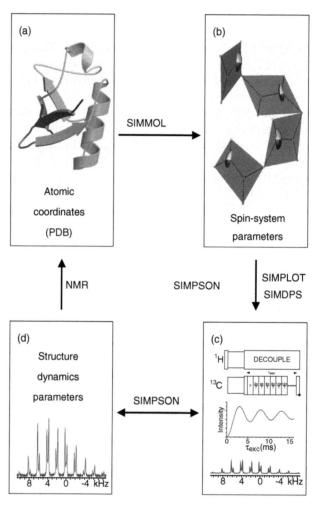

FIGURE 11.1 Schematic representation of some elements involved in numerical simulation of solid-state NMR experiments with specific attention to protein structures. (a) Atomic coordinates and graphical representation of these. (b) Spin-system relevant elements and their parameters, such as peptide planes and chemical shielding tensors illustrated for a fraction of the molecule. These two elements may be considered the *molecular object*. (c) A NMR pulse sequence along with numerical simulations and (d) establishment of structural parameters from NMR spectra by iterative fitting. The two latter elements are parts of the *NMR object*. Reproduced from ref. 34 with permission.

development of the first dipolar recoupling techniques in magic-angle spinning (MAS) solid-state NMR,[35,36] analysis of isolated spin pairs in biological molecules became popular in the late 1980s and has since played an important role as a method to unravel specific internuclear distances of prime importance for the description of molecular function without the need for a more demanding (and for many systems still impossible) full structural analysis.[37-42] The accuracy provided by these experiments may be as high as sub-Ångstrom level, which is considered important for high-precision structural definition of, for example, hydrophobic binding pockets with relevance for drug development. Similarly, on the small spin-system level, quadrupolar metal nuclei in the catalytic sites of metalloenzymes and ^{17}O or ^{2}H as probes to information on hydrogen bonding have achieved attention in the last years, governed by new technologies such as multiple-quantum MAS[43] and a quadrupolar version of the Carr-Purcell-Meiboom-Gill (QCPMG).[44] Since the late 1990s, major attention has been devoted to the study of the larger spin systems encountered in uniformly ^{13}C-, ^{15}N-, or ^{13}C, ^{15}N-labeled biomolecules. The development of methods for such studies has largely followed two paths: one based on rotating powders[7] and one, in the concern of membrane proteins, on uniaxially oriented systems[6] although hybrids[45] between the two have also been presented.

In the design of solid-state NMR methods for analysis of small and large spin systems in biological macromolecules, it is instructive to consider the modular construction of these molecules and the relation to NMR measurables, as illustrated in Figure 11.2 by a membrane protein structure. In the protein structure (Figure 11.2a, here embedded in a phospholipid membrane), the various amino acids are linked by covalent bonds, leading to the skeleton of the molecule defined by chains (Figures 11.2b, 11.2d) of linked "rigid" peptide plane structures (Figure 11.2c), which in combination with attached side chains and potential ligated smaller molecules/ions, may be regarded as our *molecular object*. Considering this object, the NMR entry to structural information goes through the properties of the NMR active nuclear spin species being present in the various residues either naturally or by isotope enrichment. It is well established from liquid-state NMR that properties such as the isotropic chemical shift, reflecting the local electronic environment of the nuclei, are sensitive toward which residue is considered and to this residue's participation in a given secondary structure. For partially oriented systems and, obviously to a much larger extent, immobile molecules, the chemical shifts are influenced by anisotropic effects. This implies that the chemical shift highly depends on the orientation of the chemical shielding tensor relative to the external magnetic field. In immobilized molecules, the dipole–dipole coupling interactions, averaged by molecular motion for rapidly reorienting molecules in liquids, also enter the scene with full effect. This effect depends on the internuclear distance and the orientation of the internuclear vector relative to the external magnetic field. Finally, anisotropic effects from electric field gradients may be probed directly for quadrupolar nuclei such as ^{39}K and ^{67}Zn, often encountered as metal ions in biological macromolecules.

The establishment of experimental methods and data-handling procedures in liquid-state NMR has benefited greatly from the simple "modular" properties of polypeptides and their NMR parameters. The same applies to a large extent when anisotropic interactions are added to the picture. It turns out that the anisotropic NMR interactions are very similar from residue to residue, at least when considering the peptide backbone. This aspect is visualized in Figures 11.2b–11.2d, which, for a single peptide plane to a larger α-helical structure, illustrates the peptide planes in gray with the atoms replaced by typical ^{1}H, ^{13}C, and ^{15}N chemical shielding tensors attached in the form of ellipsoids, with the length of the three axes reflecting the size of the principal

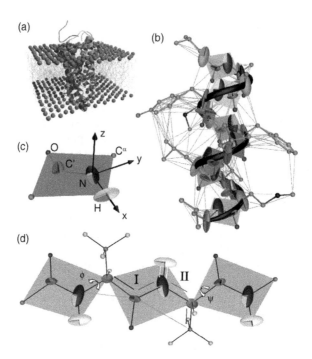

FIGURE 11.2 Graphical illustration of a membrane protein structure and fragments of this with indication of solid-state NMR relevant peptide planes, chemical shielding tensors, and specific through-space internuclear distances. The graphics have been made using SIM-MOL.[34] (a) The rhodopsin structure of Palczewski et al.[46] (PDB code: 1F88) embedded in a phospholipid bilayer (not structurally derived coordinates), (b) an α-helical fragment with indication of peptide planes, chemical shift tensors, and NMR relevant internuclear distances (larger than 75 Hz dipole–dipole couplings), (c) a peptide plane with definition of the peptide-plane coordinate system and backbone chemical shift tensors (see text), and (d) an extended structure of three peptide planes with indication of typical coherence transfer pathways (I: NCOCA, II: NCACB) used in multidimensional solid-state NMR experiments. Reproduced from ref. 25 with permission.

shielding tensor elements. These tensors have typical — and largely invariant — magnitudes and orientations relative to the rigid, planar structure of the peptide plane. Precisely this feature renders it relatively easy to consider anisotropic chemical shielding effects in numerical calculations of biological NMR spectroscopy. For example, it is possible to establish reasonable parameters for the chemical shielding tensors and the dipole–dipole interactions (merely depending on the distance between the nuclei and the orientation of the internuclear vector) on the basis of a known protein structure. Considered the other way around, this orientation and magnitude dependence of the anisotropic interactions makes it possible to extract structural information from these interactions by numerical simulations.

To facilitate the establishment of appropriate spin-system parameters and also provide a flexible tool for visualization of the relevant tensorial interactions for test and documentation, we recently introduced the SIMMOL computer program.[34]

SIMMOL is a flexible Tcl-controlled[47] program that based on structural coordinates in the PDB (Protein Data Bank) format,[48] allows three-dimensional (3D) visualization and parameter calculation of tensorial interactions for essentially any spin system relevant for polypeptides or proteins. This program has been used to establish the graphics in Figures 11.1 and 11.2 but is much more flexible than this, as illustrated in several recent papers.[25–54] The program's major force is its ability to establish reliable chemical shift, dipolar coupling, and quadrupolar coupling tensor parameters for multiple-spin systems as input to numerical calculations aimed at method evaluation, method development, and extraction of structural data from experimental data. For this purpose, SIMMOL takes advantage of the typical "peptide-plane" nuclear spin interactions parameters tabulated in Table 11.1. These parameters are primarily based on solid-state NMR measurements for single-crystal or powder samples of amino acids and small peptides, as well as membrane peptides oriented macroscopically in phospholipid bilayers.[34] Obviously, such typical parameters will gradually improve, and perhaps even be differentiated according to amino acid type, as more and more solid-state NMR experiments, as well as partially oriented liquid-state NMR and molecular dynamics studies, become available. The same applies to the information on side-chain tensors, which so far needs to be based on typical parameters for aliphatic, aromatic, or carboxyl group parameters with the tensors oriented as precisely as possible relative to local structure planes using various coordinate tools in SIMMOL.[34] In such work, it is useful that SIMMOL allows 3D interactive visualization of the molecular structure and the associated tensors to verify that the tensors are correctly associated with the molecular structure. The 3D graphics may also prove useful for documentation and illustration of the results from NMR studies, where the anisotropic interaction tensors typically bear the most important information about structure and dynamics.

11.2.2 THE NMR OBJECT AND SIMPSON

With a hold in the molecular structure, the spin systems, and the size and orientation of the anisotropic interaction tensors, the next element in numerical simulations is the extension of the spin system parameters into the spin dynamics of relevant NMR experiments. This first amounts to definition of the internal Hamiltonian, which here for the sake of generality will be described for a rotating sample. Static samples may easily be treated as special cases.

For a general spin system, the Hamiltonian typically contains various elements, which in the normal high-field truncated Zeeman interaction representation may be expressed as

$$H(t) = H_{rf}(t) + H_{CS}(t) + H_J(t) + H_D(t) + H_Q(t) \tag{11.1}$$

where the first term represents external manipulation by radiofrequency (rf) irradiation, and the following terms represent the internal chemical shielding, J coupling, dipole–dipole coupling, and quadrupolar coupling interactions, all of which represent contributions from the full spin system. We note that all terms may be time dependent — not only the rf irradiation in terms of a pulse sequence but also the internal

TABLE 11.1

Typical Magnitudes and Orientations of Chemical Shift, Scalar J Coupling, Dipole–Dipole Coupling, and Quadrupolar Coupling Tensors within the Amino-Acid Residue or Peptide Plane of Polypeptides[a]

Chemical Shift	δ_{iso}^{CS}	δ_{aniso}^{CS}	η^{CS}	α_{PE}^{CS}	β_{PE}^{CS}	γ_{PE}^{CS}
$^1H^N$	9.3	7.7	0.65	90	−90	90
$^{13}C^{\alpha\ b}$	50	−20	0.43	90	90	0
$^{13}C'$	170	−76	0.90	0	0	94
^{15}H	119	99	0.19	−90	−90	−17

J and Dipolar Coupling	J_{iso}^{IS}	$b_{IS}/2\pi$	r_{IS}	β_{PE}^{D}	γ_{PE}^{D}	
$^1H^{\alpha}$–$^{13}C^{\alpha}$	140	−23328	1.090	—[c]	—[c]	
$^1H^N$–^{15}N	−92	11341[d]	1.024[d]	90	0	
$^{13}C^{\alpha}$–$^{13}C'$	55	2142	1.525	90	120.8	
$^{13}C^{\alpha}$–$^{13}C^{\beta}$	35	2159	1.521	—[c]	—[c]	
$^{13}C^{\alpha}$–^{15}N	−11	988	1.458	90	115.3	
$^{13}C'$–^{15}N	−15	1305	1.329	90	57	

Quadrupolar Coupling	C_Q	η^Q	α_{PE}^{Q}	β_{PE}^{Q}	γ_{PE}^{Q}
$^2H^{Nh}$	0.210	0.15	−90	90	0
$^2H^{\alpha}$	0.168	0.10	—[c]	—[c]	—[c]
^{14}N	3.21	0.32	0	0	0
$^{17}O^{e,f}$	8.3	0.28	−90	90	0

[a] The Euler angles relate the principal axes frames (P^λ) to the peptide plane coordinate system (E) having x along N-H and z being the normal to the plane (cf. Figure 2c). Chemical shifts $\left(\delta_{iso}^{CS}, \delta_{aniso}^{CS}\right)$ are in ppm relative to TMS (^{1}H, ^{13}C) and liq. NH$_3$. Scalar J and dipolar couplings (J_{iso}^{IS} and $b_{IS}/2\pi$) are given in Hertz, the internuclear distance r_{IS} in Ångstroms, and the quadrupolar coupling in MHz. Because of axial symmetry, the dipolar coupling α_{PE} (and α_{PC}) angle can be chosen arbitrarily.

[b] The orientation of the $^{13}C^{\alpha}$ chemical shielding tensor depends on the secondary structure and may vary significantly from the given angles. We note that SIMMOL automatically calculates the bond length, dipolar coupling constants, and dipolar coupling. Ω_{PC}^{D} Euler angles are taken directly from the PDB structure without reference to the tabulated values.

[c] The Euler angles depend on the secondary structure.

[d] A somewhat lower value of $b_{IS}/2\pi$ = 9.9 kHz (corresponding to r_{IS} = 1.07 Å) is typically used for peptides oriented in uniaxially aligned phospholipid bilayers.

[e] We note that the magnitude, and in particular the orientation, of these tensors may be influenced by hydrogen bonding.

[f] We assume that the ^{17}O quadrupolar coupling tensor is oriented with its unique element Q_{zz} along the C'-O bond axis and Q_{yy} perpendicular to the peptide plane.

Hamiltonians for which the anisotropic (i.e., orientation dependent) terms acquire time dependence by sample rotation.

To maintain a simple picture, for the purpose of illustration, we here express the various terms of the internal Hamiltonian in the typical first-order form

$$H_\lambda(t) = \sum_{m=-2}^{2} \omega_{\lambda,0}^{(m)} e^{im\omega_r t} O_\lambda \qquad (11.2)$$

with $\omega_r/2\pi$ denoting the spinning frequency and O_λ the spin operator. We note that similar — albeit more complicated — expressions may readily be established for second-order interactions, for example, being relevant for the quadrupolar coupling interaction in spin $I > 1/2$ nuclei. The first-order Fourier coefficients may be written as

$$\omega_{\lambda,m'}^{(m)} = \omega_{iso}^\lambda \delta_{m,0} + \omega_{aniso}^\lambda$$

$$\left\{ D_{0,-m}^{(2)}\left(\Omega_{PR}^\lambda\right) - \frac{\eta^\lambda}{\sqrt{6}} \left[D_{-2,-m}^{(2)}\left(\Omega_{PR}^\lambda\right) + D_{2,-m}^{(2)}\left(\Omega_{PR}^\lambda\right) \right] \right\} d_{-m,m'}^{(2)}\left(\beta_{RL}\right) \qquad (11.3)$$

using the angular frequencies $\omega_{iso}^{CS} = \omega_0 \delta_{iso}^{CS}$, $\omega_{aniso}^{CS} = \omega_0 \delta_{aniso}^{CS}$, $\omega_{iso}^J = -2\pi\sqrt{3}\,J_{iso}$, $\omega_{aniso}^J = 2\pi\sqrt{6}\,J_{aniso}$, $\omega_{iso}^D = 0$, $\omega_{aniso}^D = \sqrt{6}\,b_{IS}$, $\omega_{iso}^Q = 0$, and $\omega_{aniso}^Q = 2\pi\sqrt{6}\,C_Q/[4I(2I - 1)]$ for the chemical shift, J coupling, dipolar coupling, and quadrupolar coupling interactions, respectively ($\omega_0 = -\gamma B_0$ is the Larmor frequency). The chemical shift parameters are related to the principal tensor elements according to $\delta_{iso} = 1/3\,(\delta_{xx} + \delta_{yy} + \delta_{zz})$, $\delta_{aniso} = \delta_{zz} - \delta_{iso}$, $\eta^{CS} = (\delta_{yy} - \delta_{xx})/\delta_{aniso}$, for which the elements ordered according to $|\delta_{zz} - \delta_{iso}| \geq |\delta_{xx} - \delta_{iso}| \geq |\delta_{yy} - \delta_{iso}|$. The dipolar coupling constant is defined as $b_{IS} = \gamma_I \gamma_S \mu_0/(r_{IS}^3\,4\pi)$, with r_{IS} being the internuclear distance. The relevant spin operators are given by $O_{CS} = I_{iz}$, $O_D = O_{Janiso} = 3I_{iz}I_{jz} - \mathbf{I}_i \cdot \mathbf{I}_j$ (truncated to $2I_{iz}I_{jz}$ for heteronuclear interactions), $O_{Jiso} = \mathbf{I}_i \cdot \mathbf{I}_j$ (truncated to $I_{iz}I_{jz}$ for heteronuclear interactions), and $O_Q = 3I_{iz}^2 - I_i^2$. Finally, $\delta_{m,0}$ is a standard Kronecker delta allocating the isotropic interactions to the $m = 0$ Fourier component. A more detailed description can be found in ref. 55.

The orientation of an anisotropic tensor is expressed using second-rank Wigner [$D^{(2)}$] and reduced Wigner [$d^{(2)}$] matrices describing coordinate transformations from the principal-axis frame (P^λ) to the laboratory-fixed frame (L). For proteins, it proves convenient to let the transformations further involve a peptide plane frame (E), a molecule (or crystal) fixed frame (C), and a rotor-fixed frame (R). With $\Omega_{XY}^\lambda = \{\alpha_{XY}^\lambda, \beta_{XY}^\lambda, \gamma_{XY}^\lambda\}$ denoting the Euler angles relating two frames X and Y, the transformations relating P and R may be written

$$D_{m',m}^{(2)}\left(\Omega_{PR}^\lambda\right) = \sum_{m',m''=-2}^{2} D_{m',m''}^{(2)}\left(\Omega_{PE}^\lambda\right) D_{m',m''}^{(2)}\left(\Omega_{EC}^\lambda\right) D_{m'',m}^{(2)}\left(\Omega_{CR}\right) \qquad (11.4)$$

TABLE 11.2
Second-Rank Reduced Wigner Matrix Elements $d_{m',m}^{(2)}(\beta)$[a]

m', m	-2	-1	0	1	2
-2	$\frac{1}{4}(1+c_\beta)^2$	$\frac{1}{2}(1+c_\beta)s_\beta$	$\sqrt{\frac{3}{8}}s_\beta^2$	$\frac{1}{2}(1-c_\beta)s_\beta$	$\frac{1}{4}(1-c_\beta)^2$
-1	$-\frac{1}{2}(1+c_\beta)s_\beta$	$c_\beta^2-\frac{1}{2}(1-c_\beta)$	$\sqrt{\frac{3}{8}}s_{2\beta}$	$\frac{1}{2}(1+c_\beta)-c_\beta^2$	$\frac{1}{2}(1-c_\beta)s_\beta$
0	$\sqrt{\frac{3}{8}}s_\beta^2$	$-\sqrt{\frac{3}{8}}s_{2\beta}$	$\frac{1}{2}(3c_\beta^2-1)$	$\sqrt{\frac{3}{8}}s_{2\beta}$	$\sqrt{\frac{3}{8}}s_\beta^2$
1	$-\frac{1}{2}(1-c_\beta)s_\beta$	$\frac{1}{2}(1+c_\beta)-c_\beta^2$	$-\sqrt{\frac{3}{8}}s_{2\beta}$	$c_\beta^2-\frac{1}{2}(1-c_\beta)$	$\frac{1}{2}(1+c_\beta)s_\beta$
2	$\frac{1}{4}(1-c_\beta)^2$	$-\frac{1}{2}(1-c_\beta)s_\beta$	$\sqrt{\frac{3}{8}}s_\beta^2$	$-\frac{1}{2}(1+c_\beta)s_\beta$	$\frac{1}{4}(1+c_\beta)^2$

[a] $c_\beta = \cos\beta$, $s_\beta = \sin\beta$.

where we defined the Wigner rotations as

$$D_{m',m}^{(2)}(\Omega) = e^{-im'\alpha} d_{m',m}^{(2)}(\beta) e^{-im\gamma} \tag{11.5}$$

using the reduced Wigner elements in Table 11.2. The term R is related to L by a Wigner rotation using $\alpha_{RL} = \omega_r t$, $\beta_{RL} = \tan^{-1}\sqrt{2}$, and $\gamma_{RL} = 0$ for a spinning sample, whereas $\Omega_{RL} = (0, 0, 0)$ for a static sample. These dependencies are included as intrinsic parts of the Hamiltonian defined by Equations (11.2) and (11.3). With Ω_{CR} describing the orientation of the individual crystallite relative to R (the "powder angles"), we are left with the need for specification of the most relevant transformations

$$P^\lambda \xrightarrow{\Omega_{PE}^\lambda} E \xrightarrow{\Omega_{EC}} C \tag{11.6}$$

relating the principal axis frame of the anisotropic tensors to the molecular frame C via a peptide plane fixed from E. Typical Euler angles for the first transformation are given in Table 11.1 or calculated directly from the structure, whereas the second transformation depends on the definition of C from the atomic coordinate data. Both transformations are readily handled by SIMMOL.

The next element in a numerical simulation is to define the external manipulations, which in addition to the macroscopic sample rotation and the influence from the dominating external magnetic field, involve rf irradiation as expressed by the first term in Equation (11.1). Indeed, our major capability of manipulation comes though the rf pulse sequence, which may be used to excite coherences/polarization, transfer these between spins, and alter the scaling of the influence from the various parts of the internal Hamiltonian in different periods of the experiment for example, with the purpose of monitoring an interaction selectively to obtain information about structure or dynamics. An example is dipolar recoupling, in which all anisotropic interactions to a good approximation are eliminated by fast MAS and in which rf irradiation is applied in synchrony with the sample rotation to prevent averaging of (and thereby recouple) specific dipole–dipole couplings. The recoupling may be aimed at measurement of dipole–dipole coupling interactions and thereby establish information about internuclear distances, or it may serve as a mediator for transfer of coherence or polarization from one spin to another. In the latter case, it may be desirable to completely eliminate the influence from isotropic and anisotropic chemical shielding interactions, giving very broadbanded transfer elements, or it may be relevant that the recoupling depends on the isotropic chemical shifts such that coherence/polarization transfer only occurs in certain spectral windows, as illustrated schematically by the typical ^{13}C and ^{15}N isotropic chemical shifts in Figure 11.3. For illustration, this figure also shows the corresponding powder patterns to visualize the chemical shift dispersion induced by anisotropic shielding. The truncation of anisotropic shielding effects and achievement of specificity with respect to isotropic chemical shifts are relevant (e.g., for specific transfer of magnetization from ^{15}N to $^{13}C^{\alpha}$ and not to ^{13}C in the peptide backbone, with the aim of obtaining the highest possible sensitivity for the desired transfers but also to provide unambiguous assignments in 2D correlation spectra). Similar considerations obviously apply to transfers in the side-chains to assign resonances to specific amino acid residues.

With such objectives, the last couple of decades have witnessed the introduction of a large number of pulse sequence elements[52–59] that to an increasing degree of perfection, tailor the nuclear spin Hamiltonian to the desired form. Representative elements are illustrated in Figure 11.4b, whereas Figure 11.4a shows a typical layout for a multiple-dimensional solid-state NMR experiment in which these elements may play a specific role. For each of the fundamental coherence/polarization transfer elements, we have given a couple of typical examples, noting that in many applications, the transfer elements are separated by time-incremented evolution periods during which the coherences under free precession or under influence from specific re-/decoupling may encode the frequency dimensions of the multiple-dimensional experiment with specific information about structure and dynamics.

The first proton to low-γ nuclei coherence transfer (here $^{1}H \rightarrow ^{15}N$) is typically conducted by Hartmann–Hahn cross-polarization (CP),[60] often ramped variants of this experiment,[61] or potentially adiabatic cross-polarization.[62] A more recent experiment for such purposes is the PRESTO pulse sequence exploiting symmetry-based recoupling of the ^{1}H-X dipolar interaction.[63] The most typical element for $^{15}N \rightarrow ^{13}C$ coherence transfer is the so-called double cross-polarization experiment (DCP),[64] potentially applied off-resonance and referred to as SPECIFIC,[65] or using the more robust i, DCP[66] or GATE[67] pulse sequences. Depending on the desired robustness toward variations in the chemical shift and chemical shift anisotropies, powerful alternatives can be various CN and RN experiments,[68] potentially applied under Lee-Goldburg[69] off-resonance conditions[70] serving to maximize the dipolar scaling factor and simultaneously decouple the ^{13}C-^{13}C dipolar interactions. Also depending on the specific requests on frequency selectivity versus broadbandedness, a large number

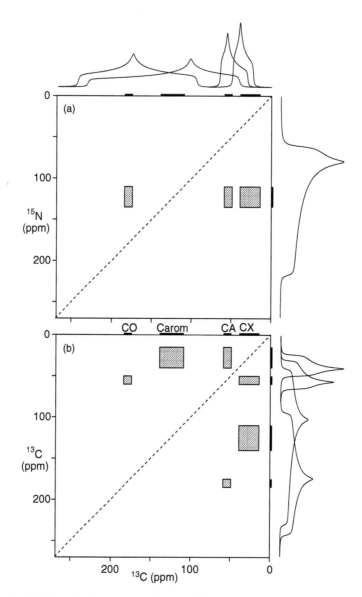

FIGURE 11.3 Schematic illustration of ^{13}C and ^{15}N isotropic chemical shift ranges (in the square diagrams) and anisotropic chemical shielding (powder patterns at the boundary of the diagrams). The hatched areas illustrate the desired specificity of the dipolar recoupling experiments used for coherence transfer in multidimensional correlation experiments.

of different elements may be proposed for the homonuclear $^{13}C^{13}C$ dipolar recoupling, among which we list HORROR (homonuclear rotary resonance)[71] and the adiabatic variant DREAM,[72] as well as a whole arsenal of different symmetry-based CN and RN recoupling experiments[59,73–76] and permuted variants of these such as

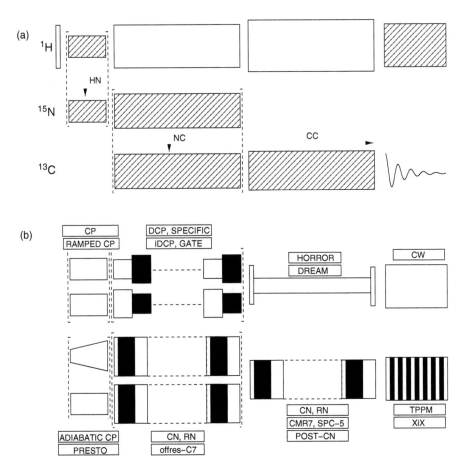

FIGURE 11.4 Schematic illustration of (a) a multidimensional solid-state NMR experiment for assignment and structure determination in uniformly (or extensively) ^{13}C,^{15}N isotope-labeled proteins along with (b) typical pulse sequence elements for re- and decoupling of specific dipole–dipole interactions in MAS experiments. The elements in (b) correspond to the hatched boxes in panel (a).

CMR7,[77] POST-C7,[78] and SPC-5.[79] Finally, for decoupling of dominating dipolar coupling interactions with protons during the detection period, as indicated in Figure 11.4a, or during the low-transfer elements or precession periods, the most popular sequences are standard continuous wave decoupling or the time-modulated TPPM[80] or XiX[81] pulse sequences. We note that these elements are only a few representative cases — dozens of different elements with different properties have been proposed over the years. Which element to use highly depends on the given application, the nuclear spins systems, the isotope labeling pattern, and the instrumental conditions in terms of static field, rf, and sample spinning properties. We note that the two latter conditions not only may be dictated by the available instrumentation but also may take into consideration the stability of the biological samples in terms of centrifugation and sample heating.

The availability of hundreds of different pulse sequence elements, often demonstrated only for very specific cases; the fact that most of these cases have been designed analytically for single- or two-spin systems but often find their major applications in larger spin systems; and the potential risk of destructive interplay between different pulse sequence elements form some of the arguments for developing numerical simulation programs, which in a fast and reliable manner, can analyze the performance under given conditions. Over the last few years, several "general" programs have been proposed, probably starting with the ANTIOPE program[82] and, later, much more flexible programs such as GAMMA,[83] SIMPSON,[55] BlochLib,[84] and SPINEVOLUTION.[85] Whereas all these packages have their stronger and weaker facets, we will here concentrate our description on the SIMPSON program which has gained high popularity and is used routinely in a large number of laboratories worldwide for evaluation, design, and data analysis purposes.

The SIMPSON software package[55] is designed to operate essentially as a computer spectrometer. Through a flexible Tcl-based[47] interface and a large number of typical commands, SIMPSON essentially allows simulations of all kinds of NMR experiments with a programming workload similar to what it takes to program a pulse sequence on a NMR spectrometer. It is precisely this flexibility that has been a key element in making SIMPSON popular among "spin-engineers" as well as more "application-oriented" solid-state NMR spectroscopists. It enables relatively simple analysis of complex pulse sequences, which earlier — in practice — was restricted to a few spin engineer research groups writing special programs associated with their development of new pulse sequences. This is obviously acceptable for development, but it is important that the users of the pulse sequences by themselves can compare the performance of different pulse sequences in their specific context. This facilitates the choice of the best experiments for practical applications.

Through a Tcl interface, SIMPSON controls all the delicate parts of a solid-state NMR simulation, including propagation of the time-dependent Hamiltonians to monitor the evolution of the density operator in course of the experiment, typically including many coherence/polarization transfer steps and several evolution periods. This is performed under consideration of finite rf pulse irradiation and different response from the many crystallites (i.e., molecules oriented differently relative to the external field). In short, the density operator is propagated from a given initial state $\rho(0)$ through time according to

$$\rho(t) = U(t,0)\rho(0)U^{\dagger}(t,0) \qquad (11.7)$$

where $U(t,0)$ — in the case where dissipative processes such as relaxation can be ignored — is the unitary propagator responsible for the spin dynamics in the period from 0 to t. This term, $U(t,0)$, is related to the Hamiltonian according to

$$U(t,0) = \hat{T} \exp\left\{-i \int_{0}^{t} H(t')\,dt'\right\} = \prod_{j=0}^{n-1} \exp\left\{-iH(j\Delta t)\Delta t\right\} \qquad (11.8)$$

where the first expression describes the formally correct integration of the Hamiltonian, with the Dyson time-ordering operator $\hat{T}$ being relevant for Hamiltonians containing noncommuting components, whereas the last expression represents the standard "numerical" reformulation into a simple time-ordered product with the time steps being taken in sufficiently small steps to ensure that the otherwise time-dependent Hamiltonian in Equation (11.1) may be considered static within each interval.

The signal for a crystallite characterized by the orientation Ω_{CR} and a detection operator (Q_{det}) may described by

$$s(t;\Omega_{CR}) = \mathrm{Tr}\{Q_{det}\rho(t;\Omega_{CR})\} \qquad (11.9)$$

typically sampled equidistantly with respect to time. In the case of a powder sample, the signal needs to be averaged over all uniformly distributed powder angles Ω_{CR} according to

$$\bar{s}(t) = \frac{1}{8\pi^2} \int_0^{2\pi} d\alpha_{CR} \int_0^{\pi} d\beta_{CR} \sin(\beta_{CR}) \int_0^{2\pi} d\gamma_{CR} s(t;\Omega_{CR}) \qquad (11.10)$$

where we, for the sake of generality, assumed averaging over the full sphere. Typically, the integrals are replaced by summations using different digitization schemes. We note that in numerous cases, the intrinsic symmetry of the orientation dependence allows reduction of the averaging to one-half or one-quarter of the sphere.[86,87]

The delicate aspect of numerical calculations in solid-state NMR is to appropriately address the time propagation of the density operator, exploiting all possible periodicities, as well as the powder averaging, using a minimum number of crystallites to ensure adequate representation of all uniformly distributed orientations of the molecules. These elements, effectively corresponding to a loop over four variables — three crystallite angles α_{CR}, β_{CR}, γ_{CR}, and the time t, are illustrated in Figure 11.5 with the upper coordinate system reflecting handling of the spatial part of the calculation and the lower the rf-relevant spin part of the calculation — both may be variables in time. Using appropriate methods such as γ-COMPUTE,[88–91] it is possible, in special cases with periodic rf irradiation, to reduce the dimensionality of the problem by concatenating the time t and γ_{CR} dependencies, both corresponding to a rotation around the axis of the sample spinner. The averaging over the two remaining crystallite angles, α_{CR} and β_{CR}, may be conducted very efficiently using methods such as the Zaremba, Conroy, Wolfsberg,[92–94] REPULSION,[95] or Lebedev.[87] In combination, efficient treatment of the time propagation and the powder averaging — while preserving the high flexibility of pulse sequence elements — have been fundamental for the development of fast numerical software with practical effect for biological solid-state NMR.

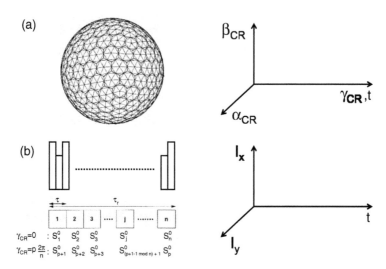

FIGURE 11.5 Illustration of powder averaging, which for stationary samples only involves averaging over the α and β Euler angles (left), for example, using REPULSION,[93] whereas for the rotating sample it also involves the third Euler angle γ typically treated in concert with the time evolution (right) either directly or using methods such as γ-COMPUTE[88] with the specific propagators $\overset{0}{S}i$ reproduced and used as illustrated in (b, left). (b) Illustration of the handling of rf pulse sequences (left) expressed as dependencies of the spin operators I_x and I_y as function of time (right) potentially handled using methods such as γ-COMPUTE. Figures reproduced from refs. 95 and 88 with permission.

11.3 EXAMPLES OF NUMERICAL SIMULATIONS IN BIOLOGICAL SOLID-STATE NMR

Numerical simulations have an effect on biological solid-state NMR on many levels. Using flexible, easily programmable software such as SIMPSON[55] and SIMMOL,[34] it is quite easy to use computer simulations for the exploration of the spectral signatures of different nuclear spin interactions (and combinations of these) under different experimental conditions. A second area of application could be pulse sequence evaluation of the individual pulse sequence elements or of the entire pulse sequence. A third aspect of relevance could be the use of numerical simulation software as an active ingredient in the design of new pulse sequences with specific characteristics. Finally, simulation software may be combined with iterative fitting procedures to extract information about the internal nuclear spin Hamiltonian — and thereby molecular structure and dynamics — from experimental spectra. These four facets of numerical simulations will be addressed in the following sections.

11.3.1 EXPLORATION OF THE INTERNAL HAMILTONIAN

One of the most fascinating aspects of solid-state NMR spectroscopy is the ability to extract structural information from various anisotropic nuclear spin interactions. This includes extraction of information about internuclear distances and dihedral angles from dipole–dipole couplings as well as information about the electronic surroundings of the nuclei from chemical shielding and quadrupole coupling parameters. Numerical simulations prove useful not only to extract such information from experimental spectra — as will be demonstrated in a subsequent section — but also for visualization and exploration of the effects of different interactions in the NMR spectra. These effects often depend critically on external parameters such as field and spinning. Under the right condition, spectra from very simple NMR experiments may provide information about several interaction tensors and their relative orientations. In biology-related solid-state NMR, one of the simplest and yet quite remarkable observations is the ability to recouple dipole–dipole interactions by adjusting the sample spinning frequency to a submultiple of the isotropic chemical shift difference between the components of abundant (naturally or isotope labeled) spin-pair systems. This macroscopic interference with the dynamics of the nuclear spins, referred to as rotational resonance,[35,96] is illustrated in Figure 11.6 by numerical simulations of solid-state NMR powder spectra for the ^{13}C-^{13}C spin-pair of Zn-acetate.

A very sensitive probe to information about the electronic structure is the quadrupolar coupling interaction that often along with anisotropic shielding, may be measured for quadrupolar nuclei. In fact, most of the NMR active nuclei in the periodic table are quadrupolar nuclei, which with few exceptions, can be studied relatively easily by solid-state NMR spectroscopy. Focusing on biology, two examples come immediately into mind. The first is metal nuclei such as ^{35}Mg, ^{39}K, ^{43}Ca, and ^{67}Zn, which potentially complicated by low resonance frequencies, low abundance, and large quadrupolar moments, may be studied to obtain information about the ligation of cations in the catalytic sites of metalloproteins. The second is oxygen and deuterium, for which the quadrupolar coupling interactions may provide important insight into structurally important features such as hydrogen bonding. Addressing here the latter example, Figure 11.7 shows a series of solid-state NMR spectra reflecting ^{17}O in the nucleic acid base guanine, as recently studied experimentally by Wu and coworkers.[97] In this case, it is of interest to establish information about the combined effects from the quadrupolar coupling and anisotropic chemical shielding interactions, under the consideration that the quadrupolar coupling constant is relatively large. This implies that the information most straightforwardly is extracted from the second-order broadened central transition. To establish information about both effects, it is instructive to note that the shielding interaction is proportional to the external field, whereas the second-order quadrupolar coupling is inversely proportional to the field. Also, it is worth noting that MAS typically reduces the line width of the second-order powder pattern by a factor of close to two, leading to more sensitive spectra than is typically recorded for static samples. With these arguments, Figures 11.7a and 11.7b show MAS powder spectra calculated for ^{17}O in guanine under conditions of 500 and 900 MHz instrumentation, respectively, with the former using a sample spinning frequency of 14.5 kHz to separate the spinning sidebands from the center band, whereas the latter achieves the same using 8 kHz spinning. In both cases, the upper spectrum reflects the quadrupolar coupling interaction alone, and the lower spectrum also includes effects from anisotropic shielding using the parameters determined by Wu et al.[98] It is clear that the 900 MHz spectra

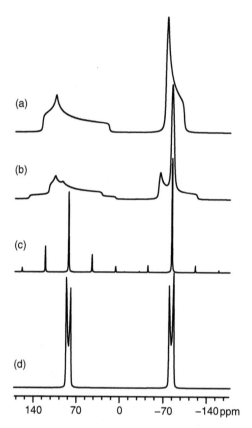

FIGURE 11.6 Numerical simulations of single-pulse (ideal rf) ^{13}C solid-state NMR powder spectra for the ^{13}C-^{13}C spin-pair in ^{13}C-labeled Zn-acetate at 9.4 T. Static spectra calculated without (a) and with (b) active dipole–dipole coupling between the spins. (c) and (d) MAS spectra corresponding to sample spinning frequencies of $\omega_r/2\pi$ = 3.8 kHz and 16.6 kHz, respectively, under consideration of all spin-pair parameters. The latter spectrum corresponds to the n = 1 rotational resonance, where $\omega_r/2\pi = \delta_{iso}^{CO} - \delta_{iso}^{CH3}$. The simulations used the parameters $b_{IS}/2\pi$ = −2150 Hz, Ω_{CR}^{IS} = {0,0,0}, J = 49 Hz, δ_{iso}^{CH3} = 8300 Hz, δ_{aniso}^{CH} = −2373 Hz, η^{CH3} = 0.017, Ω_{CR}^{CH3} = {0,0,0}, δ_{iso}^{CO} = 8300 Hz, δ_{aniso}^{CO} = −8197 Hz, η^{CH3} = 0.34, and Ω_{CR}^{CH3} = {0,90°,0}.[95]

(122 MHz for ^{17}O) is most sensitive to the combined effects (and thereby the magnitude and relative orientation of the two tensors) from the quadrupole coupling and anisotropic chemical shielding interactions, which may provide important information about hydrogen bonding to the nucleic acid oxygens.

11.3.2 EVALUATION PULSE SEQUENCE ELEMENTS

Typically, the pulse sequences used for biological solid-state NMR are composed of several building blocks, as illustrated in Figure 11.4, taking care of specific needs such as coherence transfer between spins and evolution under specific parts of the Hamiltonian to extract information about structure and dynamics. The vast majority

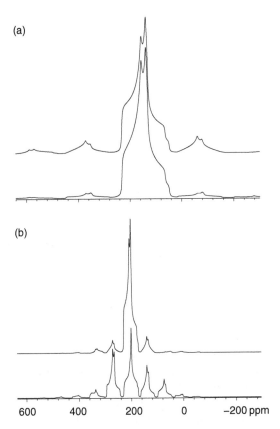

FIGURE 11.7 ^{17}O MAS NMR spectra calculated for the oxygen in guanine corresponding to (a) 500 MHz (67.8 MHz for ^{17}O) and (b) 900 MHz (122.0 MHz for ^{17}O) instrumentation. The spectra assumed spinning frequencies of 14.5 kHz (a) and 8 kHz (b). The upper spectra reflect the quadrupolar coupling interaction alone, whereas both anisotropic shielding and quadrupolar couplings are considered for the lower spectra. The calculations used the parameters δ_{iso} = 230 ppm, δ_{aniso} = 220 ppm, η^{CS} = 0.5, Ω_{PC}^{CS} = {5°, 87°, 67°}, C_Q = 7.1 MHz, η^Q = 0.8, and Ω_{PC}^Q = {0,0,0}.[98]

of these elements have been designed using a combination of intuition and analytical evaluations using effective (or average) Hamiltonian theory.[30–33] For practical purposes, this has been accomplished for one- or two-spin systems, often representing a considerable simplification relative to the large spin systems encountered in uniformly ^{13}C,^{15}N-labeled proteins (cf. Figure 11.2b).

One example could be the design of efficient schemes for the recoupling of homonuclear dipole–dipole interactions under MAS conditions. A relatively simple approach is to use a continuous rf irradiation with an amplitude half the spinning frequency, and the rf carrier frequency set at the mean isotropic chemical shifts of the two spins. In cases with modest chemical shielding anisotropy, this will ensure recoupling of the dipolar coupling between the two spins. This simple method, referred to as HORROR,[71] introduced the important concept of γ-encoded (i.e., γ_{CR}-independent) recoupling. Through a reduced dependency on the crystallite orientations, γ-encoding increases the efficiency of coherence transfer between the two spins from the typical 50% to about 73%. Furthermore, HORROR recouples the dipolar interaction with a relatively high scaling factor on the dipolar interaction, which translates into a relatively short pulse sequence element, typically being less susceptible to signal loss caused by relaxation than longer sequences. As a major drawback, however, HORROR is quite sensitive to chemical shift effects, implying that it typically works only for spin pairs characterized by relatively small chemical shielding anisotropies and confined in relatively narrow chemical shift regions. These problems motivated the design of the more advanced sevenfold symmetric C7 recoupling sequence,[73] its permutation offset stabilized variant POST-C7,[78] the CMR7[77] and SPC-5[79] sequences, and a large number of more general CN and RN recoupling experiments,[74–76] all of which suppress the influence from chemical shifts to a higher order. This improvement is typically achieved at the expense of the need for much stronger rf irradiation, lower dipolar scaling factors, and thereby, longer sequences.

Although the analytical formalism — not being repeated here — gives a reasonable picture of the performance of the recoupling phenomena for design purposes, it may still be difficult to evaluate the individual recoupling method and its comparison to other methods for $^{13}C \rightarrow ^{13}C$ coherence transfer under experimentally relevant conditions. Small changes in the experimental conditions or in the isotropic/anisotropic interaction parameters for a given application may readily cause otherwise "ideal" experiments to be useless, and vice versa. To illustrate this aspect, Figure 11.8 shows 2D contour plots for the $^{13}C^{\alpha} \rightarrow ^{13}C^{\beta}$ and $^{13}C' \rightarrow ^{13}C^{\alpha}$ coherence transfers obtained by HORROR under conditions of 4–40 kHz sample spinning for a typical peptide backbone using 400-MHz NMR equipment. These transfers are of interest for the establishment of solid-state NMR analogs to the NCACB and NCOCA experiments known from liquid-state NMR.[28] More specifically, the contour plots, also depending on the position of the ^{13}C rf carrier, have been calculated for a $^{13}C^{\alpha}-^{1}H^{\alpha}-^{13}C^{\beta}-^{1}H^{\beta}-^{13}C'$ heteronuclear five-spin system in threonine, with the geometry and the anisotropic interaction parameters established using SIMMOL for the Thr7 residue in ubiquitin.[99,100] The contour plots reveal some interesting aspects of recoupling in MAS solid-state NMR. First, a "first-order" pulse sequence such as HORROR is by no means the ideal choice for cases with large anisotropic shielding at low spinning speeds. For smaller anisotropies, however, the pulse sequence is promising considering essentially all relevant measures. It is an extremely simple pulse sequence to implement — it is only continuous rf with an amplitude being half the spinning speed; it is γ-encoded and, thereby, increases the sensitivity by about 50% relative to standard experiments; it has the highest dipolar scaling factor presented so far for a γ-encoded experiment; and it can easily be combined with efficient ^{1}H decoupling under essentially all conditions. Addressing the latter point and to facilitate direct comparison to our subsequent analysis of C7 recoupling, the contour plots have been calculated assuming 120-kHz continuous-wave decoupling to eliminate effects from the two protons in the five-spin system. Substantially lower decoupling will be adequate — the curves are essentially indistinguishable from those obtained for a three-spin system with the carbons alone.

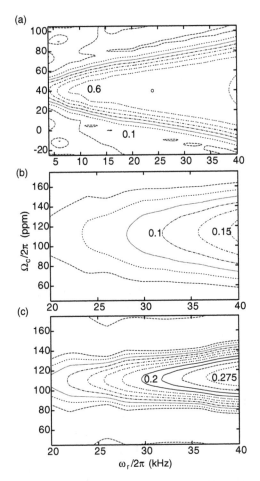

FIGURE 11.8 Contour plots mapping the efficiency of (a) $^{13}C^{\alpha}\rightarrow{}^{13}C^{\beta}$ and (b,c) $^{13}C'\rightarrow{}^{13}C^{\alpha}$ coherence transfer (relevant for NCACB and NCOCA experiments) calculated for the HORROR sequence without (a,b) and with (c) phase-alternation of the rf phase after each complete two-rotor period cycle as function of the sample spinning frequency (ω_r) and the ^{13}C carrier frequency (Ω_c). The calculations assumed the rf field strength of the HORROR irradiation to be half the spinning frequency, a decoupling rf field strength of 120 kHz, and Larmor frequencies corresponding to a 400-MHz instrument. The calculations were performed for the $^{13}C^{\alpha}-{}^1H^{\alpha}-{}^{13}C^{\beta}-{}^1H^{\beta}-{}^{13}C'$ five-spin system of Thr7 in ubiquitin,[99,100] using the following parameters. Shifts (δ_{iso}, δ_{anise} η^{CS}-, Ω_{CS}^{PC}): $^{13}C^{\alpha}$: 50 ppm, -20 ppm, 0.43, 13.5°, 106.3°, 172.4°; $^{13}C^{\beta}$: 30 ppm, -12 ppm, 0.77, 0°, 90°, 0°; $^{13}C'$: 170 ppm, -76 ppm, 0.9, 20.7°, 77.1°, -101.4°; $^1H^{\alpha}$: 4 ppm, 7.7 ppm, 0.6, 0°, 101.6°, -46.5°; $^1H^{\beta}$: 3 ppm, 5 ppm, 0.3, 0°, 155.9°, 169.3°. Dipole–dipole and J coupling ($b_{IS}/2\pi$, Ω_{IS}^{PC}, J): $^{13}C^{\alpha}-{}^{13}C$: -2125 Hz, 0°, 101.2°, 101.3°, 35 Hz; $^{13}C^{\alpha}-{}^{13}C'$: 2146 Hz, 0°, 126.2°, -22.7°, 55 Hz; $^{13}C^{\beta}-{}^1H^{\alpha}$: -23855 Hz, 0°, 101.6°, -146.5°, 140 Hz; $^{13}C^{\alpha}-{}^1H^{\beta}$: -3077 Hz, 0°, 126.8°, 115.1°, –; $^{13}C^{\beta}-{}^{13}C'$: -494 Hz, 0°, 76.0°, 126.3°, –; $^{13}C^{\beta}-{}^1H^{\alpha}$: -3081 Hz, 0°, 87.8°, -106.0°, –; $^{13}C^{\beta}-{}^1H^{\beta}$: -24004 Hz, 0°, 155.9°, 169.3°, 140 Hz; $^{13}C'-{}^1H^{\alpha}$: -3105 Hz, 0°, 71.3°, -176.9°, –; $^{13}C'-{}^1H^{\beta}$: -1408 Hz, 0°, 97.9°, 132.6°, –; $^1H^{\alpha}-{}^1H^{\beta}$: -8760 Hz, 0° H6.4°, 85.8°, –.

Considering the contour plots individually, Figure 11.8a describes the $^{13}C^\alpha \rightarrow ^{13}C^\beta$ transfer for which HORROR in the present case of 400-MHz instrumentation provides more than 60% transfer efficiency when the carrier is placed between the isotropic chemical shifts of the two relevant spins using a spinning frequency of 8 kHz and above. The method becomes increasingly broadbanded (and obviously also more robust to more dominant tensorial interactions as, for example, encountered at higher fields) on increasing the sample spinning frequency. It should be noted that HORROR is not only simple and provides ample room for decoupling without approaching Hartmann–Hahn match to the protons (i.e., $\omega_f^C \leq 3\omega_{rf}^H$), but it also possess the very important feature that the transfer profiles are sufficiently narrowbanded so the dipolar interaction to the carbonyl ^{13}C is not effective, which often is the case for more broadbanded pulse sequences. Addressing a chemical-shift-anisotropy-wise more complicated case, Figure 11.8b shows the corresponding $^{13}C' \rightarrow ^{13}C^\alpha$ coherence transfer efficiencies for HORROR. Clearly, in this case HORROR is not living up to the same standards as for the $^{13}C^\alpha \rightarrow ^{13}C^\beta$ transfer. Even on 40 kHz spinning, corresponding to a rf field strength of 20 kHz on the carbon channel, the efficiency is not higher than about 15% using HORROR. Using a slightly modified — and not γ-encoded — variant of HORROR, with the rf phase being reversed on conclusion of every second rotor period, the efficiencies may be increased to about 30%. At first sight this is not remarkable, but as we will see in the following example, it may not be as bad after all, as efficient recoupling under conditions of medium to high sample spinning frequencies using more chemical-shift-compensated approaches often faces severe problems because they are not compatible with ^1H decoupling — or in an alternative approach without decoupling does not prevent distribution of coherences to "undesired" spins. In addition to this come considerations such as sample heating by strong rf irradiation, which is not the case for HORROR or variants of this experiment, even at 40–50 kHz spinning.

Using the same spin system and similar experimental conditions, it is interesting to explore the performance of more broadbanded recoupling experiments such as the C7-type of pulse sequences. A large arsenal of symmetry-based recoupling experiments[73–76] has been proposed that improves the chemical shift compensation relative to the very simple HORROR experiment at the expense of significantly higher rf field strengths (typically 3.5–10 times the spinning frequency) and significantly lower scaling factors on the recoupled dipolar Hamiltonian. Typically, the scaling factor is reduced by more than a factor of two, thereby effectively making the pulse sequence two to three times longer and, thereby, potentially more susceptible toward relaxation effects. At relatively low spinning speeds, where ^1H decoupling is feasible, pulse sequences like C7 work reasonably well, as illustrated in Figure 11.9 for $^{13}C^\alpha \rightarrow ^{13}C^\beta$ and $^{13}C' \rightarrow ^{13}C^\alpha$ in the threonine five-spin system described above. For the first of these two transfers, efficiencies exceeding 60% are obtained for the transfer to the ^{13}C spin, whereas 45% is achieved for the $^{13}C' \rightarrow ^{13}C^\alpha$ transfer at low spinning speeds. The discrepancy is ascribed to competitive transfer to the $^{13}C^\beta$ spin in the latter case, indicating one of the problems using broadband recoupling experiments for "band-selective" transfer in multiple-spin systems, a problem that increases at higher spinning speeds and is present even to a larger extent for the permutation-stabilized POST variants.[78] For increasing spinning frequencies, the

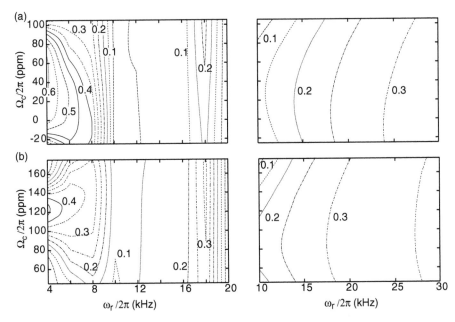

FIGURE 11.9 Contour plots mapping the efficiency of (a) $^{13}C^{\alpha} \rightarrow {}^{13}C$ and (b) $^{13}C' \rightarrow {}^{13}C_{\alpha}$ coherence transfer (relevant for NCACB and NCOCA experiments) calculated for the C7 recoupling pulse sequence under different spinning conditions with 120-kHz ^{1}H decoupling (left) and without decoupling (right). The simulations correspond to 9.4 T and the same spin system parameters as in Figure 11.8.

efficiency of both transfers degrade dramatically as a result of interference with the chosen 120-kHz decoupling. It is well known that many of the symmetrized recoupling experiments may by themselves contribute ^{1}H decoupling, as recently explored by Hughes et al.[101] for various CN and POST-CN recoupling experiments without decoupling. In this manner, the spinning speed is not limited to being 3N times lower than the available decoupling rf field strength. These aspects are explored in the right-side panels in Figure 11.9, which demonstrate efficiencies of 30–35% for the two transfers at 30-kHz spinning, using C7. We have explored other recoupling experiments, including POST-C7, which display similar recoupling profiles. The simulations may support the fact that heteronuclear effects may be partially truncated by spinning and irradiation at the carbon channel but still reinforce the fact that a major problem with the broadband recoupling experiments is undesired spin diffusion between ^{13}C spins, which here renders the performance of C7 similar to, and not much better than, HORROR at high spinning speeds, despite the use of significantly higher rf field strengths. In addition, the sequences offer slower coherence build up as a result of a lower dipolar scaling factor. It should be noted that these conclusions depend on the spin systems and the external field strengths scaling the chemical shift effects. Our major point is that homo- and heteronuclear multiple-spin effects are highly relevant and that numerical simulations may be helpful in this regard.

Numerical simulation in biological solid-state NMR — and thereby evaluation of experiment performance — is not limited to spin-1/2 nuclei. In many biological macromolecules, metal cations play an important role for the biological activity (e.g., as catalytic reagent or activator through the induction of conformational changes). Often these cations are NMR active quadrupolar nuclei such as ^{39}K, ^{67}Zn, and ^{43}Ca, which by no means are trivial cases from an NMR point of view. First, these nuclei often have a relatively low gyromagnetic ratio (γ), causing problems with the sensitivity and severe probe ringing effects. Second, the relevant spin isotope is often far from abundant, implying that isotope labeling may be an issue, in particular when addressing the occurrence in large biomolecules diluting the molar presence of these metal ions in a typical sample. Third, often the low-γ quadrupolar metal nuclei have very strong quadrupolar coupling interactions, implying not only that the first-order spectra are very broad (typically not detectable) but even the second-order spectra for the $\frac{1}{2} \rightarrow -\frac{1}{2}$ central transition may extend over tens to hundreds of kilohertz. Although fast MAS has proven useful when coping with the case in which the first-order broadening is the only problem, as often seen for nuclei such as ^{23}Na, ^{27}Al, ^{51}V,[86,102–104] the second-order broadening problem is more difficult to handle.

The second-order quadrupolar interactions are influenced not only by second-rank tensors, which may be averaged by MAS, but by fourth-rank tensors, which also may be averaged by spinning, although at an angle different from the magic-angle. For relatively small second-order effects, advanced methods such as dynamic-angle spinning,[105] double-rotation,[106] switched-angle spinning,[107] multiple-quantum MAS (MQ-MAS),[43] and satellite-transition MAS[108] may be helpful in resolving second-order broadened resonances, typically via a high-resolution dimension in two-dimensional spectra. For larger quadrupolar couplings exceeding $C_Q/4I(2I-1)$ > 0.5–1 MHz, where C_Q is the quadrupolar coupling constant and I the spin-quantum number, it is difficult to use these methods, in practice leaving us with the broad second-order powder patterns. Even in cases with only one site, or cases in which overlapping patterns from different sites may be resolved, the broad second-order patterns are a limiting factor simply from a sensitivity point of view.

To improve the sensitivity of quadrupolar NMR for nuclei with large quadrupolar coupling interactions, we have over the last decade worked on the use of spin-echo methods that alone, or in combination with MAS, split up the broad second-order powder patterns into spin-echo sidebands. The method, taking advantage of detecting the free-induction decays in the delays between the refocusing pulses in a QCPMG experiment, produces sideband spectra very similar to those typically encountered in MAS experiments in the presence of large chemical-shielding anisotropies or first-order quadrupolar coupling effects.[44] This applies even in the case of static samples,[44,109–110] and the combination with MAS[111] may be used to reduce the second-order effects to about 60% before splitting the powder pattern into spin-echo sidebands. We note that the same technology may be used for first-order powder patterns as well (e.g., the first-order quadrupolar effects of ^{2}H to provide high sensitivity for static powder samples or to probe effects from molecular dynamics through broadening of the spin-echo sidebands).[112] In addition, combinations with multiple-quantum MAS have been proposed.[113,114] In all cases, the separation between the sidebands in the spectra are determined by the separation between the spin-echo refocusing pulses and the actual sampling rate used between them.

It is well known that the effects of finite rf pulses may be a major concern in solid-state NMR spectroscopy. Finite rf pulses have also been a major motivator for establishing numerical simulation software, simply because these effects — being simultaneously influenced from the rf as well as from internal parts of the Hamiltonian in the course of the rf pulses — are extremely difficult to evaluate analytically to an extent that precise data can be extracted from experimental spectra. In addition, these effects may dramatically influence the outcome of even the simplest NMR experiments. Obviously, the concern of finite rf pulse effects increases with the increasing size of the internal interactions, probably being most debated in the case of quadrupolar interactions, which may be in the megahertz range, whereas typical rf field strengths are in the tens to hundreds of kilohertz regime. However, as sample heating for biological samples becomes an increasing problem, the wish to use rf field strength of similar magnitude as typical chemical shielding and dipolar coupling interactions increases. Taking the QCPMG experiment as a case in which huge quadrupolar coupling effects may interfere with the action of hundreds of pulses as an example, it is interesting to note that theory predicts regimes of rf field strengths that will cause almost ideal behavior, whereas others will lead to substantial finite rf pulse effects. What matters here is that whenever the magnitude of an internal interaction (here taking into account the scaling effects from the orientation-dependent part of the Hamiltonian) comes close to the rf field strength, then the finite pulse effects come into play.

Considering quadrupolar nuclei, it has been clear for many years that interference between the first-order quadrupolar interaction and the rf becomes notable in the regime

$$0.2 < |\omega_Q/\omega_{rf}| < 3 \qquad (11.11)$$

where $\omega_Q = 2\pi C_Q/4I(2I - 1)$ is the maximum quadrupolar splitting frequency, whereas ω_{rf} is the rf field strength — both in angular frequencies. In the so-called hard pulse regime below 0.2, the pulse lengths are adjusted as typically encountered in NMR where anisotropic effects are negligible. In the soft pulse regime over 3, however, the pulse lengths are adjusted as if the rf field strength were $\omega_{rf}(I - 1/2)$ instead of just ω_{rf}. In the case of large quadrupolar coupling constants, the relevant regime for avoiding finite rf pulse effects is the soft pulse regime — the question is now, what happens then with the second-order quadrupolar coupling effects? From theory,[111] it may be estimated that interference from the second-order quadrupolar coupling may occur when

$$\omega_Q^2/(\omega_0\omega_{rf}) > 0.1 \qquad (11.12)$$

where ω_0 denotes the angular Larmor frequency. To illustrate these effects and illustrate the use of numerical simulations in relation to QCPMG experiments for quadrupolar nuclei, Figures 11.10a and 11.10b give contour plots, with light areas illustrating finite pulse regimes that should be avoided, as a function of C_Q and v_{rf}

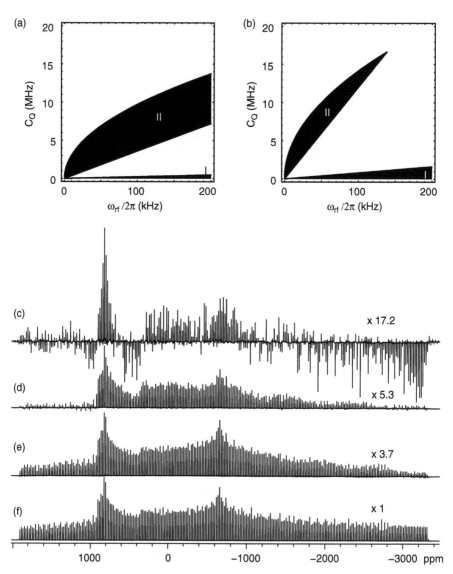

FIGURE 11.10 Recommended regimes of rf field strength relative to the quadrupolar coupling constant for (a) ^{87}Rb ($I = 3/2$) and (b) ^{67}Zn ($I = 5/2$) at 9.4 T. The black areas marked I and II represent hard- (adjust pulses using $\omega_{rf}/2\pi$) and soft-pulse (adjust pulses using $[I + 12]\omega_{rf}/2\pi$) regimes. (c–f) QCPMG spectra calculated for ^{67}Zn in zinc diimidazole diacetate at 9.4 T, assuming (f) ideal rf fields, (e) 50-kHz rf field strength (soft-pulse), as well as 200 kHz using the (d) soft and (c) hard condition for adjustment of the pulse lengths. The numbers to the right indicate the vertical scale factor used in representation of the spectra. The spectra were calculated for a QCPMG sequence with $\tau_1 = \tau_2 = \tau_3 = \tau_4 = 25$ µs, $\tau_a = 2$ ms (400 points), $C_Q = 8.2$ MHz, $\eta_Q = 0.62$, and $\delta_{iso} = 155$ ppm.[107]

$= \omega_{rf}/2\pi$ for spin $I = 3/2$ and spin $I = 5/2$ nuclei explored for ^{87}Rb and ^{67}Zn, respectively, for a 400-MHz spectrometer. The black areas represent the "ideal-behaving" hard- (marked by I) and soft-pulse (marked by II) regimes. The field dependence is introduced via the second-order term, giving rise to the curved upper boundary of regime II — we note that without consideration of the second-order effect, the "first-order" soft-pulse region would extend from the straight line at the lower side of region II to the vertical axis of the coordinate system.

To substantiate the importance of these considerations, Figure 11.10 also contains a series QCPMG sideband spectra calculated for ^{67}Zn in zinc diimidazole diacetate[109] in a 9.4-T magnet (25.02 MHz for ^{67}Zn). Zinc diimidazole diacetate is a good mimic for Zn in a biological macromolecule. The calculations are performed under consideration of ideal rf pulses (Figure 11.10f) as well as finite rf pulses in the soft-pulse regime using an rf field strength of 50 kHz (Figure 11.10e) and in the forbidden region using a field strength 200 kHz with the rf pulses adjusted using $\omega_{rf}(I - 1/2)$ (Figure 11.10d) and ω_{rf} (Figure 11.10c). Considering the overall envelope of the spin-echo sideband envelope as well as the scaling of the spectral intensity (indicated to the right in the figure), it is clearly evident that the best spectra are obtained using modest, carefully adjusted rf field strengths, as predicted by region II in Figure 11.10b, rather than using the frequently encountered "let's give the spins what we have" approach.

As a final and distinct example of experiment evaluation, we address numerical simulations of oriented-sample solid-state NMR experiments for large-membrane proteins. In this case, we do not consider the technical performance of the pulse sequence but, rather, focus on numerical simulations as a means to explore how the sample should ideally be prepared in terms of isotope labeling to provide spectra from which the structural information can be extracted. Before doing so, it is relevant to address the fundamental difference between spectra from static-oriented samples and spectra resulting from MAS experiments in terms of structure determination. The MAS approach is conceptionally very similar to the approach used in biological liquid-state NMR. The resonance positions in multiple-dimensional spectra — often ^{13}C and ^{15}N isotopes in enriched samples — are typically determined by isotropic chemical shifts that may show dependencies on residue type as well as secondary structure.[115] Tertiary structure information in terms of internuclear distances and torsion angles is typically determined from the intensity of such "isotropic" peaks encoded through periods of evolution under influence of relevant anisotropic interactions. Considering oriented samples, the resonance positions are determined not so much by the isotropic chemical shifts but, instead, by anisotropic chemical shifts and dipolar couplings. This implies that on assignment — which may be a big challenge — the structure information is more or less automatically available from the line positions (rather than the intensities). The practical availability of this information obviously depends on the resolution of the spectra, as discussed on the basis of computer simulations in several papers.[25,34,49–55] With reference to these accounts, and the fact that the so-called PISEMA[116] and ^{1}H-^{15}N chemical shift correlation experiments are among the most suitable methods for the study of oriented ^{15}N-labeled membrane proteins, Figure 11.11 shows a series of 2D spectra calculated for the aquaporin water-channel

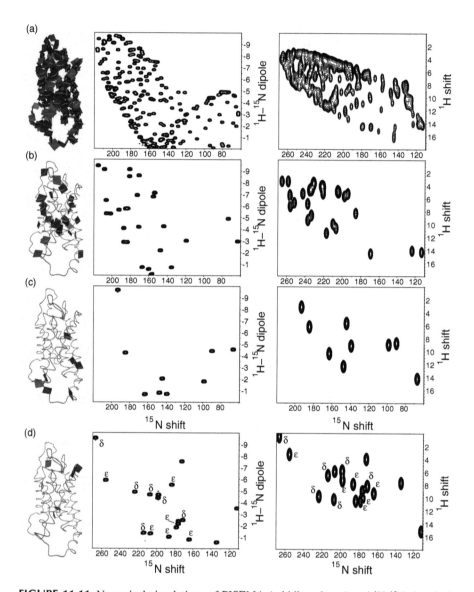

FIGURE 11.11 Numerical simulations of PISEMA (middle column) and ^{1}H-^{15}N chemical shift correlated (right column) spectra calculated for aquaporin-1 on basis of a recent 2.2-Å x-ray diffraction structure (PDB coordinates: 1J4N),[117] using SIMPSON and SIM-MOL. The different panels reflect different ^{15}N-labeling patterns, from uniform (a), alanine (b), arginine (c), and histidine (d) labeling. The simulations assumed 80 kHz rf irradiation at the protons during the FSLG cycles[118] of the two experiments. The figure is reproduced from ref. 52 with permission.

protein AQP-1 oriented as it may be arranged in planar lipid bilayers. From these spectra, calculated using typical resonance line widths as observed for smaller oriented peptides, it is evident that the two sets of 2D experiments do not provide sufficient resolution for uniformly labeled samples of the size of the aquaporins. In contrast, it appears that important structural information may be extracted from different residue-specific ^{15}N-labeled samples, as illustrated by the spectra corresponding to alanine, arginine, and histidine labeling.

11.3.3 Design of Pulse Sequence Elements

Numerical simulations may obviously not only be used to evaluate pulse sequences designed using analytical tools but also be used directly as a tool for designing new pulse sequences with specific properties. With the relevant nuclear spin systems steadily increasing in size as a result of more extensive isotope labeling and with the ideals moving gradually toward uniform ^{13}C and ^{15}N labeling to extract all information from one sample, the complexity of the experiments providing optimum resolution and sensitivity inevitably increases. Many spins, powder or partially oriented samples, the wish for frequency specific coherence transfers, the need that the samples are not heated excessively by extreme rf irradiation, and the wish for the shortest possible pulse sequences to reduce effects from relaxation impose extreme demands on the spin engineer. These demands may to some extent be handed over to computer optimizations, provided the right strategies can be programmed with sufficient efficiency. Nonetheless, numerical design of NMR experiments has so far been relatively sparse.

Numerical design of optimum pulse sequences may concern optimization of efficiency for a given coherence/polarization transfer process, on the level of the density operator $\rho(t)$, or it may concern maximization or minimization of the scaling on certain internal nuclear spin interactions, such as a specific types of dipole–dipole couplings, while leaving the linear shift terms influenced as little as possible — on the level of the Hamiltonian $H(t)$. The optimizations, typically attributing influence through the rf irradiation, should ideally be given full flexibility to the pulse sequence or be assisted by rigid analytical ideas that certain Hamiltonians or external modulations will provide the desired result.

A first entry to optimization of a given coherence/polarization transfer, or the scaling factor for a given Hamiltonian, could be to establish a theoretical upper limit on the relevant coefficient to be sure that the optimization heads toward the upper limit and to avoid using time on optimizing toward efficiencies that are not possible to reach from a theoretical point of view. Such analysis may be performed within the framework of the unitary bounds on spin dynamics,[119–121] which not only provide a number for the highest achievable efficiency but actually — under the assumption that dissipative processes can be ignored — provide the unitary operators leading to this efficiency. With the only requirement that the propagator is unitary, however, it may be difficult to translate this propagator into a practically feasible experiment. Furthermore, it may be that limitations in the available external manipulations and the nuclear spin system do not allow us to access the unitary bound efficiency but, rather, a lower efficiency, conforming with specific conditional bounds imposed by symmetry or limitations in external manipulations.[122,123]

For these reasons, the most valuable output of the unitary bounds is probably the information about what can be achieved, thereby setting a target for optimizations carried out by analytical (e.g., using the energy level diagrams) or numerical means. Taking the latter approach, we have successfully applied nonlinear computer optimization[124] based on typical pulse sequence elements (rf pulses, free precession periods, isotropic mixing, planar mixing, etc.) in arbitrary order to establish pulse sequences as a complement to analytical derivations. This work, primarily addressing spin-state- and coherence-order-selective pulse sequences in liquid-state NMR,[125–127] has been successful in the sense that the known optimum transfer could be reached. However, because of the many degrees of freedom being constrained into available pulse sequence elements, the required permutations, and the limitations of standard optimization procedures, it is not possible to deduce whether the solutions actually are the most efficient ones, providing the goal in shortest possible time.

In solid-state NMR, it is possible to take a similar approach, keeping in mind that the spin systems are often larger and the fundamental couplings mediating coherence transfer may display dependence on the orientation of the molecule. Provided it is possible by analytical means to establish a functional dependency for the optimization with a relatively small number of variables, it is possible to use standard methods for nonlinear optimization[124] in the design process. For example, such procedures have been used for the design of solid-state NMR multiple-pulse experiments for decoupling of homonuclear dipolar interactions between protons.[128–130]

We here take a different approach that offers substantially higher degrees of freedom in the design process (many more variables) and is less dependent on the establishment of a reasonable (and unfortunately often biased) analytical simplification of the optimization problem before its numerical solution. The key words are optimal control theory.[131,132] Having been developed and extensively used in economic sciences for optimization of investments, and later in engineering for optimization of instrumental performance, optimal control theory is an ideal vehicle for the experiment design described above. Because of these attractive features, optimal control theory has over the last few years gradually found its way into optimization of experiments in optical spectroscopy,[133] magnetic resonance imaging,[134,135] liquid-state NMR,[136–138] and most recently, solid-state NMR.[139] With attention to the last application — as it is most closely related to the topic of this chapter — we recently combined optimal control theory and SIMPSON for automated design of solid-state NMR experiments with optimum performance. The input for such optimizations, apart from the normal input to the SIMPSON "computer-spectrometer," is the initial operator, the destination operator, the desired time that should be used for the given transfer, the time increments, and potential constraints on the rf fields such as peak values, average rf, and rf inhomogeneity. With this information, the optimal control version of SIMPSON by itself should find the best experiments without the need for the expertise of a spin engineer with hands-on expertise on analytical optimization. One can say that this tool — if successful — may move part of the technique development to the practical NMR spectroscopist. This may give the hope that the optimizations are performed on the most relevant systems in a manner conforming with the large efforts now devoted to sample preparation.

Equipped with such tools, it is possible to optimize the efficiency of any coherence/polarization transfer process, being practically very important, as loss of signal in several of such processes during a typical multiple-dimensional experiment may translate into a very significant increase in experiment time or put excessive demands on the amount of sample required for a given experiment. Both problems are critical issues in biological solid-state NMR, in which it may be problematic — or at least expensive — to get sufficient amounts of labeled proteins. It may also be problematic for another point of view; namely, it could be that the required amount of sample (including solvent, membranes, cryo protections, surfactants, etc.) would not fit into the small sample volumes required for fast sample spinning, or the sample may not like intense rf peak powers translating into heat. The latter problem applies in particular to samples that contain a relatively high concentration of ions, which may result from using optimal conditions for preparation of micro- or nanocrystalline samples.

We here address the $^{15}N \rightarrow {}^{13}C$ coherence transfer as being of fundamental importance for essentially all solid-state NMR experiments on uniformly ^{13}C, ^{15}N-labeled proteins. For example, the $^{15}N \rightarrow {}^{13}C_\alpha$ and $^{15}N \rightarrow {}^{13}C'$ transfers are key elements in solid-state NMR variants of the NCACB (or NCX) and NCOCA experiments, respectively, for sequential and in-residue assignment of the ^{13}C, ^{15}N resonances in multiple-dimensional spectra. When considering these transfers, one of the most popular experiments has been the DCP experiment.[64] This experiment is extremely simple to implement (involving only continuous wave irradiation on two channels, with their rf field strengths differing by one or two spinning frequencies), and it is γ-encoded, implying transfers up to about 73%, matching favorably with the about 50% offered by non-γ-encoded experiments. DCP has only two disadvantages, with the most critical one being susceptibility to rf inhomogeneity and the other being relatively narrow rf and offset ranges for efficient transfer. The experiment is also sensitive to chemical shielding anisotropies. These disadvantages may, at the expense of lower efficiency, to some extent be alleviated by using a ramped (i.e., linearly increasing)[61] or adiabatically varying[62] rf field on one of the two channels, which serves to broaden the recoupling resonance condition, however, not with the result of a full solution to the sensitivity loss problem.[54]

Using SIMPSON with a module for optimal control optimization as the tool, it is possible to address the above problems with the aim of finding improved pulse sequences. In particular, it proves relevant to develop pulse sequences with much higher robustness toward rf inhomogeneity — as calculations reveal that a relatively modest 5% Lorentzian or equivalent 9% Gaussian inhomogeneity profile (which is typical for 4-mm rotors, and even larger inhomogeneities are often found for other coil diameters) in practice reduces the sensitivity of DCP by a factor of two. The same loss is often actively induced by the common use of ramped rf fields. The critical issue is that the condition for the two rf fields

$$\omega_{rf}^S = \omega_{rf}^I + n\omega_r, \quad n = \pm 1, \pm 2 \tag{11.13}$$

matches unfavorably with a typical rf inhomogeneity profile, in which a majority of the isochromats will have a similar relative change in the rf fields on the two channels, being more compatible with a $\omega_{rf}^{S} = \omega_{rf}^{I}$ match condition. Partial compensation for this drawback may be established using approaches similar to those recently presented for the iDCP[66] and GATE[67] experiments.

Under consideration of a 5% Lorentzian rf inhomogeneity on the two channels and even larger rf inhomogeneities extending over a square profile going from 80 to 120% of the nominal rf field strengths on the two channels, optimal control theory provides us with a large number of pulse sequences with improved rf inhomogeneity characteristics, higher transfer efficiencies than can be provided by γ-encoded experiments (exceeding 73% transfer efficiency and approaching 100% of the much longer adiabatic experiments), while maintaining similar off-resonance characteristics as the DCP experiments. A representative pulse sequence is given in Figure 11.12, along with an experimental comparison of the transfer efficiencies for $^{15}N \rightarrow ^{13}C_{\alpha}$ transfer in a powder of glycine spinning 10 kHz, as well as graphs illustrating the sensitivity of the optimal control DCP experiments (^{OC}DCP) relative to DCP, ramped DCP, and adiabatic DCP under various rf inhomogeneity conditions.[139] The experimental spectra reveal a gain of 53% using ^{OC}DCP instead of DCP, thereby serving to demonstrate that the numerical design of solid-state NMR experiments is relevant. We note that the proposed ^{OC}DCP pulse sequence uses only 25–30% of the rf power used for the DCP experiment. It is anticipated that similar gains may be achieved for other vital transfers used as elements in multiple-dimensional solid-state NMR experiments, allowing us to foresee that optimal control may have a major effect in the future on our ways to design pulse sequences.

11.3.4 EXTRACTION OF STRUCTURAL PARAMETERS

A major aspect of numerical simulations in relation to biological solid-state NMR concerns extraction of parameters for the nuclear spin interactions from experimental spectra. These parameters may be translated into internuclear distances (dipole–dipole coupling interactions), information about the local electronic environments of the nuclei (chemical shielding and quadrupole coupling interactions), and molecular backbone and side-chain torsion angles (J couplings, chemical shifts, dipole–dipole couplings). This information may serve as entry to molecular dynamics calculation software for the final calculation of molecular structure and dynamics. In this manner, simulation software such as SIMPSON in combination with SIMMOL[34] may serve as an interface between experimental spectra and the more general structure calculation procedures — an interface that needs considerable development relative to that commonly used for liquid-state NMR. Obviously, numerical simulations and iterative fitting may also be applied to distinguish and quantify different elements in multiple-component samples. To facilitate any of these needs, SIMPSON has been equipped with the minimization toolbox originally developed at CERN as the MINUIT program.[140] This combination[50] introduces a very powerful tool for all kinds of minimization including grid scans, Monte Carlo, SIMPLEX, and steepest-descent gradient-based procedures[124] for nonlinear optimization, as described previously.

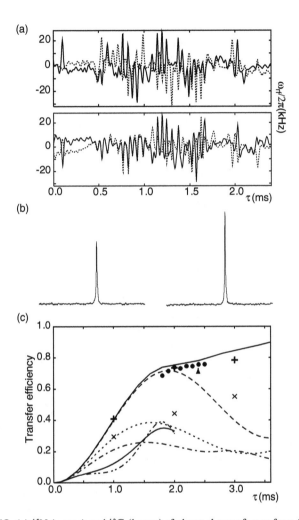

FIGURE 11.12 (a) ^{15}N (upper) and ^{13}C (lower) rf channel waveforms for a 2.4-ms optimal control variant of the DCP experiment (^{OC}DCP), designed using SIMPSON with optimal control procedures. (b) Experimental comparison of ^{15}N→^{13}C coherence transfers obtained for a powder of ^{13}C$_2$,^{15}N-labeled glycine spinning 10 kHz in a 400-MHz instrument using DCP (left; 35 and 25 kHz rf on the ^{15}N and ^{13}C channels and ^{OC}DCP (right: 10.4 and 9.5 kHz on the ^{15}N and ^{13}C channels). (c) Numerical comparison of ^{15}N→^{13}C coherence transfer efficiencies obtained using optimal control pulse sequences (solid line: without rf inhomogeneity; solid circles: with 5% Lorentzian inhomogeneity), DCP (dashed line: without inhomogeneity; dotted and dot-dashed: with 5% and 10% Lorentzian rf inhomogeneity, respectively), ramped DCP (dot-dot-dash and dot-dot-dot-dash: ramped DCP with 5% and 10% Lorentzian rf inhomogeneity, respectively), and adiabatic DCP (+: without, and ×: with 5% Lorentzian rf inhomogeneity). The arrow indicates the pulse sequence in panel (a). Reproduced with permission from ref. 137.

With attention to essentially all the numerical simulations shown in the preceding part of this chapter — ^{13}C–^{13}C rotational resonance, ^{17}O second-order quadrupolar MAS NMR, HORROR or C7 recoupling, ^{67}Zn QCPMG NMR of metallo proteins, PISEMA experiments on large residue–specific or uniformly isotope-labeled proteins — may be combined straightforwardly with iterative fitting procedures to extract the relevant nuclear spin interaction parameters. Such applications may take advantage of SIMPSON and SIMMOL, following a direct path between spectra and structure, as illustrated in Figure 11.1. Because we already demonstrated the numerical part of such evaluations, we will not here repeat any of applications of this type but will, instead, restrict ourselves to one single example in which various important elements of data analysis are combined.

Our example will be structural analysis of the antimicrobial ionophore alamethicin using oriented-sample flat-coil NMR spectroscopy.[141] Using this setup, alamethicin oriented in planar DMPC phospholipid bilayers gives rise to 2D ^{15}N-^{1}H dipolar coupling versus ^{15}N chemical shift spectra, as shown in Figure 11.13a for alamethicin with the backbone amide ^{15}N labeled at the Ala^6, Val^9, and Val^{15} residues. The positions of the resonances in this spectrum reflect the orientation of each of the three involved peptide planes (cf. Figure 11.2c) relative to the magnetic field. Actually, each set of observed chemical shifts and dipolar couplings typically corresponds to a number of possible orientations, as illustrated by the black areas in the $\{\alpha_{EL},\ \beta_{EL}\}$ restriction plots shown for Ala^6 in Figure 11.13b. The restriction plots for the two parameters and their intersection (all calculated using SIMPSON procedures) provide detailed structural information about the individual peptide planes. On assumption of a regular secondary structure, such as an α-helix, this information may be translated into information about the conformation of the overall peptide in the membrane. In this case, all three peptide plane constraints are compatible with a transmembrane arrangement of the helices.

In an attempt to extend the structural constraints obtained by solid-state NMR for alamethicin with the three given isotope labels, Figure 11.13c shows a modelled structure obtained by molecular dynamics. This structure links the three peptide planes and provides us with the possibility of calculating 2D spectra for peptides with different molecular tilt (τ) and rotational pitch (ρ) angles and numerically comparing these with the experimental spectrum. The discrepancy (or match) between the calculated and experimental spectra may be represented in an root mean square deviation (RMSD) plot, as illustrated in Figure 11.13d. In this specific case, the best match is obtained for an average tilt angle of $\tau = 11°$ and a rotational pitch angle $\rho = 230°$. This geometry is compatible with multimeric ion channels with a hydrophilic lining in the pore interior, as illustrated in top and side views in Figure 11.13e. This study, and later results for the more challenging 7 TM protein bacteriorhodopsin,[25,54,142] reinforces the value of flexible numerical simulation tools that combine data/parameters from different sources to provide an overall picture of the structure.

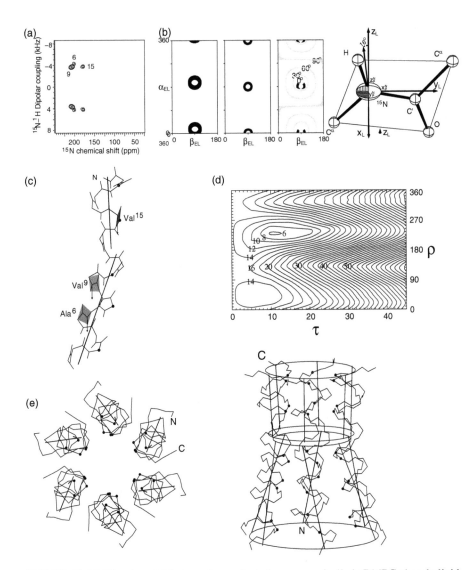

FIGURE 11.13 The alamethicin ionophore oriented macroscopically in DMPC phospholipid bilayers and studied by flat-coil solid-state NMR.[141] (a) Two-dimensional ^{15}N-^{1}H dipolar coupling versus ^{15}N chemical shift spectrum. (b) Restriction plots for Ala^6 expressing possible constraints on the peptide-plane orientation relative to the magnetic field (and thereby the phospholipid bilayer). (c) Molecular dynamics structure of alamethicin. (d) RMSD between experimental and simulated 2D spectra as function of the average helix tilt angle and the rotational pitch calculated on basis of the molecular dynamics structure in panel (c). The best orientation is compatible with the multimeric ion channels shown in panel (e). This figure is reproduced from ref. 141 with permission.

11.4 CONCLUSION

In conclusion, we have described some elementary aspects of numerical simulations with particular attention to the overall utility and need of such methods for the future development of biological solid-state NMR spectroscopy. Throughout the chapter, I have highlighted the importance of controlling the spin dynamics in relatively large spin systems to high precision, as well as emphasizing the importance of having flexible and powerful simulation software available for experiment interpretation, design, and evaluation. This is one of the key elements in the continued fast progress of biological solid-state NMR spectroscopy.

ACKNOWLEDGMENTS

I acknowledge financial support from the Danish National Research Foundation, Danish Biotechnological Instrument Centre (DABIC), the Danish Natural Science Research Council (SNF), and Carlsbergfondet. Mads Bak, Jimmy T. Rasmussen, Robert Schultz, and Thomas V. are acknowledged for collaborations and discussions.

REFERENCES

1. Haeberlen, U., *High-Resolution NMR in Solids. Selective Averaging,* Academic Press, New York, 1976.
2. Mehring, M., *Principles of High Resolution NMR in Solids*, Springer, New York, 1983.
3. Gerstein, B.C. and Dybowski, C.R., *Transient Techniques in NMR of Solids. An Introduction to Theory and Practice*, Academic Press, Orlando, FL, 1985.
4. Schmidt-Rohr, K. and Spiess, H.W., *Multidimensional Solid-State NMR and Polymers*, Academic Press, London, 1996.
5. McDowell, L.M. and Schaefer, J., High-resolution NMR of biological solids, *Curr. Opin. Struct. Biol.*, 6, 624–629, 1997.
6. Opella, S.J., NMR and membrane proteins, *Nat. Struct. Biol.,* 4, 845–848, 1997.
7. Griffin, R.G., Dipolar recoupling in MAS spectra of biological solids, *Nat. Struct. Biol.,* 5, 508–512, 1998.
8. de Groot, H.J.M., Solid-state NMR spectroscopy applied to membrane proteins, *Curr. Opin. Struct. Biol.*, 10, 593–600, 2000.
9. Tycko, R., Solid-state NMR as a probe of amyloid fibril structure, *Curr. Opin. Chem. Biol.*, 4, 500–506, 2000.
10. McDermott, A., Polenova, T., Böckmann, A., Zilm, K.W., Paulsen, E.K., Martin, R.W. and Montelione, G.T., Partial NMR assignment for uniformly (^{13}C, ^{15}N)-enriched BPTI in the solid state, *J. Biomol. NMR,* 16, 209–219, 2000.
11. Valentine, K.G., Liu, S.F., Marassi, F.M., Veglia, G., Opella, S.J., Ding, F.X., Wang, S.H., Becker, J.M. and Naider, F., Structure and topology of a peptide segment of the 6th transmembrane domain of the *Saccharomyces cerevisiae* alpha-factor receptor in phospholipid bilayers, *Biopolymer*, 59, 243–256, 2001.
12. Wang, J., Kim, S., Kovacs, F. and Cross, T.A., Structure of the transmembrane region of the M2 protein H+ channel, *Prot. Sci.*, 10, 2241–2250, 2001.

13. Castellani, J., van Rossum, B., Diehl, A., Schubert, M., Rehbein, K. and Oschkinat, H., Structure of a protein determined by solid-state NMR spectroscopy, *Nature*, 420, 98–102, 2002.
14. Rienstra, C.M., Tucker-Kellogg, L., Jaroniec, C.P., Hohwy, M., Reif, B., McMahon, M.T., Tidor, B., Lozano-Perez, T. and Griffin, R.G., De novo determination of peptide structure with solid-state magic-angle spinning NMR spectroscopy, *Proc. Natl. Acad. Sci. USA*, 99, 10260–10265, 2002.
15. Petkova, A.T., Ishii, Y., Balbach, J.J., Antzutkin, O.N., Leapman, R.D., Delaglio, F. and Tycko, R., A structural model for Alzheimer's beta-amyloid fibrils based on experimental constraints from solid-state NMR, *Proc. Natl. Acad. Sci. USA*, 99, 16742–16747, 2003.
16. van Beek, J.D., Hess, S., Vollrath, F. and Meier, B.H., The molecular structure of spider dragline silk: folding and orientation of the protein backbone, *Proc. Natl. Acad. Sci. USA*, 99, 10266–10271, 2002.
17. Marassi, F.M. and Opella, S.J., Simultaneous assignment and structure determination of a membrane protein from NMR orientational restraints, *Prot. Sci.*, 12, 403–411, 2003.
18. Zeri, A.C., Mesleh, M.F., Nevzorov, A.A. and Opella, S.J., Structure of the coat protein in fd filamentous bacteriophage particles determined by solid-state NMR spectroscopy, *Proc. Natl. Acad. Sci. USA*, 100, 6458–6463, 2003.
19. Luca, S., White, J.F., Sohal, A.K., Filiv, D.V., van Boom, J.H., Grisshammer, R. and Baldus, M., The conformation of neutotensin bound to its G-protein coupled receptor investigated by 2D solid-state NMR, *Proc. Natl. Acad. Sci. USA*, 100, 10706–10711, 2003.
20. Park, S.H., Mrse, A.A., Nevzorov, A.A., Mesleh, M.F., Oblatt-Montal, M., Montal, M. and Opella, S.J., Three-dimensional structure of the channel-forming trans-membrane domain of virus protein "u" (Vpu) from HIV-1, *J. Mol. Biol.*, 333, 409–424, 2003.
21. Böckmann, A., Lange, A., Galinier, A., Luca, S., Giraud, N., Juy, M., Heise, H., Monstserret, R., Penin, F. and Baldus, M., Solid-state NMR sequential resonance assignments and conformational analysis of the 2×10.4 kDa dimeric form of the *Bactillus subtilus* protein Crh, *J. Biomol. NMR*, 27, 323–339, 2003.
22. Luca, S., Heise, H. and Balus, M., High-resolution solid-state NMR applied to polypeptides and membrane proteins, *Acc. Chem. Res.*, 36, 858–865, 2003.
23. Jaroniec, C.P., MacPhee, C.E., Baja, V.S., McMahon, M.T., Dobson, C.M. and Griffin, R.G., High-resolution molecular structure of a peptide in an amyloid fibril determined by magic-angle spinning NMR spectroscopy, *Proc. Natl. Acad. Sci. USA*, 101, 711–716, 2004.
24. Igumenova, T.I., McDermott, A.U., Zilm, K.W., Martin, R.W., Paulson, E.K. and Wand, A.J., Assignments of carbon NMR resonances for microcrystalline ubiquitin, *J. Am. Chem. Soc.*, 126, 6720–6727, 2004.
25. Nielsen, N.C., Malmendal, A. and Vosegaard, T., Techniques and applications of NMR to membrane proteins, *Mol. Membr. Biol.*, 21, 129–141, 2004
26. Wüthrich, K., *NMR of Proteins and Nucleic Acids*, Wiley, New York, 1986.
27. Ernst, R.R., Bodenhausen, G. and Wokaun, A., *Principles of Nuclear Magnetic Resonance in One and Two Dimensions*, Clarendon Press, Oxford, 1987.
28. Cavanagh, J., Fairbrother, W.J., Palmer, A.G., III, and Shelton, N.J., *Protein NMR Spectroscopy: Principles and Practice*, Academic Press, San Diego, 1996.
29. Venter, J.C., Adams, M.D., Myers, E.W. et al., The sequence of the human genome, *Science*, 291, 1304–1351, 2001.

30. Haeberlen, U. and Waugh, J.S., Coherent averaging effects in magnetic resonance, *Phys. Rev.*, 175, 453–467, 1968.
31. Hohwy, M. and Nielsen, N.C., Systematic design and evaluation of multiple-pulse experiments in nuclear magnetic resonance spectroscopy using a semi-continuous Baker-Campell-Hausdorff expansion, *J. Chem. Phys.*, 109, 3780–3791, 1998.
32. Untidt, T.S. and Nielsen, N.C., Closed solution to the Baker-Campbell-Hausdorff problem: exact effective Hamiltonian theory for analysis of nuclear-magnetic-resonance experiments, *Phys. Rev. E*, 65, 021108-1–021108-17, 2003.
33. Siminovitch, D., Untidt, T.S. and Nielsen, N.C., Exact effective Hamiltonian theory. II: Expansion of matrix functions and entangled unitary exponential operators, *J. Chem. Phys.*, 120, 51–66, 2004.
34. Bak, M., Schultz, R., Vosegaard, T. and Nielsen, N.C., Specification and visualization of anisotropic interaction tensors in polypeptides and numerical simulations in biological solid-state NMR, *J. Magn. Reson.*, 154, 28–45, 2002.
35. Raleigh, D.P., Levitt, M.H. and Griffin, R.G., Rotational resonance solid-state NMR, *Chem. Phys. Lett.*, 146, 71–76, 1988.
36. Gullion, T. and Schaefer, J., Rotational-echo double resonance NMR, *J. Magn. Reson.*, 81, 196–200, 1989.
37. Creuzet, F., McDermott, A.E., Gebhard, R., van der Hoef, K., Spijker-Assink, M.B., Herzfeld, J., Lugtenburg, J., Levitt, M.H. and Griffin, R.G., Determination of membrane protein structure by rotational resonance NMR: bacteriorhodopsin, *Science*, 251, 783–786, 1991.
38. McDowell, L.M., Lee, M.S., McKay, R.A., Anderson, K.S. and Schaefer, J., Intersubunit communication in tryptophan syntase by carbon-13 and fluorine-19 REDOR NMR, *Biochemistry*, 35, 3328–3334, 1996.
39. Smith, S.O. and Bormann, B.J., Determination of helix-helix interactions in membranes by rotational resonance NMR, *Proc. Natl. Acad. Sci. USA*, 92, 488–491, 1995.
40. Wang, J., Kalazs, Y.S. and Thompson, L.K., Solid-state REDOR NMR distance measurements at the ligand site of a bacterial chemotaxis membrane receptor, *Biochemistry*, 36, 1699–1703, 1997.
41. Yang, J., Gabrys, C.M. and Weliky, D.P., Solid-state nuclear magnetic resonance evidence for an extended β strand conformation of the membrane bound HIV-1 fusion peptide, *Biochemistry*, 40, 8126–8137, 2001.
42. Nishimura, K., Kim, S., Zahng, L. and Cross, T.A., The closed state of a H+ channel helical bundle combining precise orientational and distance restraints from solid-state NMR, *Biochemistry*, 41, 13170–13177, 2002.
43. Frydman, L. and Harwood, J.S., Isotropic spectra of half-integer quadrupolar spins from bidimensional magic-angle spinning NMR, *J. Am. Chem. Soc.*, 117, 5367–5368, 1995.
44. Larsen, F.H., Jakobsen, H.J., Ellis, P.D. and Nielsen, N.C., Sensitivity-enhanced quadrupolar-echo NMR of half-integer quadrupolar nuclei. Magnitudes and relative orientation of chemical shielding and quadrupolar coupling tensors, *J. Phys. Chem A*, 101, 8597–8606, 1997.
45. Glaubitz, C. and Watts, A., Magic angle-oriented sample spinning (MAOSS): a new approach toward biomembrane studies, *J. Magn. Reson.*, 130, 305–316, 1998.
46. Palczewski, K., Kumasaka, T., Hori, T., Behnke, C.A., Motoshima, H., Fox, B.A., Trong, I.L., Teller, D.C., Okada, T., Stenkamp, R.E, Yamamoto, M. and Miyano, M., Crystal structure of rhodopsin: a G protein-coupled receptor, *Science*, 289, 739–745, 2000.
47. Ousterhout, J.K., *Tcl and Tk Toolkit*, Addison-Wesley, Reading, MA, 1994.

48. Berman, H.M., Westbrook, J., Feng, Z., Gilliland, G., Bhat, T.N., Weissig, H., Shindy-alov, I.N. and Bourne, P.E., The protein data bank, *Nucl. Acids Res.*, 28, 235–242, 2000.

49. Bak, M., Schultz, R. and Nielsen, N.C., Homo- and heteronuclear dipolar recoupling under magic-angle spinning conditions, in *Perspectives on Solid-State NMR in Biology*, Kiihne, S. and de Groot, H.J.M., Eds., Kluwer Academic, Dordrecht, 2001.

50. Vosegaard, T., Malmendal, A. and Nielsen, N.C., The flexibility of SIMPSON and SIMMOL for numerical simulations in solid and liquid-state NMR spectroscopy, *Chem. Monthly*, 133, 1555–1574, 2002.

51. Vosegaard, T. and Nielsen, N.C., Towards high-resolution solid-state NMR on large uniformly ^{15}N- and [^{13}C,^{15}N]-labeled membrane proteins in oriented and lipid bilayers, *J. Biomol. NMR*, 22, 225–247, 2002.

52. Bjerring, M., Vosegaard, T., Malmendal, A. and Nielsen, N.C., Methodological development of solid-state NMR for characterization of membrane proteins, *Concepts Magn. Reson.*, 18A, 111–129, 2003.

53. Vosegaard, T.V. and Nielsen, N.C., Improved pulse sequences for pure exchange solid-state NMR spectroscopy, *Magn. Reson. Chem.*, 42, 285–290, 2004.

54. Sivertsen, A.C., Bjerring, M., Kehlet, C.T., Vosegaard, T.V. and Nielsen, N.C., Numerical simulations in biological solid-state NMR spectroscopy, *Ann. Rep. NMR Spectrosc.*, 54, 243–293, 2005.

55. Bak, M., Rasmussen, J.T. and Nielsen, N.C., SIMPSON: A general simulation program for solid-state NMR spectroscopy, *J. Magn. Reson.*, 147, 296–330, 2000.

56. Bennett, A.E., Griffin, R.G. and Vega, S., in *NMR Basic Principles and Progress*, Diehl, P., Fluck, E., Gunther, H. and Kosfeld, R., Eds., Springer, Berlin, Vol. 33, pp. 1–77.

57. Dusold, S. and Sebald, A., Dipolar recoupling under magic-angle spinning conditions, *Ann. Rep. NMR Spectrosc.*, 41, 185–264, 2000.

58. Baldus, M., Correlation experiments for assignment and structure elucidation of immobilized polypeptides under magic-angle spinning, *Progr. NMR Spectrosc.*, 41, 1–47, 2002.

59. Levitt, M.H., Symmetry-based pulse sequences in magic-angle spinning solid-state NMR, in *Encyclopedia of NMR*, Wiley, Chichester, 2002, pp. 165–196.

60. Pines, A., Gibby, M.G. and Waugh, J.S., Proton-enhanced nuclear induction spectroscopy. A method for high resolution NMR of dilute spins in solids, *J. Chem. Phys.*, 56, 1776–1777, 1972.

61. Metz, G., Hu, W. and Smith, S.O., Ramped-amplitude cross polarization in magic-angle-spinning NMR, *J. Magn. Reson. A*, 110, 219–227, 1994.

62. Hediger, S., Meier, B.H. and Ernst, R.R., Adiabatic passage Hartmann-Hahn cross polarization in NMR under magic-angle sample spinning, *Chem. Phys. Lett.*, 240, 449–456, 1995.

63. Zhao, X., Hoffbauer, W., Schmedt Auf Der Gunne, J. and Levitt, M.H., Heteronuclear polarization transfer by symmetry-based recoupling sequences in solid-state NMR, *Solid State Nucl. Magn. Reson.*, 26, 57–64, 2004.

64. Schaefer, J., Stejskal, E.O., Garbow, J.R. and McKay, R.A., Quantative determination of the concentrations of ^{13}C-^{15}N chemical bonds by double cross-polarization NMR, *J. Magn. Reson.*, 59, 150–156, 1984.

65. Baldus, M., Petkova, A.T., Herzfeld, J. and Griffin, R.G., Cross polarization in the tilted frame: assignment and spectral simplification in the heteronuclear spin systems, *Mol. Phys.*, 95, 1197–1207, 1998.

66. Bjerring, M. and Nielsen, N.C., Solid-state NMR heteronuclear coherence transfer using phase and amplitude modulated rf irradiation at the Hartmann-Hahn sideband conditions, *Chem. Phys. Lett.*, 382, 671–678, 2003.

67. Bjerring, M., Rasmussen, J.T., Krogshave, R.S. and Nielsen, N.C., Heteronuclear coherence transfer in solid-state nuclear magnetic resonance using a γ-encoded transferred echo experiment, *J. Chem. Phys.*, 119, 8916–8926, 2003.

68. Brinkmann, A. and Levitt, M.H., Symmetry principles in the nuclear magnetic resonance of spinning solids: heteronuclear recoupling by generalized Hartmann-Hahn sequences, *J. Chem. Phys.*, 115, 357–384, 2001.

69. Lee, M. and Goldburg, W.I., Nuclear-magnetic-resonance line narrowing by a rotating rf field, *Phys. Rev.*, 140A, 1261–1271, 1965.

70. Bjerring, M. and Nielsen, N.C., Solid-state NMR heteronuclear dipolar recoupling using off-resonance symmetry-based pulse sequences, *Chem. Phys. Lett.*, 370, 496–503, 2003.

71. Nielsen, N.C., Bildsøe, H., Jakobsen, H.J. and Levitt, M.H., Double-quantum homonuclear rotary resonance: efficient dipolar recovery in magic-angle spinning nuclear magnetic resonance, *J. Chem. Phys.*, 101, 1805–1812, 1994.

72. Verel, R., Ernst, M. and Meier, B.H., Adiabatic dipolar recoupling in solid-state NMR: the DREAM sequence, *J. Magn. Reson.*, 150, 81–99, 2001.

73. Lee, Y.K., Kurur, N.D., Helmle, M., Johannessen, O.G., Nielsen, N.C. and Levitt, M.H., Efficient dipolar recoupling in the NMR of rotating solids. A sevenfold symmetric radiofrequency pulse sequence, *Chem. Phys. Lett.*, 242, 304–309, 1995.

74. Brinkmann, A., Eden, M. and Levitt, M.H., Synchronous helical pulse sequences in magic-angle spinning nuclear magnetic resonance: application to double quantum spectroscopy, *J. Chem. Phys.*, 112, 8539–8554, 2000.

75. Carravetta, M., Edén, M., Zhao, X., Brinkmann, A. and Levitt, M.H., Symmetry principles for the design of radiofrequency pulse sequences in the nuclear magnetic resonance of rotating solids, *Chem. Phys. Lett.*, 321, 205–221, 2000.

76. Zhao, X., Carravetta, M., Madhu, P.K. and Levitt, M.H., Symmetry-based pulse sequences in solid-state NMR and applications to biological systems, *Rec. Res. Develop. Biophys.*, 2, 121–146, 2003.

77. Rienstra, C.M., Hatcher, M.E., Mueller, L.J., Sun, B., Fesik, S.W. and Griffin, R.G., Efficient multispin homonuclear double-quantum recoupling for magic-angle-spinning NMR: $^{13}C^{13}$-C correlation spectroscopy of U-^{13}C-erythromycin A, *J. Am. Chem. Soc.*, 120, 10602–10612, 1998.

78. Hohwy, M., Jakobsen, H.J., Edén, M., Levitt, M.H. and Nielsen, N.C., Broadband dipolar recoupling in the nuclear magnetic resonance of rotating solids: a compensated C7 pulse sequence, *J. Chem. Phys.*, 108, 2686–2694, 1998.

79. Hohwy, M., Rienstra, C.M., Jaroniec, C.P. and Griffin, R.G., Fivefold symmetric homonuclear dipolar recoupling in rotating solids: application to double quantum spectroscopy, *J. Chem. Phys.*, 110, 7983–7992, 1999.

80. Bennett, A.E., Rienstra, C.M., Auger, M., Lakshmi, K.V. and Griffin, R.G., Heteronuclear decoupling in solids, *J. Chem. Phys.*, 103, 6951–6958, 1995.

81. Ernst, M., Samoson, A. and Meier, B.H., Low-power XiX decoupling in MAS NMR experiments, *J. Magn. Reson.*, 163, 332–339, 2003.

82. de Bouregas, F.S. and Waugh, J.S., ANTIOPE, a program for computer experiments on spin dynamics, *J. Magn. Reson.*, 96, 280–289, 1992.

83. Smith, S.A., Levante, T.O., Meier, B.H. and Ernst, R.R., Computer simulations in magnetic resonance. An object-oriented programming approach, *J. Magn. Reson.*, 106, 75–105, 1994.

84. Blanton, W.B., BlochLib: a fast NMR C++ tool kit, *J. Magn. Reson.*, 162, 269–283, 2003.
85. Veshtort, M. and Griffin, R.G., SPINEVOLUTION: A powerful tool for the simulation of solid and liquid state NMR experiments, Presentation at 45th Rocky Mountain NMR Conference, Denver, CO, 2004.
86. Skibsted, J., Nielsen, N.C., Bildsøe, H. and Jakobsen, H.J., Satellite transitions in MAS NMR of and quadrupolar nuclei, *J. Magn. Reson.*, 95, 88–117, 1991.
87. Edén, M. and Levitt, M.H., Computation of orientational averages in solid-state NMR by gaussian spherical quadrature, *J. Magn. Reson.*, 132, 220–239, 1998.
88. Hohwy, M., Bildsøe, H., Jakobsen, H.J. and Nielsen, N.C., Efficient spectral simulations in NMR of rotating solids. The γ-COMPUTE algorithm, *J. Magn. Reson.*, 136, 6–14, 1999.
89. Charpentier, T., Fermon, C. and Virlet, J., Efficient time propagation technique for MAS NMR simulation: application to quadrupolar nuclei, *J. Magn. Reson.*, 132, 181–190, 1998.
90. Levitt, M.H. and Eden, M., Numerical simulation of periodic nuclear magnetic resonance problems: fast calculation of carousel averages, *Mol. Phys.*, 95, 879–890, 1998.
91. Blanton, W.B., Logan, J.W. and Pines, A., Rational reduction of periodic propagators for off-period observations, *J. Magn. Reson.*, 166, 174–181, 2004.
92. Zaremba, S.K., Good lattice points, discrepancy, and numerical integration, *Ann. Mat. Pure Appl.*, 293, 4–73, 1966
93. Conroy H., Molecular Schrödinger equation. VIII. A new method for the evaluation of multidimensional integration, *J. Chem. Phys.*, 47, 5307–5318, 1967
94. Cheng, V.B., Suzukawa, H.H., Jr. and Wolfsberg, M., Investigations of a nonrandom numerical method for multidimensional integration, *J. Chem. Phys.*, 59, 3992–3999, 1973.
95. Bak, M. and Nielsen, N.C., REPULSION, a novel approach to efficient powder averaging in solid-state NMR, *J. Magn. Reson.*, 125, 132–139, 1997.
96. Levitt, M.H., Raleigh, D.P., Creuzet, F. and Griffin, R.G., Theory and simulations of homonuclear spin-pair systems in rotating solids, *J. Chem. Phys.*, 92, 6347–6364, 1990.
97. Wu, G., Dong, S., Ida, R. and Reen, N., A solid-state [17]O nuclear magnetic resonance study of nucleic acid bases, *J. Am. Chem. Soc.*, 124, 1768–1777, 2002.
98. Nielsen, N.C., Creuzet, F., Griffin, R.G. and Levitt, M.H., Enhanced double-quantum nuclear magnetic resonance in spinning solids at rotational resonance, *J. Chem. Phys.*, 96, 5668–5677, 1992.
99. Cornilescu, G., Marquart, J.S., Ottiger, M. and Bax, A., Validation of protein structure from anisotropic carbonyl chemical shifts in dilute liquid crystalline phase, *J. Am. Chem. Soc.*, 120, 6836–6837, 1998.
100. Wand, A.J., Urbauer, J.L., McEvoy, R.P. and Bieber, R.J., Internal dynamics of human ubiquitin revealed by ^{13}C-relaxation studies of randomly fractionally labeled protein, *Biochemistry*, 35, 6116–6125, 1996.
101. Hughes, C.E., Luca, S. and Baldus, M., Radio-frequency driven polarization transfer without heteronuclear decoupling in rotating solids, *Chem. Phys. Lett.*, 385, 435–440, 2004.
102. Jakobsen, H.J., Skibsted, J., Bildsøe, H. and Nielsen, N.C., Magic-angle spinning NMR spectra of satellite transitions for quadrupolar nuclei in solids, *J. Magn. Reson.*, 85, 173–180, 1989.

103. Kristensen, J.H., Bildsøe, H., Jakobsen, H.J. and Nielsen, N.C., Deuterium quadrupole couplings from least-squares computer simulations of ^{2}H MAS NMR spectra, *J. Magn. Reson.*, 92, 443–453, 1991.

104. Skibsted, J., Nielsen, N.C., Bildsøe, H. and Jakobsen, H.J., ^{51}V MAS NMR spectroscopy: determination of quadrupole and anisotropic shielding tensors, including the relative orientation of their principal-axis systems, *Chem. Phys. Lett.*, 188, 405–412, 1992.

105. Llor, A. and Virlet, J., Towards high-resolution NMR of more nuclei in solids: sample spinning with time-dependent spinner axis angle, *Chem. Phys. Lett.*, 152, 248–253, 1998.

106. Chmelka, B.F., Mueller, K.T., Pines, A., Stebbins, J., Wu, Y. and Swanziger, J.W., Oxygen-17 NMR in solids by dynamic-angle spinning and double rotation, *Nature*, 339, 42–43, 1989.

107. Shore, J.S., Wang, S., Taylor, R.E., Bell, A.T. and Pines, A., Determination of quadrupolar and chemical shielding tensor elements and the relative orientation of the principal axis systems using two-dimensional NMR spectroscopy, *J. Chem. Phys.*, 105, 9412–9420, 1996.

108. Gan, Z.H., Isotropic NMR spectra of half-integer quadrupolar nuclei using satellite transitions and magic-angle spinning, *J. Am. Chem. Soc.*, 122, 3242–3243, 2000.

109. Larsen, F.H., Lipton, A.S., Jakobsen, H.J., Nielsen, N.C. and Ellis, P.D., ^{67}Zn QCPMG solid-state NMR studies of zinc complexes as models for metallo proteins, *J. Am. Chem. Soc.*, 121, 3783–3784, 1999.

110. Larsen, F.H., Skibsted, J., Jakobsen, H.J. and Nielsen, N.C., Solid-state QCPMG NMR of low-γ quadrupolar metal nuclei in natural abundance, *J. Am. Chem. Soc.*, 122, 7080–7086, 2000.

111. Larsen, F.H., Jakobsen, H.J., Ellis, P.D. and Nielsen, N.C., High-field QCPMG-MAS NMR of half-integer quadrupolar nuclei with large quadrupole couplings, *Mol. Phys.*, 95, 1185–1195, 1998.

112. Larsen, F.H., Jakobsen, H.J., Ellis, P.D. and Nielsen, N.C., Molecular dynamics from H-2 quadrupolar Carr-Purcell-Meiboom-Gill solid-state NMR spectroscopy, *Chem. Phys. Lett.*, 292, 467–473, 1998.

113. Vosegaard, T., Larsen, F.H., Jakobsen, H.J., Ellis, P.D. and Nielsen, N.C., Sensitivity-enhanced multiple-quantum MAS NMR of half-integer quadrupolar nuclei, *J. Am. Chem. Soc.*, 119, 9055–9056, 1997.

114. Larsen, F.H. and Nielsen, N.C., Effect of finite rf pulses and sample spinning speed in multiple-quantum magic-angle apinning (MQ-MAS) and multiple-quantum Carr-Purcell-Meiboom-Gill magic-angle spinning (MQ-QCPMG-MAS) nuclear magnetic resonance of half integer quadrupolar nuclei, *J. Chem. Phys A*, 103, 10825–10832, 1999.

115. Luca, S., Filippov, D.V., van Boom, J.H., Oschkinat, H., de Groot, H.J. and Baldus, M., Secondary chemical shifts in immobilized peptides and proteins: a qualitative basis for structure refinement under magic-angle spinning, *J. Biomol. NMR*, 20, 325–331, 2001.

116. Wu, C.H., Ramamoorthy, A. and Opella, S.J., High-resolution heteronuclear dipolar solid-state NMR spectroscopy, *J. Magn. Reson. A*, 109, 270–272, 1994.

117. Sui, H., Han, B.G., Lee, J.K., Walian, P. and Jap, B.K., Structural basis of water-specific transport through the AQP1 water channel, *Nature*, 414, 872–878, 2001.

118. Bielecki, A., Kolbert, A.C. and Levitt, M.H., Frequency-switched pulse sequences: homonuclear decoupling and dilute spin NMR in solids, *Chem. Phys. Lett.*, 155, 341–346, 1989.

119. Sørensen, O.W., Polarization transfer experiments in high-resolution NMR spectroscopy, *Progr. NMR Spectrosc.*, 21, 503–569, 1989.
120. Stoustrup, J., Schedletzky, O., Glaser, S.J., Griesinger, C., Nielsen, N.C. and Sørensen, O.W., Generalized bound on quantum dynamics: efficiency of unitary transformations between non-hermitian states, *Phys. Rev. Lett.*, 74, 2921–2924, 1995.
121. Glaser, S.J., Schulte-Herbruggen, T., Sieveking, K., Schedletzky, O., Nielsen, N.C., Sørensen, O.W. and Griesinger, C., Unitary control in quantum ensembles: maximizing signal intensity in coherent spectroscopy, *Science*, 280, 421–424, 1998.
122. Nielsen, N.C., Schulte-Herbruggen, T. and Sørensen, O.W., Bounds on spin dynamics tightened by permutation symmetry. Application to coherence transfer in I_2S and I_3S spin systems, *Mol. Phys.*, 85, 1205–1216, 1995.
123. Untidt, T.S., Glaser, S.J., Griesinger, S.J. and Nielsen, N.C., Unitary bounds and controllability of quantum evolution in NMR spectroscopy, *Mol. Phys.*, 96, 1739–1744, 1999.
124. Press, W.H., Teukolsky, S.A., Vetterling, W.T. and Flannery, B.P., Numerical recipes in C: the art of scientific computing, Cambridge University Press, Cambridge, 1993.
125. Nielsen, N.C., Thøgersen, H. and Sørensen, O.W., A systematic strategy for design of optimum coherent experiments applied to efficient interconversion of double- and single-quantum coherences in nuclear magnetic resonance, *J. Chem. Phys.*, 105, 3962–3968, 1996.
126. Untidt, T.S., Schulte-Herbruggen, T., Luy, B., Glaser, S.J., Griesinger, C., Sørensen, O.W. and Nielsen, N.C., Design of NMR pulse experiments with optimum sensitivity: coherence-order-selective transfer in I_2S and I_3S spin systems, *Mol. Phys.*, 95, 787–796, 1998.
127. Untidt, T.S., Schulte-Herbruggen, T., Sørensen, O.W. and Nielsen, N.C., Nuclear magnetic resonance coherence-order and spin-state-selective correlation in I_2S spin systems, *J. Phys. Chem. A*, 103, 8921–8926, 1999.
128. Liu, H., Glaser, S.J. and Drobny, G.P., Development and optimization of multipulse propagators: applications to homonuclear spin decoupling in solids, *J. Chem. Phys.*, 93, 7543–7560, 1990.
129. Sakellariou, D., Lesage, A., Hodgkinson, P. and Emsley, L., Homonuclear dipolar decoupling in solid-state NMR using continuous phase modulation, *Chem. Phys. Lett.*, 319, 253–260, 2000.
130. Lesage, A., Sakellariou, D., Hediger, S., Elena, B., Charmont, P., Steuernagel, S. and Emsley, L., Experimental aspects of proton NMR spectroscopy in solids using phase-modulated homonuclear dipolar decoupling, *J. Magn. Reson.*, 63, 105–113, 2003.
131. Pontryagin, L., Boltyanskii, B., Gamkredlidze, R. and Mishchenko, E., *The Mathematical Theory of Optimal Processes*, Wiley-Interscience, New York, 1962.
132. Bryson Jr, A. and Ho, Y.C., *Applied Optimal Control*, Hemisphere, Washington, DC, 1975.
133. Peirce, A.P., Dahleh, M. and Rabitz, H., Optimal control of quantum mechanical systems: existence, numerical approximations and applications, *Phys. Rev. A*, 37, 4950–4964, 1988.
134. Conolly, S., Nishimura, D. and Macovski, A., Optimal control solution to the magnetic resonance selective excitation problem, *IEEE Trans Med Imag, MI_5*, 106–115, 1986.
135. Rosenfeld, D. and Zur, Y., Design of adiabatic selective pulses using optimal control theory, *Magn. Reson. Med.*, 36, 401–409, 1996.
136. Reiss, T.O., Khaneja, N. and Glaser, S.J., Time optimal coherence-order-selective transfer of in-phase coherence in heteronuclear IS spin systems, *J. Magn. Reson.*, 154, 192–195, 2002.

137. Khaneja, N., Reiss, T., Luy, B. and Glaser, S.J., Optimal control of spin dynamics in the presence of relaxation, *J. Magn. Reson.*, 162, 311–319, 2003.
138. Skinner, T.E., Reiss, T.O., Luy, B., Khaneja, N. and Glaser, S.J., Application of optimal control theory to the design of broadband excitation pulses for high resolution NMR, *J. Magn. Reson.,* 163, 8–15, 2003.
139. Kehlet, C.T., Sivertsen, A.C., Bjerring, M., Reiss, T.O., Khaneja, N., Glaser, S.J. and Nielsen, N.C., Improving solid-state NMR dipolar recoupling by optimal control, *J. Am. Chem. Soc.*, 126, 10202–10203, 2004.
140. James, F. and Ross, M., MINUIT: a system for function mimimization and analysis of the parameter errors and correlations, *Comput. Phys. Commun.*, 10, 343–367, 1975.
141. Bak, M., Bywater, R.P., Hohwy, M, Thomsen, J.K., Adelhorst, K., Jakobsen, H.J., Sørensen, O.W. and Nielsen, N.C., Conformation of alamethicin in oriented phospholipid bilayers determined by ^{15}N solid-state nuclear magnetic resonance, *Biophys, J.*, 81, 1684–1698, 2001.
142. Kamihira, M., Vosegaard, T., Mason, A.J., Straus, S.K., Nielsen, N.C. and Watts, A., Structural constraints of bacteriorhodopsin in purple membranes determined by oriented-oriented sample solid-state NMR spectroscopy, *J. Struct. Biol.*, 149, 7–16, 2005.

12 Chemical Shift Tensors of Nucleic Acids: Theory and Experiment

John Persons, Paolo Rossi, and Gerard S. Harbison

CONTENTS

ABSTRACT We examine the experimental and computational evidence for the effect of conformation on the chemical shielding tensors of ^{31}P in the phosphodiester linkage and ^{13}C in the ribose/deoxyribose moieties of nucleic acids. The phosphodiester group has considerable flexibility and a relatively flat potential surface, resulting in a diversity of conformations in nucleic acids; however, theory shows that the chemical shielding tensors of the low-energy conformations are very similar, which accounts for the very small chemical shift dispersion encountered experimentally. Calculations also show that the orientation of the phosphate chemical shielding tensor is invariant within a couple of degrees over all conformations. In contrast, the ribose/deoxyribose ^{13}C chemical shielding tensors vary dramatically as a function of several conformational variables, and the major task of data analysis is integrating a large number of measurable isotropic chemical shifts and shift tensor elements into a unique conformational solution. We present computational data showing that the full chemical shielding tensors carry additional information not obtainable from isotropic chemical shifts alone.

KEY WORDS: *RNA, chemical shielding, tensor, ribose*

12.1 INTRODUCTION

In modern biomolecular nuclear magnetic resonance (NMR), solution- and solid-state experimental techniques and concepts are rapidly converging. In solution-state

NMR, the use of anisotropic media to determine orientational information relative to a molecular frame,[1] as well as the determination of relative tensor orientations by cross-correlated relaxation,[2] have made it essential to know the magnitude and orientations of the chemical shielding tensor (CST) of biologically relevant prosthetic groups. Conversely, magic-angle spinning and the use of very high fields have made it possible to obtain solution-like resolution in the solid state while using recoupling methods to retain anisotropic information.[3] In this chapter, we examine what is currently understood about the effect of conformation on shielding tensors in DNA and RNA, integrating experimental and computational results. Because conformation is our interest, we focus on the sugar and phosphate tensors; DNA and RNA bases are rigid. Readers interested in nucleic acid base tensors are referred to measurements of their ^{13}C principal values[4] and ^{15}N principal values[5] and to a variety of more sophisticated chemical shift/dipolar correlation[6] and computational methods[7] that have established the magnitudes and orientations of the CSTs in the standard DNA and RNA bases.

12.2 ^{31}P CHEMICAL SHIELDING TENSORS OF PHOSPHODIESTERS

The disappointingly limited dispersion of ^{31}P chemical shifts in DNA and RNA has been appreciated since the first solution NMR studies of oligomers (reviewed in ref. 8). Hopes that inclusion of the anisotropic part of the CST would increase this dispersion have, in general, not been borne out. The earliest ^{31}P powder patterns of polynucleotides,[9] which were done under unspecified conditions of humidity and electrolyte content and therefore of limited biological relevance, showed little difference between the ^{31}P tensor principal values between the four RNA homopolymers; principal values of $\sigma_{11} = 88 \pm 5$, $\sigma_{22} = 26 \pm 4$, and $\sigma_{33} = 110 \pm 4$ (with respect to 85% H_3PO_4) were obtained not just for the homopolymers but also for tRNA and salmon sperm DNA. Calf-thymus DNA and intact fd bacteriophage fell within the same range.[10] In our hands, calf-thymus DNA prepared with a sodium counterion, with about 0.5 M excess NaCl per base pair, and incubated at 79% relative humidity for several weeks (which causes most of the DNA to be in the A form) gave a slightly reduced chemical shielding anisotropy (CSA) at room temperature (Table 12.1); the span of the tensor from B-form DNA, with a lithium counterion and stored at 66% humidity, was reduced still further. It is likely, however, that in both these materials,

---

TABLE 12.1
Chemical Shielding Tensor Principal Values of Sodium-A-DNA and Lithium-B-DNA, Obtained at Room Temperature and High Humidity

	σ_{11} (ppm)	σ_{22} (ppm)	σ_{33} (ppm)	σ_i (ppm)	$\Delta\sigma$ (ppm)
A-Na⁺-DNA	83	18	104	1	154
B-Li⁺-DNA	70	12	85	1	126

the reduction in the CSA was a result of dynamical averaging and not an intrinsic property of the shielding tensor, particularly as a careful study of frozen cyclic adenosine monophosphate (AMP) solutions[11] indicated that such phosphodiester motions persist down to 248 K, even in that more constrained sample. The same study showed that hydrogen bonding to the phosphodiester also reduces the anisotropy, although it is doubtful that this has much biological relevance, as phosphodiesters in DNA and RNA are invariably exposed to water or hydrogen bonded to a bound protein.

The phosphodiester ^{31}P CST orientation was established by Herzfeld et al.,[12] using single-crystal solid-state NMR of barium diethyl phosphate. The least-shielded element σ_{11} lies approximately along the vector connecting the two phosphodiester oxygens, the most-shielded element σ_{33} lies along the vector between the two other oxygens, and the intermediate element lies orthogonal to these two. The principal values of barium diethyl phosphate are very similar to those of DNA, indicating that this is indeed a valid model compound. This orientation was used by Nall et al.[13] to interpret the ^{31}P spectra of oriented DNA fibers, showing them to be consistent with the standard structural model for A-DNA. Subsequently, we showed[4] that the orientation of the phosphodiester moiety in A and B-DNA could be determined using two-dimensional magic-angle spinning (MAS) spectroscopy[14,15] of oriented DNA samples, and that using the same technique,[16] proflavin intercalation induced a change in that orientation, leading to an "extended" conformation about that linkage. Levitt's group has subsequently used the same technique[17,18] to study the effect of other drugs on phosphodiester conformation. Recently, solution NMR of DNA oligomers in anisotropic media[19] has exploited the anisotropy of the ^{31}P chemical shift in a similar way, and relaxation studies in isotropic solution[20] have examined the cross-correlation between the C–H dipolar coupling and the ^{31}P CSTs; both of these techniques rely on the assumption of a ^{31}P chemical shielding tensor invariant in magnitudes and orientation.

The reasons for the experimental near-invariance of the isotropic ^{31}P chemical shift in DNA and RNA, which limits its usefulness in structural studies and in generating resolution in multidimensional solution NMR, and the (largely assumed) invariance of the anisotropic part of the shielding, which permits quantitative interpretation of oriented DNA spectra, can most easily be explored computationally. It is, in fact, rather surprising that so little computational effort has been devoted to this problem. Liang et al.[21] used Hartree-Fock methods, with the very small 6–31*G basis set, to study the torsional minima of dimethylphosphate (DMP), the simplest meaningful fragment of a phosphodiester linkage. The researchers found that the global minimum of DMP has C_2 symmetry and an approximate g^+g conformation about the P–O bonds. Numbering the atoms as they would be numbered in DNA, with θ_1 being the C3OPO and θ_2 the OPOC5 torsion angles, this conformation has $(\theta_1, \theta_2) = (\pm 75.1°, \pm 75.1°)$. Those authors also located a local minimum with C_2 symmetry, near but not at the tt conformation, at $(\theta_1, \theta_2) = (\pm 156.9°, \pm 156.9°)$, and one with C_1 symmetry, near canonical gt, at $(\theta_1, \theta_2) = (\pm 76.8°, \pm 189.4°)$. Ab initio computations at a similar level were used to assign the Raman spectrum of barium diethyl phosphate.[22] Finally, writers of an early paper[23] used gauge-invariant atomic orbital methods, with a very small basis set and no attempt to account for electron correlation, to examine differences between two conformers of DMP, the absolute shieldings obtained were highly unrealistic, and so it is doubtful how useful the work is; in any case, it found the differences between conformers to be negligible.

With a larger 6-311++G(2d,p) basis set and the B3LYP density functional to account for correlation, we have used the program GAMESS[24] to reexamine the energetics of the dimethylphosphate anion. We have also used the same basis set and function in conjunction with gauge-invariant atomic orbital methods and the program Gaussian 03[25] to probe the effect of conformation on the chemical shift. In addition to the g^+g global minimum, we find three, rather than two, additional local minima. These are depicted[26] in Figure 12.1, and the torsion angles and energies for the minima, along with calculated ^{31}P chemical shieldings, are shown in Table 12.2. The gt (Figure 12.1b) conformation lies only 5.6 kJ/mol above g^+g, and it is closer to the canonical torsion angles than the earlier work would have predicted. A new g^+g^+ conformation with C_s symmetry was detected (Figure 12.1c), and finally, the tt conformation (Figure 12.1d) was found to have rigorous C_{2v} symmetry; in the previous work, the C_{2v} conformation lay at a saddle point near a C_2-symmetric local minimum.

The relatively small energy difference between gt and g^+g indicate that the former should have a significant Boltzmann population at room temperature. However, these two conformations have only a 4-ppm isotropic chemical shift difference and a 9-ppm difference in chemical shielding anisotropies. In fact, as can be seen from Table 12.1, only the tt conformation, which is comparatively high in energy, differs substantially in chemical shift. It appears also that the major differences in chemical shifts reside in the σ_{11} element.

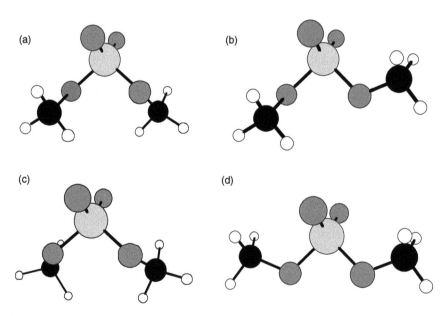

FIGURE 12.1 The four minimum energy conformations of the dimethylphosphate anion, computed as described in the text. (a) g^+g^- (C_2), (b) gt (C_1), (c) g^+g^+ (C_s), (d) tt (C_{2v}).

TABLE 12.2
Energies, Optimized Torsion Angles, and Chemical Shielding Tensor Principal Values, Computed as Described in the Text, for the Four Minimum Energy Conformations of the Dimethylphosphate Anion, a Model Compound for the DNA and RNA Phosphodiester Linkage

Conformation	E (kJ/mol)	θ_1°	θ_2°	σ_{11} (ppm)	σ_{22} (ppm)	σ_{33} (ppm)	σ_i (ppm)	$\Delta\sigma$ (ppm)
C_2	0	73.6	73.6	474.6	228.1	163.2	288.7	278.9
C_1	5.6	71.6	182.2	464.5	223.6	165.8	284.6	269.8
C_s	9.8	88.9	88.9	469.8	236.3	164.3	290.1	269.5
C_{2v}	13.5	180.0	180.0	449.9	228.1	169.0	279.6	255.2

In terms of orientation, the CST principal axes lie rigorously along the symmetry axes of the system in the C_{2v} conformation. In the C_2 conformation, the σ_{22} principal axis is constrained to the C_2 axis by symmetry; the σ_{11} lies within 1° of the unesterified O-O axis. This angle is less than the deviation of the two pairs of O-O axes from orthogonality. Similarly, the unconstrained axes in the C_s form fall approximately 2° off the C_{2v} symmetry axes formed by the phosphorus and the two unesterified oxygens; and the completely unconstrained axes of the C_1 form all lie less than 2° from these same reference frame axes. Thus, density-functional theory (DFT) calculations suggest that the usual assumption of an invariant ^{31}P CST orientation in DNA is entirely justified.

Both A-DNA and B-DNA tend to favor the gg^+ conformation that DFT finds to be lowest in energy (and is actually a mirror image of Figure 12.1a). Crystallographic studies show that torsion angles in A-DNA are in fact tightly clustered about the $(\theta_1, \theta_2) = (-73.6°, -73.6°)$ minimum. This is consistent with the well-known rigidity of the A-DNA structure and with the relatively non–motionally averaged room-temperature ^{31}P powder pattern. Interestingly, B-DNA, although it tends also to a gg^+ structure, shows a much larger scatter of torsion angles; most of the scatter in B-DNA phosphodiester conformations can be generated from gg^+ by synchronous opposite torsional motions about the P–O bonds. It is tempting to assign the reduced anisotropy of the B-DNA ^{31}P powder pattern, which we have observed in lithium DNA samples even at low humidity, to excursions of the phosphodiester group along this torsional trajectory. In any case, the evidence from DFT studies is that significant changes in the conformation about the phosphodiester group have relatively minor effects on the ^{31}P NMR spectrum, as might be expected from our experimental experience, and that therefore the assumption of an invariant phosphodiester shielding tensor is well founded.

12.3 CHEMICAL SHIELDING TENSORS OF RIBOSE/DEOXYRIBOSE

The major conformational variables of the ribose or deoxyribose moieties in DNA and RNA are the two parameters defining the pseudorotation — the pucker angle ϕ and the pucker amplitude q — and the torsion angles between 5 carbon and ring (the exocyclic angle γ) and the base and the ring (the glycosidic angle χ). The sugar pucker falls into two broad classes S and N, and the potential minima for the two classes fall near C2-*endo* and C3-*endo*, respectively. The exocyclic torsion angle has the usual approximate threefold potential expected for a bond between two sp$_3$ carbons, whereas the glycosidic angle can be divided very roughly into *syn* and *anti* configurations, in which the "business end" of the base (i.e., the Watson-Crick hydrogen bonding edge) faces toward or away from the sugar.

The first useful correlation of deoxyribose conformation with chemical shift was in fact established by solid-state CPMAS NMR; in 1989, Santos et al.[27] showed that A-DNA (in which most of the sugars have an N-type pucker) and B-DNA (in which all of the sugars are S-type) differ markedly in the isotropic ^{13}C chemical shifts at the 3 and 5 positions. This correlation was applied to RNA by Varani and Tinoco[28] and was widely adopted as a structural probe, although applicability of the shift correlation to ribonucleotides was not established in the initial work. Using two-dimensional NMR of oriented A- and B-DNA,[29] we were able to determine weak two-dimensional sideband intensities from the C3, C4, and C5 carbons, arising from the small CSAs of those carbons; then, using the assumption that the most-shielded tensor element was along the C–O bond, as determined by Grant and coworkers for sucrose,[30] we were able to establish the orientation of the C3, C4, and C5 bonds relative to the fiber axis in both A- and B-DNA and to show that it was in agreement with canonical structures.

DFT calculations by Sitkoff and Case[31] correlated the CSA of C1, C3, and C4 of deoxythymidine with the pseudorotation angle, which controls the sugar pucker. The calculations showed a considerable decrease in the chemical shielding aniso-tropy from N to S. Unfortunately, the work lacks computational detail — it is unclear how the pucker angle was adjusted and, specifically, whether the rest of the molecule was allowed to relax with the pseudorotation angle fixed — and as some pseudo-rotations have large internal steric effects, allowing the molecule to readjust after fixing the pseudorotation angle is probably necessary to achieve relevant results. Dejaegere and Case[32] were more explicit about the methods used to fix the pseudo-rotation angle — four-ring dihedrals were fixed (which constrains the pucker angle and amplitude) and the structure optimized before chemical shift calculation. The authors were able to reproduce computationally the experimental effects of ring pucker on the C3 and C5 chemical shifts; they also found that the C1 and C3 CSAs were influenced by ring pucker (see discussion pp. 311–313).

Dissatisfied with the haphazard approach of looking at individual chemical shift changes as a result of single perturbations, we conducted a more extensive study of the effect of global conformation on chemical shifts in RNA nucleotides and nucle-osides,[33] showing that the effect of sugar pucker and of the exocyclic angle can quite clearly be determined and the effect of the different conformational change

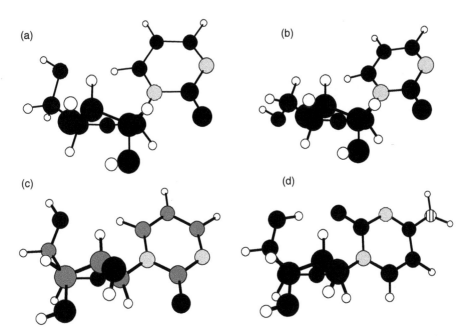

FIGURE 12.2 The four most common nucleotide conformations in RNA, computed as described in the text. (a) N-*anti-gg*, (b) N-*anti-gt*, (c) S-*anti-gg*, (d) S-*syn-gg*.

distinguished by NMR. We used a statistical technique known as discriminant analysis to find the linear combination of measured isotropic ^{13}C chemical shifts that optimally separate these different structural classes. Unfortunately, the same statistical method showed that torsion about the glycosidic angle could not be distinguished from ring pucker. This latter limitation, however, appeared from computational studies[34] to be partly a result of our inability to separately assign the signals for C2 and C3 in the solid state. Those studies also showed that DFT calculations at the B3LYP/6-311+G(2d,p) level of theory nicely reproduced experimental results.

To avoid this pitfall, all of our calculations began with known crystal structures. For the four common classes of nucleotide conformation depicted in Figure 12.2, two sets of coordinates were used; in the first, only the hydrogen positions were optimized (as these are not accurately measured by x-ray crystallography); in the second, the entire structure was optimized. The final optimized structures were all close to the crystal structures, and perhaps significantly, the agreement between computed isotropic chemical shifts and experiment was better for the fully optimized structures.

In Table 12.3, we show the principal values of the chemical shielding tensor for the four most common conformations of nucleotides. All calculations were done essentially as described in reference 34, using Gaussian 98[35] at the B3LYP/6-311+G(2d,p) level, both for optimization and for the NMR chemical shielding calculation (which used the gauge-invariant atomic orbital package). The computed shieldings were referenced to tetramethylsilane and an empirically determined correction of 4.6 ppm subtracted from all shifts. We believe that the need for this correction lies in

TABLE 12.3
Chemical Shielding Tensor Principal Values, Computed as Described in the Text, for the Four Most Common RNA Nucleotide Conformations

Conformation	σ_{11}	σ_{22}	σ_{33}	σ_i	$\dfrac{\sigma_{22}+\sigma_{33}}{2}-\sigma_{11}$
N-*anti-gg*					
C1'	75.431	100.295	117.359	97.695	33.396
C2'	67.649	84.757	90.224	80.877	19.842
C3'	34.108	74.459	112.255	73.607	59.249
C4'	44.890	98.175	119.008	87.358	63.702
C5'	26.934	80.066	87.437	64.812	56.818
N-*anti-gt*					
C1'	78.047	101.209	118.380	99.212	31.748
C2'	67.061	82.050	89.219	79.443	18.574
C3'	44.676	77.447	113.348	78.490	50.722
C4'	47.634	94.827	120.363	87.608	59.961
C5'	34.699	83.155	93.631	70.495	53.694
S-*anti-gg*					
C1'	57.379	97.950	123.120	92.817	53.156
C2'	55.999	83.429	111.606	83.678	41.519
C3'	65.075	81.071	96.412	80.853	23.667
C4'	68.469	97.504	108.729	91.567	34.648
C5'	40.326	80.887	86.653	69.288	43.444
S-*syn-gg*					
C1'	76.104	108.058	130.679	104.947	43.265
C2'	32.166	73.319	108.141	71.209	58.564
C3'	67.288	80.883	93.263	80.478	19.785
C4'	72.850	104.597	116.279	97.908	37.588
C5'	41.135	75.885	86.702	67.907	40.159

the peculiarities of the vibrational averaging of the chemical shift of tetramethylsilane. In Table 12.4 we include the principal values of the chemical shielding tensor for five less-common nucleotide conformations; as each conformation was represented by a single crystal structure, we used the heavy atom positions from that structure, in every case optimizing only the hydrogen positions. The citation for the parent crystal structure is given in the table. For C2'–C5', the most-shielded principal value is likely along the C–O bond, based on expectations from the sucrose single-crystal study of Grant's group.[30] We also include the isotropic chemical shift and a quantity we call σ_d — the difference between the most-shielded principal value and the average of the two least-shielded principal values. This quantity is not quite the same as the standard chemical shift anisotropy, whose unique principal value may be the most or least shielded depending on whether σ_{22} is closer to σ_{33} or σ_{11}, but it reflects the difference between the chemical shifts parallel and perpendicular to the C–O bond.

TABLE 12.4
Chemical Shielding Tensor Principal Values, Computed as Described in the Text, for Five Less Common RNA Nucleotide Conformations, with Heavy Atom Positions Derived from the References Cited

Conformation	σ_{11}	σ_{22}	σ_{33}	σ_i	σ_d	Reference
N-syn-gg						40
C1′	78.269	103.295	124.367	101.977	35.562	
C2′	56.847	79.100	92.361	76.102	28.884	
C3′	35.401	69.814	107.625	70.946	53.319	
C4′	46.521	95.195	119.714	87.143	60.934	
C5′	27.302	78.425	87.117	64.281	55.469	
S-anti-gt						41
C1′	52.583	96.714	120.896	90.064	56.222	
C2′	51.541	69.890	109.357	76.929	38.083	
C3′	61.721	84.312	87.055	77.696	23.963	
C4′	60.893	91.994	110.051	87.646	40.130	
C5′	42.307	77.156	95.792	71.751	44.167	
S-syn-gt						42
C1′	54.221	94.482	116.743	88.482	51.392	
C2′	37.527	75.170	114.668	75.788	57.392	
C3′	58.947	84.781	87.332	77.020	27.110	
C4′	68.860	92.452	111.014	90.775	32.873	
C5′	39.361	81.679	93.164	71.401	48.061	
S-syn-tg						43
C1′	66.417	100.413	118.752	95.194	43.166	
C2′	36.943	75.935	104.527	72.469	53.288	
C3′	51.780	76.932	100.213	76.309	36.793	
C4′	66.949	99.721	109.579	92.083	37.701	
C5′	24.505	77.041	96.947	66.164	62.489	
S-anti-tg						44
C1′	52.636	97.450	118.176	89.421	55.177	
C2′	50.401	76.009	108.035	78.148	41.621	
C3′	58.981	80.279	95.829	78.363	29.073	
C4′	62.125	99.442	113.843	91.803	44.518	
C5′	22.713	76.097	100.139	66.316	65.405	

It can be seen immediately from the data that the σ_d values of C2′ and C3′ distinguish between S and N puckers in an unambiguous way, irrespective of the glycosidic angle; in all nine conformations, the C2′ σ_d is less than 30 ppm for N puckers and over 35 ppm for S puckers. The C3′ σ_d shows almost exactly opposite behavior — over 50 ppm for N puckers, under 40 ppm for S puckers. It is tempting to attribute these effects to an increase in chemical shift anisotropy resulting from steric crowding on the *endo* side of the sugar ring, which primarily affects C3′ in N conformers and C2′ in S conformers. Our data also confirm Dejaegere and Case's observation of a difference in the C1′ CSA between N and S. In any case, this is clearly a circumstance in which knowledge of the full CST gives conformational information unavailable from the isotropic chemical shift alone. Although the data are more limited, we also note that the unusual *tg* exocyclic conformation is accompanied by an unusually high C5′ CSA, which may be diagnostic and useful, as the isotropic shifts in this conformer are not distinctive.

One final remark: although these chemical shift anisotropies are not large, it is not difficult to obtain a reasonably accurate estimate of their magnitude. At a ^{13}C frequency of 150 MHz, a 50-ppm tensor gives measurable MAS sideband intensities at spinning frequencies of 3 KHz and below. Whereas such slow MAS can clutter the spectrum with sidebands from the DNA bases, they are still in a regime in which they can be suppressed by sequences such as TOSS[36] or SELTICS[37] (two-dimensional sequences such as 2D-TOSS[38] or 2D PASS[39] can thus be used to separate isotropic and anisotropic information and allow determination of CSAs).

ACKNOWLEDGMENTS

G.S.H. wishes to thank the National Institutes of Health for funding, under grant number R01 GM 065252.

REFERENCES

1. Tjandra, N. and Bax, A., Direct measurement of distances and angles in biomolecules by NMR in a dilute liquid crystalline medium, *Science*, 278, 1111, 1997.
2. Ghose R. and Prestegard J.H., Improved estimation of CSA-dipolar coupling cross-correlation rates from laboratory-frame relaxation experiments, *J. Magn. Reson.*, 134, 308, 1998.
3. Chan, J.C.C. and Tycko, R., Recoupling of chemical shift anisotropies in solid-state NMR under high-speed magic-angle spinning and in uniformly ^{13}C-labeled systems, *J. Chem. Phys.*, 118, 8378, 2003.
4. Tang, P., Santos, R.A. and Harbison, G.S., Two-dimensional solid-state NMR studies of the conformation of oriented A-DNA, *Adv. Magn. Reson.*, 13, 225, 1989.
5. Hu, J.Z., Facelli, J.C., Alderman, D.W., Pugmire, R.J. and Grant, D.M., ^{15}N chemical shift tensors in nucleic acid bases, *J. Am. Chem. Soc.*, 120, 9863, 1998.
6. Anderson-Altmann, K.L., Phung, C.G., Mavromoustakos, S., Zheng, Z., Facelli, J.C., Poulter, C.D. and Grant, D.M., ^{15}N chemical shift tensors of uracil determined from ^{15}N powder pattern and ^{15}N-^{13}C dipolar NMR spectroscopy, *J. Phys. Chem.*, 99, 10454, 1995.
7. Czernek, J., An ab initio study of hydrogen bonding effects on the ^{15}N and ^{1}H chemical shielding tensors in the Watson-Crick base pairs, *J. Phys. Chem. A*, 105, 1357, 2001.
8. Gorenstein, D.G., Conformation and dynamics of DNA and protein-DNA complexes by ^{31}P NMR, *Chem. Rev.*, 94, 1315, 1994.
9. Terao, T., Matsui, S. and Akasaka, K., ^{31}P Chemical shift anisotropy in nucleic acids, *J. Am. Chem. Soc.*, 99, 6136, 1977.
10. Opella, S.J., Cross, T.A., DiVerdi, J.A. and Sturm, C.F., Nuclear magnetic resonance of the filamentous bacteriophage fd. *Biophys. J.*, 32, 531, 1980.

11. Gerothanassis, I.P., Barrie, P.J. and Tsanaktsidis, C., Observation of large solvent effects on the ^{31}P shielding tensor of a cyclic nucleotide, *Chem. Commun.*, 2639, 1994

12. Herzfeld, J., Griffin, R.G. and Haberkorn, R.A., ^{31}P chemical-shift tensors in barium diethyl phosphate and urea-phosphoric acid: model compounds for phospholipid head-group studies, *Biochemistry*, 17, 2711, 1978

13. Nall, B.T., Rothwell, W.P., Waugh, J.S. and Rupprecht, A., Structural studies of A-form sodium deoxyribonucleic acid: phosphorus-31 nuclear magnetic resonance of oriented fibers, *Biochemistry*, 20, 1881, 1981.

14. Harbison, G.S. and Spiess, H.W., Two-dimensional magic-angle spinning NMR of partially ordered systems, *Chem. Phys. Lett.*, 124, 128, 1986.

15. Harbison, G.S., Vogt, V.-D. and Spiess, H.W., Structure and order in partially-oriented solids: characterization by 2D-magic-angle-spinning NMR, *J. Chem. Phys.*, 86, 1206, 1987.

16. Tang, P., Juang, C.-L. and Harbison, G.S., Intercalation complex of proflavine with DNA: structure and dynamics via solid-state NMR, *Science*, 249, 70, 1990.

17. Song, Z., Antzutkin, O.N., Lee, Y.K., Shekar, S.C., Rupprecht, A. and Levitt, M.H., Conformational transitions of the phosphodiester backbone in native DNA: two-dimensional magic-angle-spinning 31P-NMR of DNA fibers, *Biophys. J.*, 73, 1539, 1997.

18. Lee, S.A., Grimm, H., Pohle, W., Scheiding, W., van Dam, L., Song, Z., Levitt, M.H., Korolev, N., Szabo, A. and Rupprecht, A., NaDNA-bipyridyl-(ethylenediamine)platinum (II) complex: structure in oriented wet-spun films and fibers, *Phys. Rev. E Stat. Phys. Plasmas, Fluids, Relat. Interdisc. Topics*, 62, 7044, 2000.

19. Wu, Z., Tjandra, N. and Bax, A., ^{31}P Chemical shift anisotropy as an aid in determining nucleic acid structure in liquid crystals, *J. Am. Chem. Soc.*, 123, 3617, 2001.

20. Richter, C., Reif, B., Griesinger, C. and Schwalbe, H. NMR Spectroscopic determination of Angles α and ζ in RNA from CH-dipolar coupling, P-CSA cross-correlated relaxation, *J. Am. Chem. Soc.*, 122, 12728, 2000.

21. Liang, C., Ewig, C.S., Stouch, T.R. and Hagler, A.T. *Ab initio* studies of lipid model systems. 1. Dimethyl phosphate and methyl propyl phosphate anions, *J. Am. Chem. Soc.*, 115, 1537, 1993.

22. Guan, Y. and Thomas, G.J. Vibrational analysis of nucleic acids. IV. Normal modes of the DNA phosphodiester structure modeled by diethyl phosphate, *Biopolymers*, 39, 813, 1996.

23. Ribas Prado, F., Giessner-Prettre, C., Pullman, B. and Daudey, J-P. *Ab initio* quantum mechanical calculations of the magnetic shielding tensor of phosphorus-31 of the phosphate group, *J. Am. Chem. Soc.*, 101, 1737, 1979.

24. Schmidt, M.W., Baldridge, K.K., Boatz, J.A., Elbert, S.T., Gordon, M.S., Jensen, J.J., Koseki, S., Matsunaga, N., Nguyen, K.A., Su, S., Windus, T.L., Dupuis, M. and Montgomery, J.A., The general atomic and molecular electronic structure system, *J. Comput. Chem.*, 14, 1347, 1993.

25. Frisch, M.J., Trucks, G.W., Schlegel, H.B., Scuseria, G.E., Robb, M.A., Cheeseman, J.R., Montgomery, Jr., J.A., Vreven, T., Kudin, K.M., Burant, J.C. et al., *Gaussian 03, Revision A.4*, Gaussian, Pittsburgh, PA, 2003.

26. Bode, B.M. and Gordon, M.S.J., MacMolPlt: a graphical user interface for GAMESS, *Mol. Graphics Modeling*, 16, 133, 1999.

27. Santos, R.A., Tang, P. and Harbison, G.S. Determination of the DNA sugar pucker using ^{13}C NMR spectroscopy, *Biochemistry*, 28, 9372, 1989.

28. Varani, G. and Tinoco, I., Jr., Carbon assignments and heteronuclear coupling constants for an RNA oligonucleotide from natural abundance carbon-13-proton correlated experiments. *J. Am. Chem. Soc.*, 113, 9349., 1991

29. Juang, C.L., Tang, P. and Harbison, G.S. Solid-state NMR of DNA, *Meth. Enzymol.*, 261, 256., 1995

30. Sherwood, M.H., Alderman, D.W. and Grant, D.M. Assignment of carbon-13 chemical-shift tensors in single-crystal sucrose, *J. Magn. Reson. A*, 104, 132, 1993.

31. Sitkoff, D. and Case, D.A. Theories of chemical shift anisotropies in proteins and nucleic acids, *Prog. NMR Spectroscopy*, 32,165, 1998.

32. Dejaegere, A.P. and Case, D.A. Density functional study of ribose and deoxyribose chemical shifts, *J. Phys. Chem. A*, 102, 5280, 1998.

33. Ebrahimi, M., Rossi, P., Rogers, C. and Harbison, G.S. Dependence of ^{13}C NMR chemical shifts on conformations of RNA nucleosides and nucleotides, *J. Magn. Reson.*, 150, 1, 2001.

34. Rossi, P. and Harbison, G.S. DFT calculation of ^{13}C chemical shifts in RNA nucleosides: structure-^{13}C chemical shift relationships, *J. Magn. Reson.*, 151, 1, 2001.

35. Frisch, M.J., Trucks, G.W., Schlegel, H.B., Scuseria, G.E., Robb, M.A., Cheeseman, J.R., Zakrzewski, V.G., Montgomery, Jr., J.A., Stratman, R.E., Burant, J.C. et al., *Gaussian 98, Revision A.7*, Gaussian, Pittsburgh PA, 1998.

36. Dixon, W.T., Schaefer, J., Sefcik, M.D., Stejskal, E.O. and McKay, R.A. Total suppression of sidebands in CPMAS carbon-13 NMR, *J. Magn. Reson.*, 49, 341, 1982.

37. Hong, J. and Harbison, G.S., Magic-angle spinning sideband elimination by temporary interruption of the chemical shift, *J. Magn. Reson. A*, 105, 128, 1993.

38. Kolbert, A.C. and Griffin, R.G., Two-dimensional resolution of isotropic and anisotropic chemical shifts in magic-angle spinning NMR, *Chem. Phys. Lett.*, 166, 87, 1990.

39. Antzutkin, O.N., Shekar, S.C. and Levitt, M.H. Two-dimensional sideband separation in magic-angle-spinning NMR, *J. Magn. Reson. A*, 115, 7, 1995.

40. Cody, V. and Kalman, T.I., Conformational analysis of 6-substituted uridine analogs. Crystal structures of uridine-6-thiocarboxamide and 6-cyanouridine, *Nucleosides Nucleotides*, 4, 587, 1985.

41. Berman, H.M., Hamilton, W.C. and Rousseau, R.J. Crystal structure of an antiviral agent 5-[N-(L-phenylalanyl)amino]uridine, *Biochemistry*, 12, 1809, 1973.

42. Mizuno, H., Kitamura, K., Miyao, A., Yamagata, Y., Wakahara, A., Tomita, K. and Ikehara, M. The structure of 8-thioxoadenosine monohydrate, *Acta Cryst. B*, 36, 902, 1980.

43. Wang, A.H.J., Dammann, L.G., Barrio, J.R. and Paul, I.C., Crystal and molecular structure of a derivative of 1,N6-ethenoadenosine hydrochloride. Dimensions and molecular interactions of the fluorescent ε-adenosine (εAdo) system, *J. Am. Chem. Soc.*, 96, 1205, 1974.

44. Yamagata, Y. and Tomita, K., Structure of 1-methyladenosine trihydrate, *Acta Cryst.*, C43, 2117, 1987.

13 Solid-State Nuclear Magnetic Resonance Studies of Alkali Metal Ions in Nucleic Acids and Related Systems

Gang Wu and Alan Wong

CONTENTS

ABSTRACT We describe a solid-state NMR approach to directly study alkali metal ions in DNA and related systems. The focus of this chapter is on the recent sold-state 23Na and 39K NMR results for characterization of ion coordination environments in a special four-stranded DNA structure known as the G-quadruplex. We also introduce a new NMR titration experiment to determine relative ion binding affinity in a site-specific manner. We demonstrate that solid-state alkali NMR can be used to localize alkali metal ions in biological systems, in a complementary fashion to the currently available X-ray crystallographic techniques.

KEY WORDS: *alkali metal ions, DNA, G-quadruplex, 23Na and 39K NMR*

13.1 INTRODUCTION

Alkali metal ions such as Na^+ and K^+ are crucial to life. The traditional view of the role that alkali metal ions play in the context of nucleic acid structures is that they simply act as counter ions. Recent discoveries from x-ray crystallographic, nuclear magnetic resonance (NMR), and molecular dynamics simulation studies have prompted new discussions regarding the function of alkali metal ions in nucleic acid structures.[1–8] Crystallography is at this time the most reliable experimental technique for localizing alkali metal ions in proteins and nucleic acids. However, direct detection of light alkali metal ions by diffraction methods is a challenging task.[9] It is particularly difficult to localize Na^+ ions, because a Na^+ ion has virtually an identical x-ray scattering power as a water molecule, making it difficult to distinguish the two species in electron density maps. In many cases, it is also possible that a binding site is shared between alkali metal ions and water. This partial occupancy problem often renders it impossible to identify Na^+ ions, even with the state-of-the-art diffraction techniques. In contrast, it is also quite difficult to obtain DNA/RNA single crystals that would diffract to subatomic resolution. In solutions, because the association between alkali metal ions and biological molecules is usually weak, ions often undergo rapid exchange between free and bound states. As a consequence, conventional alkali metal NMR cannot provide site-specific information. The magnetic relaxation dispersion method is by far the most direct NMR technique capable of providing some insightful information about DNA–ion interactions.[10–12] Another effective NMR method for studying alkali metal ions bound to nucleic acid molecules relies on the use of surrogate spin-1/2 NMR probes such as ^{15}N and ^{205}Tl.[13]

It is quite clear that new spectroscopic techniques that can yield direct information about alkali metal ion binding in nucleic acids are highly desirable. Several years ago, we proposed using solid-state NMR spectroscopy as a complementary approach to x-ray crystallography for detecting alkali metal ions in biological structures.[14] One important advantage of the solid-state NMR approach over crystallography is that samples in forms of polycrystalline powders or lyophilized solids can be directly used in the NMR experiment. In addition, solid-state NMR parameters are a sensitive measure of the local electronic structure at the ion-binding site. In the last decade, solid-state NMR methodologies for studying spin-1/2 nuclei such as ^{13}C and ^{15}N have been developed and used to determine biomolecular structures. It is perhaps time to pursue a similar degree of success in the study of alkali metal ions by solid-state NMR.

In this chapter we review the solid-state NMR methods that have been developed and used in our laboratory for studying alkali metal ions in nucleic acids and related systems. Focus will be placed on solid-state ^{23}Na and ^{39}K NMR results. This chapter is organized as follows. In Section 13.2, we give a brief description of the solid-state NMR techniques and corresponding spectral analyses for studying alkali metal ions in nucleotides. In Section 13.3, we describe solid-state NMR studies of a special type of four-stranded nucleic acids. In Section 13.4, we describe a potential application of solid-state ^{23}Na NMR in studying ion binding in double-stranded nucleic acids. Section 13.5 provides a summary and a prognosis for future directions.

13.2 SOLID-STATE NMR TECHNIQUES FOR ALKALI METAL IONS

All NMR-active alkali metal isotopes have an atomic nucleus, the nuclear spin quantum number of which is greater than 1/2 ($S > 1/2$). These nuclei are known as quadrupolar nuclei. In this chapter, we focus on ^{23}Na (natural abundance = 100%) and ^{39}K (natural abundance = 93%), which are both $S = 3/2$ nuclei. The quadrupolar nature of these alkali metal nuclei makes it necessary to use solid-state NMR techniques that are quite different from those developed for more familiar spin-1/2 nuclei such as ^{13}C and ^{15}N. Very often, the resolution in NMR spectra for quadrupolar nuclei is poor, as the quadrupole interactions are usually large (e.g., on the order of several megahertz for ^{23}Na and ^{39}K).

For half-integer (or noninteger) quadrupolar nuclei such as ^{23}Na and ^{39}K, one often detects only the central transition ($m = +1/2 \leftrightarrow m = -1/2$), which is independent of the quadrupole interaction to the first order if the language of perturbation theory is used. Under some favorable conditions, such as for quadrupolar nuclei with small quadrupole moments or at very high magnetic fields, reasonable resolution can be obtained in central-transition spectra simply by using a conventional technique known as magic-angle spinning (MAS).[15] However, the biggest problem in MAS spectra for half-integer quadrupolar nuclei is the incomplete averaging of second-order quadrupole interactions, which often results in residual line broadening much larger than the chemical shift dispersion. Under such circumstances, it is very difficult to obtain high-resolution NMR spectra from which chemical information can be extracted. In 1995, Frydman and coworkers[16,17] introduced a new solid-state NMR technique known as multiple-quantum MAS (MQMAS). Using this technique, it is possible to achieve complete removal of the second-order quadrupole interactions. Because MQMAS is a two-dimensional (2D) NMR experiment, it is more time consuming than a one-dimensional (1D) MAS experiment. However, the tremendous benefit in resolution enhancement by using MQMAS has made this technique a remarkably powerful tool in many applications.

In the discussion that follows, we use a hydrated Na salt of guanosine 5'-monophosphate, Na_2(5'-GMP)·7H_2O (in its orthorhombic form), as an example to illustrate how to extract NMR parameters from a combined analysis of 1D MAS and 2D MQMAS spectra. As seen in Figure 13.1, the ^{23}Na MAS spectrum for Na_2(5-GMP)·7H_2O does not show any recognizable feature that might be useful for extracting information about the number of Na^+ sites and the corresponding NMR parameters. In the ^{23}Na MQMAS spectrum, however, four spectral regions are clearly observed, indicating that there are four distinct Na^+ sites. From individual F_1 slice spectra, we can obtain for each Na site three ^{23}Na NMR parameters: isotropic chemical shift (δ_{iso}), quadrupole coupling constant (C_Q), and asymmetry parameter (η_Q). Finally, we can compare the simulated total MAS spectrum with the experimental spectrum. For orthorhombic Na_2(5-GMP)·7H_2O, we obtain the following data: Na1, $C_Q = 1.30$ MHz, $\eta_Q = 0.7$, $\delta_{iso} = -4.5$ ppm; Na2, $C_Q = 1.85$ MHz, $\eta_Q = 0.5$, $\delta_{iso} = -2.0$ ppm; Na3, $C_Q = 1.85$ MHz, $\eta_Q = 0.6$, $\delta_{iso} = -2.0$ ppm; Na4, $C_Q = 2.30$ MHz, $\eta_Q = 0.7$, $\delta_{iso} = -5.5$ ppm. Because the crystal structure for orthorhombic Na_2(5-GMP)·7H_2O is known,[18] an assignment of these NMR parameters to individual

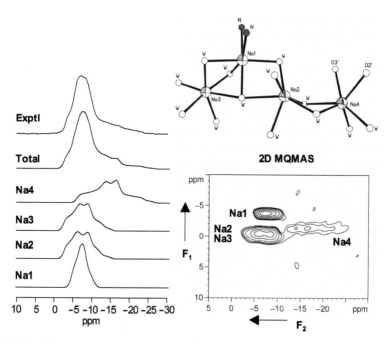

FIGURE 13.1 Experimental and simulated ^{23}Na MAS spectra for orthorhombic Na$_2$(5'-GMP)·7H$_2$O at 11.75 T (left) and partial crystal structure and two-dimensional ^{23}Na MQMAS spectrum of Na$_2$(5'-GMP)·7H$_2$O (right).

Na$^+$ sites can be made on the basis of a simple correlation between C_Q and ion-binding geometry. Na1 is assigned to the Na site with four water molecules and two hydroxyl groups from the ribose groups. Na2 and Na3 correspond to the two fully hydrated Na sites. Na4 is the site coordinated to four water molecules and two N7 nitrogen atoms from the pyrimidine moieties. A more reliable method for ^{23}Na spectral assignment is to use *ab initio* electric-field-gradient calculations.[19,20] It is important to point out that in the central-transition NMR spectra for half-integer quadrupolar nuclei, the center of mass of a peak (or a line shape) does not correspond to the true isotropic chemical shift. In this chapter, we use the symbol δ to indicate the center of mass for a peak (in parts per million) and δ$_{iso}$ for the isotropic chemical shift (also in parts per million); δ$_{iso}$ is always larger than δ. The difference between these two quantities is the isotropic second-order quadrupole shift, which approaches zero at the very high magnetic field limit.

Figure 13.2 shows ^{23}Na MAS spectra for a series of Na-nucleotides. For systems containing only a single Na$^+$ ion, spectral analysis is straightforward. For systems containing multiple Na$^+$ sites, we must use both 1D and 2D MQMAS spectra to extract ^{23}Na NMR parameters. A general procedure is similar to what has been described in the example of orthorhombic Na$_2$(5-GMP)·7H$_2$O. Several ^{23}Na MQMAS spectra for Na-nucleotides are shown in Figure 13.3. A summary of solid-state ^{23}Na NMR parameters observed for these Na-nucleotides is given in Table 13.1. The Na$^+$ ions in Na-nucleotides exhibit a variety of coordination environments. The oxygen

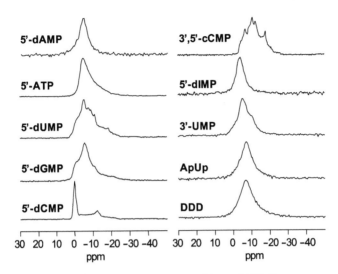

FIGURE 13.2 1D ^{23}Na MAS spectra for Na-nucleotides at 11.75 T.

ligand comes from several sources: water (W), phosphate (P), hydroxyl (S), and carbonyl groups (B). There are also several cases in which nitrogen atoms from the base are involved in the first-sphere Na$^+$ coordination. The coordination number for Na$^+$ ions varies from 5 to 6 to 7. To our knowledge, Table 13.1 represents the only collection of experimental solid-state ^{23}Na NMR data for Na-nucleotides. We antic- ipate that continuing accumulation of solid-state ^{23}Na NMR data will lead to a better understanding of the relationship between ^{23}Na NMR parameters and Na$^+$ binding structure. In addition to the work from our laboratory, there are also a few scattered reports in which solid-state ^{23}Na NMR was used to study DNA-related systems. Klinowski and coworkers studied the solid-state ^{23}Na NMR spectra of Na-DNA with and without competing species.[21] Ding and McDowell reported ^{23}Na MQMAS spec- tra for hydrated adenosine 5'-triphosphate (5'-ATP).[22] Madeddu demonstrated the importance of water content in analysis of solid-state ^{23}Na NMR spectra.[23] Frydman and coworkers recently introduced a new NMR experiment for Na$^+$ site assignment by using ^{1}H-^{23}Na dipolar interactions.[24]

As one of the low-γ quadrupolar nuclei, ^{39}K is notoriously difficult to study by NMR. The difficulty of solid-state ^{39}K NMR experiments is primarily twofold. First, the low ^{39}K NMR frequency not only makes the overall NMR sensitivity very low but also causes severe second-order quadrupole broadening. This is because the second-order quadrupole broadening is inversely proportional to the NMR frequency of the nucleus under observation. Second, the ^{39}K chemical shift range (ca. 200 ppm) usually is much smaller than the second-order quadrupole broadening, making the ^{39}K NMR spectra lack of site resolution. For these reasons, solid-state ^{39}K NMR experiments are often time consuming and produce very broad spectra at low and moderate magnetic fields (e.g., 11.75 T or lower). To date, solid-state ^{39}K NMR studies have been largely restricted to simple inorganic salts.[25]

One simple (but not necessarily easy) solution to improving NMR sensitivity is to perform solid-state ^{39}K NMR measurement at a high magnetic field. Figure 13.4

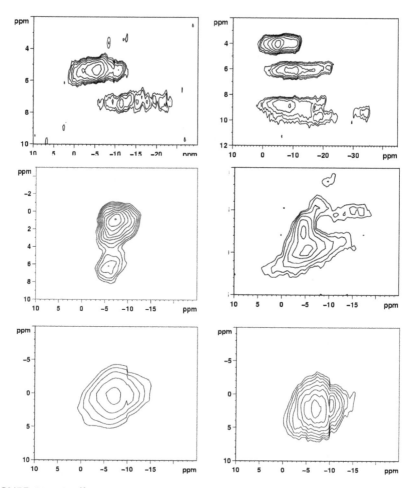

FIGURE 13.3 2D ^{23}Na MQMAS spectra for Na-nucleotides at 11.75 T.

shows the high-field (19.6 T) ^{39}K MAS NMR spectra for hydrated K$^+$ salts of adenosine 2′-monophosphate, K(2′-AMP)·1.5H$_2$O, and adenosine 5′-diphosphate, K(5′-ADP)·2H$_2$O. From an analysis of these spectra, we obtain the following ^{39}K NMR parameters: 2′-AMP, δ_{iso} = –55 ppm, C_Q = 1.85 MHz, η_Q = 0.80; 5-ADP, δ_{iso} = –105 ppm, C_Q = 2.05 MHz, η_Q = 0.25. The K$^+$ ion in 2′-AMP is coordinated to two phosphate oxygen atoms, two water molecules, and two hydroxyl groups from the ribose,[26] representing a typical environment for phosphate-bound K$^+$ ions. In comparison, the K$^+$ ion K(5′-ADP) is coordinated to seven ligands: four phosphate groups, one water, one hydroxyl, and a nitrogen atom (N3) from the adenine base.[27] The participation of N3(A) in K$^+$ coordination is quite unusual for K nucleotides. Perhaps this unusual coordination is responsible for the very shielded environment observed in K(5′-ADP), δ_{iso} = –105 ppm. Although the ^{39}K chemical-shift range is relatively small, the resolution observed in the ^{39}K MAS spectra at 19.6 T is

TABLE 13.1
Summary of Solid-State ^{23}Na NMR Parameters for Na-Nucleotides

System and Site	Na Coordination[a]	δ_{iso} (ppm)	C_Q (MHz)	η_Q
Na$_2$(5'-dAMP)·7H$_2$O				
Na1	6W	−0.3	1.6	0.7
Na$_2$(5'-ATP)·3H$_2$O				
Na1	5P, 1N	−1.5	1.9	0.6
Na2	5P, 1N	−1.5	1.8	0.5
Na3	4W, 2P	−2.5	1.2	0.7
Na4	1P, 4S	−7	2.1	0.5
Na(5'-CDP-Choline)				
Na1	1W, 1P, 2S, 1B, 1N	7	1.82	0.5
Na$_2$(3',5'-cCMP)·3H$_2$O				
Na1	2W, 1P, 2B, 1N	−8.5	1.4	0.5
Na2	2W, 2B, 1S, 1N	−1.65	2.75	0.2
Na$_2$(5'-dCMP)·7H$_2$O				
Na1	6W	1.0	0.85	1.0
Na2	5W	−2.6	2.2	1.0
Na$_2$(5'-GMP)·7H$_2$O				
(orthorhombic)				
Na1	4W, 2N	−4.55	1.3	0.7
Na2	6W	−2.25	1.8	0.6
Na3	6W	−2.25	1.8	0.5
Na4	4W, 2S	6.0	2.3	0.7
Na$_2$(5'dGMP)·4H$_2$O				
Na1	4W, 1S, 1B	0.8	1.82	0.9
Na2	2W, 2P, 1B	0.0	2.45	0.8
Na$_2$(5'-dIMP)·8H$_2$O				
Na1	4W, 2S	−1.0	1.38	0.6
Na$_2$(3'-UMP)·4H$_2$O				
Na1	6W	−1.8	2.05	0.3
Na2	3W, 1P, 3S	2.0	2.05	0.9
Na$_2$(5'-dUMP)·5H$_2$O				
Na1	3W, 1P, 1S, 1B	0.8	2.45	0.65
Na2	6W	1.1	1.75	0.95
Na3	2W, 1P, 1S, 1B	1.35	3.12	0.35
Na4	6W	2.8	2.42	0.70
ApUp				
Na1	—	≈0	≈2	—
[d(CGCGAATTCGCG)]$_2$ (DDD)				
Na1	—	≈0	≈2	—

[a] The oxygen ligands include water (W), phosphate (P), carbonyl (B), and hydroxyl (S). The nitrogen ligand is indicated by N.

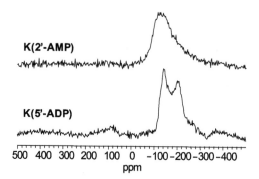

FIGURE 13.4 1D ^{39}K MAS spectra for K-nucleotides at 19.6 T. The ^{39}K chemical shift was referenced to the signal of KBr(s).

encouraging. Further studies are necessary to accumulate more solid-state ^{39}K NMR data for K-nucleotides.

13.3 ALKALI METAL ION BINDING IN G-QUADRUPLEXES

The focus of this section is on alkali metal ion binging in a special type of four-stranded nucleic acid structure known as the guanine-quadruplex (or G-quadruplex). The basic structural motif of the G-quadruplex is a guanine tetrad referred to as a G-quartet, in which four guanine molecules are linked by eight hydrogen bonds (Figure 13.5). The G-quartet model was first proposed in 1962 on the basis of x-ray

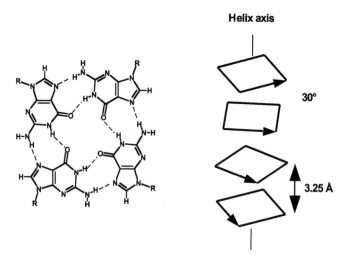

FIGURE 13.5 G-quartet (left) and helical stacking of G-quartets in 5′-GMP self-aggregates (right).

fiber diffraction data for guanylic acid.[28] Recent discoveries of the existence of the G-quartet motif in many biologically important systems such as telomeres, promoters of many genes, and sequences related to various human diseases have triggered tremendous research interest in this unusual type of nucleic acid structure. Now G-quartet structures can be found in such diverse areas as molecular biology, medicinal chemistry, supramolecular chemistry, and nanotechnology. For more detailed coverage of this exciting field, we refer readers to several excellent review articles.[29-35]

Alkali metal ions have been found to be critical for the formation, stability, and structural variation of G-quadruplexes. To date, localization of alkali metal ions in nucleic acids is largely restricted to crystallographic studies. In the next section, we review recent solid-state NMR results for direct detection of Na^+ and K^+ ions in G-quadruplex structures. Because our purpose is to validate solid-state NMR as a new method for detecting alkali metal ions, it is necessary to use crystallographic data as benchmarks for establishing solid-state NMR spectral signatures for various binding sites. For this reason, we first summarize what has been known from x-ray crystallographic studies about the mode of alkali metal ion binding in G-quadruplexes.

13.3.1 CRYSTALLOGRAPHIC DATA ON ALKALI METAL ION COORDINATION

Historically, x-ray fiber diffraction has made tremendous contributions to our understanding of the structures of nucleic acids. Not only was the original work of Watson and Crick[36] on the DNA double-helix based on x-ray fiber diffraction data, but so was the development of the G-quartet model by Gellert et al.[28] Early x-ray diffraction studies[28,37,38] have established that the 5′-GMP structure consists of stacking G-quartets in a right-handed helical fashion with 30° rotation and 3.25-Å advance between adjacent G-quartets (Figure 13.5). However, x-ray fiber diffraction data cannot yield atomic positions. The first atomic resolution crystal structure for a G-quadruplex was reported by Rich and coworkers in 1992[39] on a single crystal of *Oxytricha nova* telomere DNA repeat $d(G_4T_4G_4)$ (Oxy-1.5). The researchers observed a lose electron density, which they assigned to a K^+ ion, at the center of the Oxy-1.5 G-quadruplex structure between two G-quartets. This type of ion-binding site is in agreement with the ion-binding model first proposed by Pinnavaia and coworkers[40-42] and later confirmed by Laszlo and coworkers[43-45] using solution NMR techniques. The first crystallographic localization of Na^+ ions in a G-quadruplex was reported for $d(TG_4T)$.[46,47] As shown in Figure 13.6, $d(TG_4T)$ forms a parallel-stranded G-quadruplex. A total of seven Na^+ ions are observed to occupy the inner core of the quadruplex, thus giving a striking impression of an ion channel. The similarity between G-quadruplex structures and ion channel proteins was first pointed out by Feigon and coworkers.[48] This concept was also used by Davis and coworkers in the design of artificial ion channels.[49] As will be discussed in Section 13.3.3, the structural similarity between G-quadruplexes and ion channels will also lead to similar functions (ion selectivity) between these two classes of systems.

Table 13.2 summarizes the currently available high-resolution x-ray crystal structures for G-quadruplexes in which alkali metal ions are localized unambiguously. The G-quadruplex systems include DNA/RNA oligomers, DNA–ligand complexes,

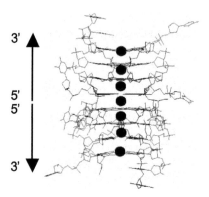

FIGURE 13.6 Na$^+$ ion binding in d(TG$_4$T) (PDB code: 352D).

TABLE 13.2
A Summary of X-Ray Crystal Structures for G-Quadruplexes

System[a]	Channel Cation[b]	Year	PDB Entry	Reference
d(G$_4$T$_4$G$_4$)	K$^+$ (1)	1992	1D59	39
d(TG$_4$T)				
S3	Na$^+$ (7)	1994	244D	46
S3	Na$^+$ (7)	1997	352D	47
[G1]$_{16}$·3K$^+$·Cs$^+$·4pic$^-$	K$^+$ (3), Cs$^+$ (1)	2000	—	49
[G1]$_{16}$·2Pb^{2+}·4pic$^-$	Pb^{2+} (2)	2000	—	50
r(UG$_4$U)	Sr^{2+} (4), Na$^+$(1)	2001	1J8G	51
d(G$_4$T$_4$G$_4$)/protein	Na$^+$ (4)	2001	1JB7	52
[G1]$_{16}$·2Ba^{2+}·4pic$^-$	Ba^{2+} (2)	2001	—	53
[G1]$_{16}$·2Sr^{2+}·4pic$^-$,	Sr^{2+} (2)	2001	—	54
[G1]$_{16}$·3Na$^+$·Cs$^+$·4pic$^-$	Na$^+$ (3), Cs$^+$ (1)	2002	—	55
d(G$_4$T$_4$G$_4$)	K$^+$ (5)	2002	1JPQ, 1JRN	56
d(TAG$_3$T$_2$AG$_3$T)-12mer	K$^+$ (2)	2002	1K8P	57
d[(AG$_3$(T$_2$AG$_3$)$_3$)]-22-mer	K$^+$ (2)	2002	1KF1	57
d(TG$_4$T)/daunomycin	Na$^+$ (3)	2003	1OOK	58
d(G$_4$T$_4$G$_4$)/acridine	K$^+$ (4)	2003	1L1H	59
(^{Br}dU)r(GAGGU)	Na$^+$ (8), Ba^{2+} (7)	2003	1J6S	60
r(U)(^{Br}dG)r(AGGU)	Na$^+$ (5)	2003	1MDG	61
r(U)(^{Br}dG)r(UGGU)	K$^+$ (3)	2003	1P79	62
d(gcGA[G]Agc)				
Low K$^+$	K$^+$ (3)	2004	1V3P	63
High K$^+$	K$^+$ (2)	2004	1V3O	63
d(TG$_4$T)				
S1	Na$^+$ (2), Tl$^+$ (5)	2004	1S45	64
S2	Na$^+$ (6), Tl$^+$ (6)	2004	1S47	64

[a] G1 = 5-*tert*-butyl-dimethylsilyl-2,3-*O*-isopropylidene guanosine.
[b] The numbers in parenthesis indicate the number of ions in the asymmetric unit.

FIGURE 13.7 Three classes of metal ion binding sites in G-quadruplex structures.

and self-assemblies of guanosine derivatives. On the basis of these crystal structures, the mode of ion binding in G-quadruplexes can be classified into three major categories (Figure 13.7).

13.3.1.1 Class I (Cavity Cations)

For this class of ion-binding sites, a cation is sandwiched between two quartets coordinating to eight donor atoms in a bipyramidal-antiprism geometry. Because the cation is inside a cavity generated by eight donor atoms, we refer to this class of cations as cavity cations. The cation may be either centered or off-centered between the two quartets. The twist angle between the two quartets may vary from 0° to 45°. Class I is the most common mode of ion binding in G-quadruplexes. In most cases, the cation is sandwiched between two G-quartets coordinating to eight O(6) atoms of guanine denoted as G_4-M^+-G_4 with typical M^+-O6(G) distances of 2.8–3.0 Å. Because monovalent cations are usually located in every cavity, the separation between two adjacent M^+ ions is approximately 3.4 Å. The electrostatic repulsion between the adjacent cations is minimized by the partially negative O6(G) atoms.

Examples also exist in which a cation is sandwiched between a G-quartet and a U-quartet $(G_4$-M^+-$U_4)$[60–62] and between a G-quartet and a T-quartet $(G_4$-M^+-$T_4)$.[64] In these cases, the cation is coordinated to four O6(G) and four O4(U/T) atoms. Both the M^+-O distance and M^+-M^+ separation are similar to those in G_4-M^+-G_4. Another subclass of class I cations is the cation sandwiched between a G-quartet and an A-quartet $(G_4$-M^+-$A_4)$. For this subclass, situations are more complicated. To date, three different types of G_4-M^+-A_4 binding have been reported, depending on the details in the A-quartet formation. If the A-quartet is formed through N6-H···N7 hydrogen bonds, the cation is coordinated to four N6 atoms of adenine, N6(A), with a distance of 3.3 Å. This is the case seen in $(^{Br}dU)r(GAGGU)$.[60] If the A-quartet is formed via N6-H···N3 hydrogen bonds (3.2 Å), the cation is then coordinated to four N1(A) atoms with much longer distances, 3.5–3.6 Å, forming a very distorted bipyramidal-antiprism geometry.[61] Another type of A-quartet is known as a water-mediated A-quartet, in which four adenine molecules are linked by two water molecules.[63] In this case, the cation is coordinated only to four O6(G) and two water molecules (not directly to adenines), denoted as G_4-M^+-$W_2(A_4)$. Under such a circumstance, although the cation is actually hexacoordinate, rather than octacoordinate, we treat it as a special case of class I. It is easy to appreciate that a cation would experience a more asymmetrical environment in G_4-M^+-A_4 than the situations in G_4-M^+-G_4, G_4-M^+-U_4, and G_4-M^+-T_4.

Class I cations also include some divalent cations such as Ba^{2+}, Pb^{2+}, and Sr^{2+}.[50,53,54] One special feature of divalent cation bindings is that they are located in every other cavity, whereas monovalent cations occupy every cavity. Therefore, the separation between two adjacent class I divalent cations in a quadruplex channel is approximately 6.5 Å. This large cation–cation separation is necessary to reduce the strong electrostatic repulsion between divalent cations.

13.3.1.2 Class II (In-Plane Cations)

Class II describes cations residing within a quartet plane. Because the distance between diagonal O(6) atoms in a G-quartet is approximately 4.5 Å, a Na^+ ion of ionic radius of 0.95 Å can fit into the central void. Cations such as K^+ (1.33 Å) and NH_4^+ (1.43 Å) are too large to fit into this site. Two subclasses of Class II cations exist according to the arrangement of axial ligands. The common type is to have one axial ligand forming a square-pyramidal geometry. Very often the fifth ligand is a water molecule (denoted as G_4-Na^+-W). For example, the end Na^+ ions in all three structures of $d(TG_4T)$[46,47,64] belong to this subclass with a Na^+-O_W distance of 2.4 Å. Similarly, a Na^+ ion can also fit into a U-quartet, U_4-Na^+-W.[51] The second subclass is to have multiple axial ligands. For example, the end Na^+ ions in a $d(G_4T_4G_4)$-protein complex are within the G-quartet plane, also coordinating to three additional ligands: two O2(T) atoms and one water.[52] Clearly, class II cations are usually found at the end of a G-quadruplex.

13.3.1.3 Class III (Loop Cations)

Some G-quadruplexes have diagonal loops. Class III describes cations residing between the end G-quartet and the loop. To date, the only occasion in which loop cations are found is in the K^+ form of $d(G_4T_4G_4)$, where two K^+ ions are located outside the end G-quartet interacting with two O2(T) atoms from the T_4 loop and two water molecules.[56] This type of loop K^+ ion can be displaced by drug molecules.[59]

13.3.2 NMR Spectral Signatures for Bound Na^+ and K^+ Ions

It becomes quite clear from Section 13.3.1 that there exists a variety of ion-binding sites in a G-quadruplex structure that may be occupied by Na^+ and K^+ ions. To establish solid-state NMR as a useful tool for detecting Na^+ and K^+ ions, the first step is to obtain NMR spectral signatures for Na^+ and K^+ ions in different binding sites. Because 5′-GMP is one of the earliest guanosine derivatives examined by x-ray fiber diffraction and solution NMR, it is a natural candidate for our solid-state ^{23}Na NMR study.[65]

Figure 13.8 shows a ^{23}Na MAS NMR spectrum for 5′-GMP. This 5′-GMP sample is prepared in its hexagonal form, in which stacking G-quartets are present.[28,37,38] This is different from the orthorhombic form of Na_2(5′-GMP)·$7H_2O$, discussed in Section 13.2. As seen in Figure 13.8, three classes of Na^+ ions are clearly

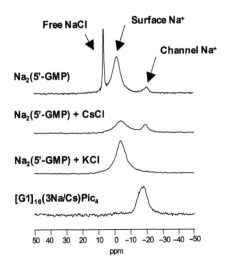

FIGURE 13.8 ^{23}Na MAS spectra of 5′-GMP and [G1]$_{16}$·3Na$^+$·Cs$^+$·4pic$^-$ at 11.75 T.

distinguished. The signal at δ 7 ppm is a result of excessive NaCl salt. The Na$^+$ ions bound to the peripheral phosphate groups (referred to as the surface Na$^+$ ions) give rise to a ^{23}Na NMR signal at approximately δ −4 ppm. The Na$^+$ ions residing inside the G-quadruplex channel (referred to as the channel Na$^+$ ions) exhibit a signal at δ −19 ppm. The above spectral assignment is confirmed by examining changes in solid-state ^{23}Na NMR spectra when CsCl and KCl salts are added to the 5′-GMP sample. When CsCl is added to the sample, the signal intensity for the surface Na$^+$ ions is significantly reduced, indicating that Cs$^+$ ions do not enter the G-quadruplex cavity but replace partially the surface Na$^+$ ions (Figure 13.8). In contrast, because K$^+$ ions are preferred over Na$^+$ ions to occupy the channel sites, the ^{23}Na NMR signal for the channel Na$^+$ ions disappears completely when KCl is added (Figure 13.8). These observations are consistent with the known affinity of the G-quadruplex formed by 5-GMP for Cs$^+$ and K$^+$ ions.[40–42] An unambiguous confirmation of our spectral assignment comes from the ^{23}Na NMR spectrum for 5-*tert*-butyl-dimethyl-silyl-2,3′-*O*-isopropylidene guanosine (G1).[55] This compound can extract Na$^+$ and Cs$^+$ ions from aqueous phase into organic phase to form a lipophilic G-quadruplex structure consisting of 16 equivalents of G1 (Figure 13.9). Because all Na$^+$ ions in [G1]$_{16}$·3Na$^+$·Cs$^+$·4pic$^-$ reside inside the G-quadruplex channel as class I ions, the observation of a ^{23}Na NMR signal centered at δ −19 ppm for this compound establishes unequivocally the spectral characteristics for channel Na$^+$ ions. At this point, we should mention that in the first solid-state ^{23}Na NMR study for G-quadruplexes, by Rovnyak et al.,[66] the reported spectral assignment was erroneous.

Although conventional 1D ^{23}Na MAS has achieved spectral separation for the major types of Na$^+$ ions in a G-quadruplex, higher resolution is needed to detect Na$^+$ ions with subtle difference. As we demonstrate in Section 13.2, MQMAS NMR is well suited for this purpose. Figure 13.10 shows 2D ^{23}Na MQMAS spectra for 5′-GMP and [G1]$_{16}$·3Na$^+$·Cs$^+$·4pic$^-$. For 5′-GMP, the 2D spectrum reveals the presence of two groups of channel Na$^+$ ions. Interestingly, one of the two spectral

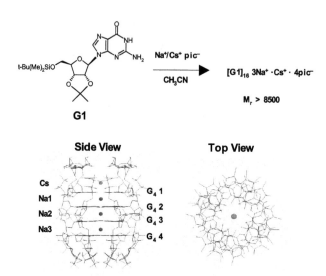

FIGURE 13.9 Self-association of **G1** (top) and crystal structure of $[G1]_{16} \cdot 3Na^+ \cdot Cs^+ \cdot 4pic^-$ (bottom). pic^- = picrate.

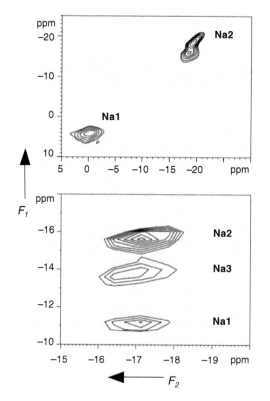

FIGURE 13.10 2D ^{23}Na MQMAS spectra of 5′-GMP (top) and $[G1]_{16} \cdot 3Na^+ \cdot Cs^+ \cdot 4pic^-$ (bottom).

regions (Na2) is tilted from the F_2 axis, indicating the presence of multiple Na^+ ions with a distribution of chemical shifts. For $[G1]_{16} \cdot 3Na^+ \cdot Cs^+ \cdot 4pic^-$, three well-resolved signals are observed, in agreement with the crystal structure.[55] Furthermore, because the sample is in polycrystalline form, the three signals exhibit line shapes parallel to the F_2 axis, a spectral feature associated with well-defined Na^+ coordination environment. $[G1]_{16} \cdot 3Na^+ \cdot Cs^+ \cdot 4pic^-$ is a nice example illustrating the remarkable resolution improvement of 2D MQMAS over 1D MAS spectra. In this case, the resolution limit in the F_1 axis is less than 0.5 ppm. From an analysis of both 1D and 2D spectra, we obtain the following ^{23}Na NMR parameters: for 5'-GMP, surface Na^+ ions, $C_Q \approx 1.6$ MHz, $\eta_Q \approx 1$, $\delta_{iso} \approx 3$ ppm; channel Na^+ ions, $C_Q \approx 1.1$ MHz, $\eta_Q \approx 1$, $\delta_{iso} \approx -18$ ppm. For $[G1]_{16} \cdot 3Na^+ \cdot Cs^+ \cdot 4pic^-$, Na1, $\delta_{iso} = -12.8 \pm 0.2$ ppm, $C_Q = 1.65 \pm 0.05$ MHz, $\eta_Q = 0.60 \pm 0.05$; Na2, $\delta_{iso} = -16.5 \pm 0.2$ ppm, $C_Q = 1.35 \pm 0.05$ MHz, $\eta_Q = 0.80 \pm 0.05$; Na3, $\delta_{iso} = -15.0 \pm 0.2$ ppm, $C_Q = 1.70 \pm 0.05$ MHz, $\eta_Q = 0.60 \pm 0.05$.

We have also obtained a ^{23}Na MAS NMR spectrum for a Na^+ salt of polyinosinic acid, poly(rI) (Figure 13.11). Poly(rI) is a high–molecular weight RNA homopolymer (>100 kDa) and is known to form four-stranded helix in both solution and solid states.[67–69] Our ^{23}Na NMR observation proves that the Na^+ ion binding in poly(rI) is similar to that observed in G-quadruplexes. That is, each Na^+ ion in poly(rI) is sandwiched between two adjacent hypoxanthine-quartets and belongs to class I.

After obtaining the ^{23}Na NMR spectral signatures for Na^+ ions in G-quadruplexes, we now apply this NMR approach to determining the number and coordination environment of Na^+ ions in a DNA dodecamer, $d(G_4T_4G_4)$ (Oxy-1.5).[70] Oxy-1.5 is perhaps one of the most studied telomeric oligonucleotides. In solution, the Na^+ form of Oxy-1.5 adopts a symmetric foldback quadruplex structure consisting

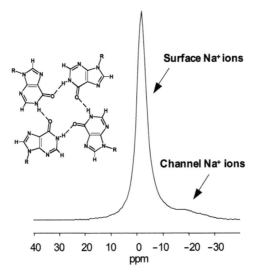

FIGURE 13.11 1D ^{23}Na MAS spectrum of a Na salt of poly(rI) (molecular weight > 100 kDa) at 11.75 T.

of four stacked G-quartets and two diagonal thymine loops.[71] The crystal structure of the K+ form of Oxy-1.5 showed that the overall structure of Oxy-1.5 in the solid state is identical to that found in solution. Feigon and coworkers also examined the structural details between various forms of Oxy-1.5 in solution and concluded that the nature of monovalent cations present in the solution (Na+, K+, and NH$_4$+) does not affect the overall fold of Oxy-1.5.[72] However, there is also strong evidence indicating that the details of ion binding in Oxy-1.5 may be different, depending on the type of ions present. For example, five K+ ions were found in the K+ form of Oxy-1.5,[56] whereas only three NH$_4$+ ions were observed in the NH$_4$+ form.[73] Because the crystal structure for the Na+ form of Oxy-1.5 has not been reported, the question regarding the mode of Na+ binding in Oxy-1.5 has remained unanswered.

Figure 13.12 shows solid-state ^{23}Na NMR spectra for Oxy-1.5. Both 1D and 2D spectra for Oxy-1.5 are similar to those observed for 5'-GMP. The most important feature in the 2D MQMAS spectrum for Oxy-1.5 is the fine spectral feature associated with the channel Na+ ions. In particular, the channel Na+ ions give rise to two distinct NMR signals with a peak volume ratio of approximately 2:1. In the dimeric structure of Oxy-1.5, there are four G-quartets stacking on top of one another, creating three coordination pockets inside the quadruplex channel. The simplest interpretation for the ^{23}Na NMR observation is that the signal with the double intensity is associated with two Na+ ions in the symmetry-related outer pockets, and

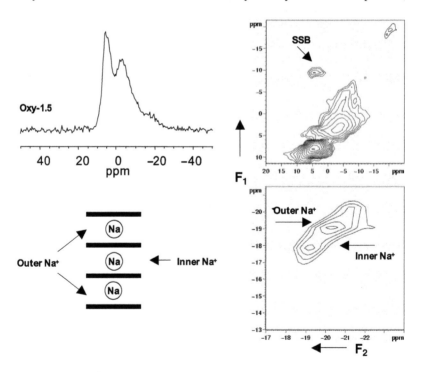

FIGURE 13.12 1D ^{23}Na MAS (left) and 2D MQMAS (right) spectra of the Na form of d(G$_4$T$_4$G$_4$). SSB = spinning sideband.

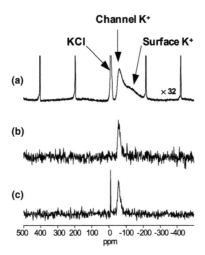

FIGURE 13.13 1D ^{39}K MAS spectra for 5'-GMP (a), $[G1]_{16}\cdot3K^+\cdot Cs^+\cdot4pic$ (b), and guanosine/KCl (c) at 19.6 T. The ^{39}K chemical shift was referenced to the signal of KBr(s).

the less intense signal is attributed to a Na$^+$ ion in the central pocket, in analogy to the NH$_4^+$ coordination in the NH$_4^+$ form of Oxy-1.5.[73] Analyses of the 1D and 2D spectra yield the following ^{23}Na NMR parameters for the channel Na$^+$ ions in Oxy-1.5: outer site, C_Q = 1.2 ± 0.2 MHz, η_Q = 0.4 ± 0.2 and δ_{iso} = −19.0 ± 0.2 ppm; central site, C_Q = 0.9 ± 0.2 MHz, η_Q = 1.0 ± 0.2 and δ_{iso} = −17.5 ± 0.2 ppm. These NMR parameters are consistent with those observed for the channel Na$^+$ ions in $[G1]_{16}\cdot3Na^+\cdot Cs^+\cdot4pic^-$ mentioned earlier, indicating a similar Na$^+$ coordination between the two systems. In $[G1]_{16}\cdot3Na^+\cdot Cs^+\cdot4pic^-$, each of the channel Na$^+$ ions is coordinated to eight O6(G) atoms from two adjacent G-quartets, having a bipyramidal antiprism coordination geometry with an averaged Na-O distance of 2.81 Å. As also seen from Figure 13.12, each of the two signals for the channel Na$^+$ ions exhibits a line shape parallel to the F_2 axis, indicating that the coordination environment for the channel Na$^+$ ions is well defined. In contrast, the signal from the surface Na$^+$ ions exhibits a diffuse line shape, indicating that the chemical environment for the surface Na$^+$ ions is not homogeneous.

Now we turn our attention to solid-state ^{39}K NMR for G-quadruplexes. To obtain the ^{39}K NMR signature for K$^+$ ions bound to a G-quadruplex structure, we examine three guanosine derivatives: 5'-GMP, G1, and guanosine.[74] Similar to the case in ^{23}Na NMR, the ^{39}K MAS NMR spectrum for 5'-GMP shown in Figure 13.13 also exhibits three groups of signals: δ–9 ppm (free KCl salt), δ –45 ppm (channel K$^+$ ions), and δ –120 ppm (surface K$^+$ ions). The ^{39}K MAS spectrum of $[G1]_{16}\cdot3K^+\cdot Cs^+\cdot4pic^-$ exhibits a peak centered at δ –45 ppm with a line width of approximately 1.0 kHz. Again, because all K$^+$ ions in this system are inside the channel,[49] the spectral assignment for channel K$^+$ ions (class I) is unambiguous. The detailed features in the line shape indicate an overlap of several central-transition powder spectra. However, in the absence of ^{39}K MQMAS data, it is not possible to extract accurate values of C_Q for the three crystallographically distinct K$^+$ sites.

Guanosine is also known to form a viscous gel in aqueous solution in the presence of KCl, indicating formation of a highly ordered molecular assembly. An early x-ray fiber diffraction study confirmed that the guanosine aggregates have a G-quadruplex structure in which two adjacent G-quartets are separated by 3.4 Å with a twist of 45° forming a continuous helix.[75] The ^{39}K MAS spectrum of guanosine shows a signal identical to that for $[G1]_{16}\cdot 3K^+\cdot Cs^+\cdot 4pic^-$. In addition, a very sharp peak is observed at δ–9 ppm, which arises from a small excess of KCl. This was confirmed by examining a solid sample of pure KCl salt. These observations indicate that, in the self-aggregates of guanosine, the K^+ ions reside exclusively inside the G-quadruplex channel in a similar fashion as those in $[G1]_{16}\cdot 3Na^+\cdot Cs^+\cdot 4pic^-$. Analyses of the ^{39}K MAS NMR spectra yield the following ^{39}K NMR parameters: 5'-GMP, $\delta_{iso} \approx -40$ ppm and $C_Q < 1.3$ MHz; $[G1]_{16}\cdot 3K^+\cdot Cs^+\cdot 4pic^-$, $\delta_{iso} = -42$ ppm and $C_Q < 0.7$ MHz; guanosine, $\delta_{iso} = -45$ ppm and $C_Q \approx 0.7$–0.8 MHz.

As mentioned in Section 13.2, the K^+ coordination environment in K(2'-AMP) is typical of phosphate-bound K^+ ions. Indeed, the ^{39}K NMR signal observed for the surface K^+ ions in 5'-GMP is in excellent agreement with that for K(2'-AMP). It is expected that ^{39}K MAS spectra for G-rich oligonucleotides would resemble that observed for 5'-GMP. However, it is important to point out that the observed ratio between the signal areas for the surface and channel K^+ ions is approximately 3:2 in the ^{39}K MAS spectrum for 5'-GMP, which is much smaller than 8:1, a ratio expected for a 5'-GMP self-assembly saturated with K^+ ions. As demonstrated by Laszlo and coworkers,[43–45] the G-quadruplex channel strongly favors K^+ ions, whereas the doubly charged phosphate group of 5'-GMP prefers Na^+ ions over K^+ ions. As the sample of 5'-GMP was prepared in the presence of both K^+ and Na^+ ions, the G-quadruplex channel is saturated by K^+ ions, but there is still a considerable amount of Na^+ ions that are bound to the phosphate groups. This selective affinity for the channel site results in a reduced number of surface K^+ ions in the 5'-GMP sample. A more detailed discussion on ion affinity will be presented in the next section.

As noted in Section 13.3.1, the K^+ coordination environment in the G-quadruplex channel is very similar to that of the K^+ ions recently observed inside the selectivity filter of the KcsA K^+ channel protein.[76–78] It is likely that these two types of K^+ ions have very similar ^{39}K NMR parameters. Although KcsA (~70 kDa) is considerably larger than the largest system examined here, $[G1]_{16}\cdot 3K^+\cdot Cs^+\cdot 4pic^-$ (8.7 kDa), our results indicate that the ^{39}K NMR sensitivity at 19.6 T is adequate for studying K^+ channel proteins.

13.3.3 DETERMINATION OF SITE-SPECIFIC ION AFFINITY

As discussed in the previous section, the two major types of alkali metal ion binding sites in an oligonucleotide G-quadruplex (class I channel and surface sites) have very different chemical environments. As a consequence, these binding sites must also exhibit different affinities for various alkali metal ions. In solution, because alkali metal ions often undergo rapid exchange between all possible binding sites, it is generally difficult to obtain site-specific information about ion affinity by conventional spectroscopic techniques. For this reason, although extensive

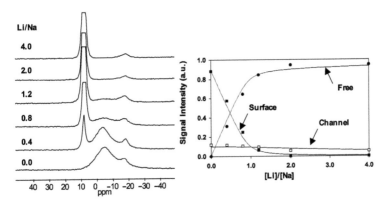

FIGURE 13.14 Experimental [23]Na MAS spectra for 5′-GMP samples as a function of the amount of Li[+] and Na[+] ions.

thermodynamic data on G-quadruplex stability are available in the literature,[79,80] none of the previous studies has been able to address the question of alkali metal ion affinity for a G-quadruplex structure in a site-specific manner.

We have recently developed a solid-state NMR method for determining relative ion affinity for different binding sites simultaneously.[81] The method is based on a titration experiment in which competitive equilibria between Na[+] and M[+] (M = Li, K, Rb, NH$_4$, and Cs) ions are monitored by ^{23}Na NMR. Because Na[+] ions at different binding sites exhibit clearly resolved signals, following individual ^{23}Na NMR peak intensities as a function of the added M[+] salt would yield a thermodynamic equilibrium constant (K_{eq}) for each of the binding competitions between Na[+] and M[+] ions. Figure 13.14 shows the results from a titration experiment in which a 5′-GMP sample is titrated with Li salt. From these data, equilibrium constants are determined for the channel and surface sites (Table 13.3). Figure 13.15 depicts the dependence of

---

TABLE 13.3
Thermodynamic Properties (K_{eq} and $\Delta G°$) Determined for the G-Quadruplex Structure Formed by 5′-GMP Self-Assembly at 25°C[a]

	Li[+]	Na[+]	K[+]	NH$_4$[+]	Rb[+]	Cs[+]
Ionic Radius (Å)	0.76	0.95	1.33	1.43	1.48	1.69
Channel Site						
K_{eq}	0.3(1)	1	25(5)	20(2)	1.8(1)	0.05(5)
$\Delta G°$ (kcal mol[-1])	0.7(2)	0	−1.9(4)	−1.8(2)	−0.3(1)	1.8(4)
Surface Site						
K_{eq}	40(10)	1	0.015(10)	9(2)	0.15(5)	0.20(5)
$\Delta G°$ (kcal mol[-1])	−2.2(1)	0	2.5(6)	−1.3(2)	1.1(3)	0.9(4)

[a] The numbers in parenthesis indicate the errors.

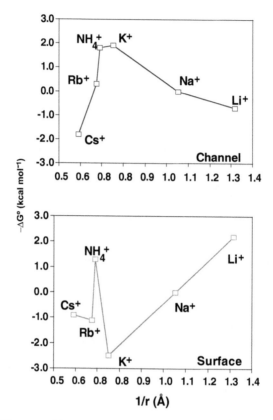

FIGURE 13.15 Diagrams of free energy difference versus reciprocal ionic radius for binding of monovalent cations to 5′-GMP structure.

$\Delta G°$ on the reciprocal ionic radius for the two ion binding sites in 5′-GMP. These data indicate that the affinity of monovalent cations for the G-quadruplex channel cavity site follows the order of Eisenman sequence V.[82] The channel cavity site of the G-quadruplex structure prefers K^+ and NH_4^+ over Na^+, whereas Cs^+ is much less favored than Na^+. We obtain the following ion affinity sequence for the channel cavity site (Class I): $K^+ > NH_4^+ > Rb^+ > Na^+ > Li^+ > Cs^+$. This ion affinity sequence is in agreement with the qualitative ranking first reported for 5′-GMP by Pinnavaia et al.:[40–42] $K^+ > Rb^+, Na^+ \gg Li^+, Cs^+$. By studying the melting temperatures of Br^8-guanosine gels in various salt solutions, Chantot and Guschlbauer[83] also observed a stability sequence: $K^+ \gg Rb^+ > NH_4^+ > Na^+ > Cs^+ > Li^+$. Several telomeric DNA sequences show a similar trend in their cation-induced stability, $K^+ > Na^+ > Cs^+$.[29] However, the sequence of ion affinity for the 5′-GMP surface site is drastically different from that for the channel cavity site. According to the observed $\Delta G°$ values, the order of cation affinity for the 5′-GMP surface site is $Li^+ > NH_4^+ > Na^+ > Cs^+ > Rb^+ > K^+$. These results clearly indicate that the overall stability of a G-quadruplex structure depends only on the affinity of the monovalent cations present in solution for the quadruplex cavity site.

Because no site-specific ion affinity data have been reported with other techniques, it is not possible at this time to make an independent evaluation of the thermodynamic data derived from our solid-state NMR experiments. However, a qualitative examination for the affinity difference between Na^+ and K^+ ions for G-quadruplexes is possible. From the solid-state ^{23}Na NMR titration experiment, the free-energy difference between K^+ and Na^+ binding to the G-quadruplex cavity site is determined to be -1.9 kcal mol^{-1}. Raghuraman and Cech[84] have analyzed the folding of $d(T_4G_4)_4$ (Oxy-4) in the presence of 50 mM NaCl and 50 mM KCl at 37°C. Their estimated $\Delta G°$ value for selective binding of K^+ over Na^+ was -0.8 kcal mol^{-1}, assuming that three cations are sandwiched between the 4 G-quartets in folded Oxy-4 quadruplex. Scaria et al.[85] have also studied the thermal melting behavior of $d(G_3T_4G_3)$ in the presence of salt. They showed that at 25°C, the free energy of the quadruplex was 4.2 kcal mol^{-1} lower in 100 mM KCl than in 100 mM NaCl. Assuming that two cations are involved in $d(G_3T_4G_3)$, $\Delta G°(Na^+\rightarrow K^+)$ was approximately -2.1 kcal mol^{-1}. Balagurumoorthy and Brahmachari[86] have determined thermodynamic data for cation binding stability in both intramolecular and hairpin dimer structures by performing van't Hoff analyses of CD melting profiles. For $d(T_2AG_3)_4$, the authors obtained a free-energy difference of -2.7 kcal mol^{-1} between K^+ and Na^+ binding, which corresponds to a value of -1.3 kcal mol^{-1}, if two cations are assumed to reside between the three G-quartets. For the hairpin dimer formed by $d(G_3T_2AG_3)$, they determined a smaller value, -1.1 kcal mol^{-1}. Again, assuming two cations are bound to the three G-quartets, $\Delta G°(Na^+\rightarrow K^+)$ is -0.5 kcal mol^{-1}.

There have also been several theoretical studies on the ion-induced stability of various G-quadruplexes. For example, Ross and Hardin[87] performed FEP/MD calculations on the relative affinity of the antiparallel $d(T_2G_4)_4$ quadruplex for K^+ and Na^+ ions. Using two sets of ionic Lenard-Jones "6–12" van der Waals parameters, the authors obtained a preferential binding of Na^+ over K^+ by 4.3 kcal mol^{-1} per cation. This is clearly inconsistent with the experimental results. Recently, Gu and Leszczynski[88] reported *ab initio* calculations for two G_4-M^+-G_4 models (M = K and Na). At the HF/6–311G(d,p)//HF/6–31G(d,p) level, their calculations predicted a free-energy difference of -5.9 kcal mol^{-1} between K^+ and Na^+ binding, correctly reproducing the order of cation selectivity. Meyer et al.[89] have also examined the binding of metal ions to G-quartets. They reported that the interaction energy difference between Na^+-G_4 and K^+-G_4 is approximately 36 kcal mol^{-1}. After taking into consideration the hydration free-energy difference between Na^+ and K^+, Meyer et al.[89] predicted that Na^+-G_4 would be about 18 kcal mol^{-1} lower in energy than K^+-G_4, indicating that Na^+ is preferred over K^+ by a G-quartet in aqueous solution. This prediction is clearly inconsistent with the experimental findings. It should be pointed out that in the model of Meyer et al., the cation is assumed to be in the G-quartet plane (class II). Although class II cations have been observed, as discussed in Section 13.3.1, it is safe to conclude that in-plane binding to a G-quartet does not contribute significantly to the stability of the G-quadruplex structure. However, it is possible that class II binding at the termini of a G-quadruplex may contribute partially to structural polymorphism between sodium and potassium forms.

Another striking feature of the data presented in Figure 13.15 is that the affinity sequence observed for the G-quadruplex channel site is remarkably similar to that for K$^+$ ion channel proteins.[82] The structural basis for the observed similarity in cation selectivity becomes quite clear after MacKinnon and coworkers published the first high-resolution crystal structure for a K$^+$ ion channel protein (KcsA) from the bacterium *Streptomyces lividans*.[78] Four K$^+$ binding sites were identified inside the selectivity filter of KcsA channel, each coordinating to eight carbonyl oxygen atoms from four signature sequences, Thr[75]-Val[76]-Gly[77]-Tyr[78]. At each binding site, the K$^+$ ion resides near the center of a bipyramidal antiprism with a mean K-O distance of 2.85 Å. This type of cation coordination environment is remarkably similar to those found in G-quadruplex structures, as already discussed in Section 13.3.1.

13.4 ALKALI METAL ION BINDING IN DOUBLE-STRANDED NUCLEIC ACIDS

Alkali metal ion binding is also important in double-stranded nucleic acids. One interesting area that has drawn considerable attention recently is related to the question of whether Na$^+$ ions are located in the minor groove region of B-DNA. The long-standing view about the minor groove in B-DNA is that the region is filled with structured water molecules forming a "spine of hydration."[90] However, a molecular dynamics simulation study by Beveridge and coworkers indicates that the minor groove is strongly electronegative and may be a potential binding site for Na$^+$ ions.[4] Now experimental results from x-ray crystallography[91–94] and NMR[10–12,95] all support the picture that monovalent cations can indeed occupy (maybe partially) the minor groove of A tract DNA. Although crystallographic evidence clearly shows that partially dehydrated K$^+$, Rb$^+$, Cs$^+$, and Tl$^+$ ions are coordinated to the ApT step of the minor groove, direct detection of Na$^+$ ions continues to be a challenge. The following statement by Egli describes the situation adequately: "Suffice it to say that it is basically impossible to locate Na$^+$ ions with tetrahedral coordination geometry and partial occupancies in electron density maps of virtually any resolution without resorting to metal ions 'mimicking' Na$^+$ in combination with anomalous scattering techniques" (p. 282).[8] However, because the bound Na$^+$ ions in the narrow minor groove are partially dehydrated, the heavier alkali metal ions used in anomalous scattering experiments must exhibit different affinities for the binding site. Our discussions in Section 13.3.3 clearly demonstrate this point. Because of this discrepancy in binding affinity, to what extent heavier alkali metal ions can "mimic" Na$^+$ is uncertain. For this reason, it is always desirable to be able to detect Na$^+$ ions directly. Solid-state ^{23}Na NMR can be used to detect Na$^+$ ions in the minor groove of B-DNA. As we demonstrated in Section 13.3.2, the challenge is to establish solid-state ^{23}Na NMR signature for such Na$^+$ ions. Our previous data on Na-ionophore complexes may be useful in this respect[96] because the minor groove of B-DNA can also be considered as an ionophore.[97] In addition, a miniduplex, [r(AU)]$_2$, may also be used as a model,[97] because the Na$^+$ ion binding in this compound is identical to that observed in B-DNA. Once the Na$^+$ binding site is identified, it is possible to carry out affinity measurement as we described in Section 13.3.3.

13.5 CONCLUDING REMARKS

We have demonstrated that solid-state alkali metal NMR is a viable new method for detecting Na^+ and K^+ ions in nucleic acids and related systems. This approach offers new possibilities for monitoring Na^+ and K^+ ions in biological processes. The complementary nature of solid-state NMR to crystallography is also attractive. The benefit of the solid-state NMR approach ranges from easy sample preparation, freedom from cryogenic artifacts, and additional information about the local electronic environment at the binding site.

It is also important to address the question of whether the NMR sensitivity is sufficient for detecting Na^+ or K^+ ions in large biomolecular systems. We have shown that the ^{23}Na NMR signal from a single Na^+ ion in a G-quadruplex of ca. 9 kDa molecular weight can be detected with confidence at a moderate magnetic field strength, 11.75 T. The very high spectral resolution (less than 1 ppm) observed in ^{23}Na MQMAS spectra strongly indicates that solid-state ^{23}Na NMR holds great promise to become a practical technique for obtaining detailed information about Na^+ binding sites. Meanwhile ^{39}K NMR sensitivity at magnetic fields greater than 19.6 T is sufficient for 1D MAS experiments; however, it is more challenging to obtain high-quality MQMAS spectra for ^{39}K-nucleotides. As very high magnetic fields are becoming available to the NMR community, including resistive magnets and hybrid magnets at 40 T or higher, the ^{23}Na and ^{39}K NMR sensitivity should be sufficient for studying large molecular systems. Active research is under way in our laboratory.

Although we have focused on ^{23}Na and ^{39}K NMR in this chapter, extension to other alkali metal ions is straightforward. As alkali metal ions are abundant in both proteins and nucleic acids, solid-state alkali metal NMR may find useful applications. One of the future directions is related to solid-state NMR studies of ion binding/transporting in ion channel proteins. An important advantage of solid-state NMR is the feasibility of studying ion channel proteins in their membrane-bound state. For many years, structural determination for membrane proteins has been a major driving force for the development of new solid-state NMR methodologies. If information about ion coordination and dynamics can be obtained directly by solid-state alkali metal NMR, it may be possible to link the structural aspect of ion channel proteins to their function (e.g., ion transport).

ACKNOWLEDGMENTS

Our research described here was supported by the Natural Sciences and Engineering Research Council (NSERC) of Canada. We wish to thank Professor Jeffery Davis (University of Maryland) and Dr. Zhehong Gan (National High Magnetic Field Laboratory) for collaborations.

REFERENCES

1. Shui, X., McFail-Isom, L., Hu, G.G. and Williams, L.D., The B-DNA dodecamer at high resolution reveals a spin of water on sodium, *Biochemistry*, 37, 8341, 1998.
2. Basu, S., Rambo, R.P., Strauss-Soukup, J., Cate, J.H., Ferré-D'Amaré, A.R., Strobel, S.A. and Doudna, J.A. A specific monovalent metal ion integral to the AA platform of the RNA tetraloop receptor, *Nat. Struct. Biol.*, 5, 986, 1998.
3. McFail-Isom, L., Sines, C.C. and Williams, L.D., DNA structure: cations in charge? *Curr. Opin. Struct. Biol.*, 9, 298, 1999.
4. Young, M.A., Jayaram, B. and Beveridge, D.L. Intrusion of counterions into the spin of hydration in the minor groove of B-DNA: fractional occupancy of electronegative pockets, *J. Am. Chem. Soc.*, 119, 59, 1997.
5. Cheatham, T.E. and Kollman, P.A., Molecular dynamics simulation of nucleic acids, *Ann. Rev. Phys. Chem.*, 51, 435, 2000.
6. Hud, N.V., Sklenár, V. and Feigon, J., Localization of ammonium ion in the minor groove of DNA duplexes in solution and the origin of DNA A-tract bending, *J. Mol. Biol.*, 286, 651, 1999.
7. Hud, N.V. and Polak, M., DNA-cation interactions: the major and minor grooves are flexible ionophores, *Curr. Opin. Struct. Biol.*, 11, 293, 2001.
8. Egli, M., DNA-cation interactions quo vadis? *Chem. Biol.*, 9, 277, 2002.
9. Tereshko, V., Wilds, C.J., Minasov, G., Prakash, T.P., Maier, M.A., Howard, A., Wawrzak, Z., Manoharan, M. and Egli, M. Detection of alkali metal ions in DNA crystals using state-of-the-art x-ray diffraction experiments, *Nucl. Acids Res.*, 29, 1208, 2001.
10. Halle, B. and Denisov, V.P., Magnetic relaxation dispersion studies of biomolecular solutions, *Meth. Enzymol.*, 338, 178, 2001.
11. Denisov, V.P. and Halle, B., Sequence-specific binding of counterions to B-DNA, *Proc. Natl. Acad. Sci. USA*, 97, 629, 2000.
12. Marincola, F.C., Denisov, V.P. and Halle, B., Competitive Na$^+$ and Rb$^+$ binding in the minor groove of DNA, *J. Am. Chem. Soc.*, 126, 6739, 2004.
13. Feigon, J., Butcher, S.E., Finger, L.D. and Hud, N.V., Solution nuclear magnetic resonance probing of cation binding sites on nucleic acids, *Meth. Enzymol.*, 338, 400, 2001.
14. Wu, G., Recent developments in solid state NMR of quadrupolar nuclei and applications to biological systems. *Biochem. Cell Biol.*, 76, 429, 1998.
15. Andrew, E.R., Bradbury, A. and Eades, R.G., Nuclear magnetic resonance spectra from a crystal rotated at high speed, *Nature (London)*, 182, 1659, 1958.
16. Frydman, L. and Harwood, J.S., Isotropic spectra of half-integer quadrupolar spins from bidimensional magic-angle spinning NMR, *J. Am. Chem. Soc.*, 117, 5367, 1995.
17. Medek, A., Harwood, J.S. and Frydman, L., Multiple-quantum magic-angle spinning NMR: a new method for the study of quadrupolar nuclei in solids, *J. Am. Chem. Soc.*, 117, 12779, 1995.
18. Katti, S.K., Seshadri, T.P. and Viswamitra, M.A., Structure of disodium guanosine 5-monophosphate heptahydrate, *Acta Crystallogr. Sect. B*, 37, 1825, 1981.
19. Wong, A. and Wu, G., Characterization of the pentacoordinate sodium ions in hydrated nucleoside 5′-phosphates by solid-state ^{23}Na NMR and quantum mechanical calculations, *J. Phys. Chem. A*, 107, 579, 2003.
20. Wong, A. and Wu, G., Experimental solid-state ^{23}Na NMR and computational studies of sodium ethylenediaminetetraacetaes: site resolution and spectral assignment, *Can. J. Anal. Sci. Spectrosc.*, 46, 188, 2001.

21. He, H., Klinowski, J., Saba, G., Casu, M. and Lai, A., ^{23}Na NMR studies of Na-DNA in the solid state, *Solid State Nucl. Magn. Reson.*, 10, 169, 1998.
22. Ding, S. and McDowell, C.A. ^{23}Na solid-state NMR studies of hydrated disodium adenosine triphosphate, *Chem. Phys. Lett.*, 320, 316, 2000.
23. Madeddu, M., The influence of water content on the ^{23}Na and ^{31}P NMR spectral parameters in solid Na-DNA, *Solid State Nucl. Magn. Reson.*, 22, 83, 2002.
24. Grinshtein, J., Grant, C.V. and Frydman, L., Separate-local-field NMR spectroscopy on half-integer quadrupolar nuclei, *J. Am. Chem. Soc.*, 124, 13344, 2002.
25. Smith, M.E., Recent progress in solid-state NMR of low-γ nuclei, *Annu. Rep. NMR Spectrosc.*, 43, 121, 2001.
26. Padiyar, G.S. and Seshadri, T.P., Metal-nucleotide interactions: crystal structures of alkali (Li$^+$, Na$^+$, K$^+$) and alkaline earth (Ca^{2+}, Mg^{2+}) metal complexes of adenosine 2′-monophosphate, *J. Biomol. Struct. Dyn.*, 15, 803, 1998.
27. Adamiak, D.A. and Saenger, W., Structure of the monopotassium salt of adenosine 5′-diphosphate dihydrate, KADP·2H$_2$O, *Acta Crystallgr. Sect. B*, 36, 2585, 1980.
28. Gellert, M., Lipsett, M.N. and Davies, D.R., Helix formation by guanylic acid, *Proc. Natl. Acad. Sci. USA*, 48, 2013, 1962.
29. Guschlbauer, W., Chantot, J.-F. and Thiele, D., Four-stranded nucleic acid structures 25 years later: from guanosine gels to telomer DNA, *J. Biomol. Struct. Dyn.*, 8, 491, 1990.
30. Sen, D. and Gilbert, W., Guanine quartet structures, *Methods Enzymol.*, 211, 191, 1992.
31. Williamson, J.R., G-quartet structures in telomeric DNA, *Annu. Rev. Biophys. Biomol. Struct.*, 23, 703, 1994.
32. Gilbert, D.E. and Feigon, J., Multistranded DNA structures, *Curr. Opin. Struct. Biol.*, 9, 305, 1999.
33. Keniry, M.A. Quadruplex structures in nucleic acids, *Biopolymers*, 56, 123, 2001.
34. Neidle, S. and Parkinson, G.N., The structure of telomeric DNA, *Curr. Opin. Struct. Biol.*, 13, 275, 2003.
35. Davis, J.T., G-quartets 40 years later: from 5′-GMP to molecular biology and supramolecular chemistry, *Angew. Chem. Int. Ed.*, 43, 668, 2004.
36. Watson, J.D. and Crick, F.H., Molecular structure of nucleic acids: a structure for deoxyribose acid, *Nature (London)*, 171, 737, 1953.
37. Zimmerman, S., X-ray study by fiber diffraction methods of a self-aggregate of guanosine-5′-monophosphate with the same helical parameters as poly(rG), *J. Mol. Biol.*, 106, 663, 1976.
38. Lipanov, A.A., Quintana, J. and Dickerson, R.E., Disordered single crystal evidence for a quadruple helix formed by guanosine 5′-monophosphate, *J. Biomol. Struct. Dyn.*, 3, 483, 1990.
39. Kang, C., Zhang, X., Ratliff, R., Moyzis, R. and Rich, A. Crystal structure of four-stranded *Oxytricha* telomeric DNA, *Nature (London)*, 356, 126, 1992.
40. Pinnavaia, T.J., Miles, H.T. and Becker, E.D., Self-assembled 5′-guanosine monophosphate. Nuclear magnetic resonance evidence for a regular, ordered structure and slow chemical exchange, *J. Am. Chem. Soc.*, 97, 7198, 1975.
41. Pinnavaia, T.J., Marshall, C.L., Mettler, C.M., Fisk, C.L., Miles, H.T. and Becker, E.D., Alkali metal ion specificity in the solution ordering of a nucleotide, 5′-guanosine monophosphate, *J. Am. Chem. Soc.*, 100, 3625, 1978.
42. Fisk, C.L., Becker, E.D., Miles, T.H. and Pinnavaia, T.J., Self-structured guanosine 5′-monophosphate. A ^{13}C and ^{1}H magnetic resonance study, *J. Am Chem. Soc.*, 104, 3307, 1982.

43. Borzo, M., Detellier, C., Laszlo, P. and Paris, A., ^{1}H, ^{23}Na, and ^{31}P NMR studies of the self-assembly of the 5'-guanosine monophosphate dianion in neutral aqueous solution in the presence of sodium cations, *J. Am. Chem. Soc.*, 102, 1124, 1980.

44. Detellier, C. and Laszlo, P., Role of alkali metal and ammonium cations in the self-assembly of the 5'-guanosine monophosphate dianion, *J. Am. Chem. Soc.*, 102, 1135, 1980.

45. Delville, A., Detellier, C. and Laszlo, P., Determination of the correlation time for a slowly reorienting spin-3/2 nucleus: binding of Na$^+$ with the 5'-GMP supramolecular assembly, *J. Magn. Reson.*, 34, 301, 1979.

46. Laughlan, G., Murchie, A.I.H., Norman, D.G., Moore, M.H., Moody, P.C.E., Lilley, D.M.J. and Luisi, B., The high-resolution crystal structure of a parallel-stranded guanine tetraplex, *Science (Washington, DC)*, 265, 520, 1994.

47. Phillips, K., Dauter, Z., Murchie, A.I.H., Lilley, D.M.J. and Luisi, B., The crystal structure of a parallel-stranded guanine tetraplex at 0.95 Å resolution, *J. Mol. Biol.*, 273, 171, 1997.

48. Hud, N.V., Smith, F.W., Anet, F.A.L. and Feigon, J., The selectivity for K$^+$ versus Na$^+$ in DNA quadruplexes is dominated by relative free energies of hydration: a thermodynamic analysis by ^{1}H NMR, *Biochemistry*, 35, 15383, 1996.

49. Forman, S.L., Fettinger, J.C., Pieraccini, S., Gottarelli, G. and Davis, J.T., Toward artificial ion channels: a lipophilic G-quadruplex, *J. Am. Chem. Soc.*, 122, 4060, 2000.

50. Kotch, F.W., Fettinger, J.C. and Davis, J.T., A lead-filled G-quadruplex: insight into the G-quartet's selectivity for Pb^{2+} over K$^+$, *Org. Lett.*, 2, 3277, 2000.

51. Deng, J., Xiong, Y. and Sundaralingam, M., X-ray analysis of an RNA tetraplex (UGGGGU)$_4$ with divalent Sr^{2+} ions at subatomic resolution (0.61 Å), *Proc. Natl. Acad. Sci. USA*, 2001, 98, 13665.

52. Horvath, M.P. and Schultz, S.C., DNA G-quartets in a 1.86 Å resolution structure of an *Oxytricha nova* telomeric protein-DNA complex, *J. Mol. Biol.*, 310, 367, 2001.

53. Shi, X., Fettinger, J.C. and Davis, J.T., Homochiral G-quadruplexes with Ba^{2+} but not with K$^+$: the cation programs enantiomeric self-recognition, *J. Am. Chem. Soc.*, 123, 6738, 2001.

54. Shi, X.D., Fettinger, J.C. and Davis, J.T., Ion-pair recognition by nucleoside self-assembly: guanosine hexadecamers bind cations and anions, *Angew. Chem. Int. Ed.*, 40, 2827, 2001.

55. Wong, A., Fettinger, J.C., Forman, S.L., Davis, J.T. and Wu, G., The sodium ions inside a lipophilic G-quadruplex channel as probed by solid-state ^{23}Na NMR, *J. Am. Chem. Soc.*, 124, 742, 2002.

56. Haider, S., Parkinson, G.N. and Neidle, S., Crystal structure of the potassium form of an *Oxytricha nova* G-quadruplex, *J. Mol. Biol.*, 320, 189, 2002.

57. Parkinson, G.N., Lee, M.P.H. and Neidle, S., Crystal structure of parallel quadruplexes from human telomeric DNA, *Nature (London)*, 417, 876, 2002.

58. Clark, G.R., Pytel, P.D., Squire, C.J. and Neidle, S., Structure of the first parallel DNA quadruplex-drug complex, *J. Am. Chem. Soc.*, 125, 4066, 2003.

59. Haider, S.M., Parkinson, G.N. and Neidle, S., Structure of a G-quadruplex-ligand complex, *J. Mol. Biol.*, 326, 117, 2003.

60. Pan, B., Xiong, Y., Shi, K., Deng, J. and Sundaralingam, M., Crystal structure of an RNA purine-rich tetraplex containing adenine tetrads: implications for specific binding in RNA tetraplexes, *Structure*, 11, 815, 2003.

61. Pan, B., Xiong, Y., Shi, K., Deng, J. and Sundaralingam, M., An eight-stranded helical fragment in RNA crystal structure: implications for tetraplex interaction, *Structure*, 11, 825, 2003.

62. Pan, B., Xiong, Y., Shi, K. and Sundaralingam, M., Crystal structure of a bulged RNA tetraplex at 1.1 Å resolution: implications for a novel binding site in RNA tetraplex, *Structure*, 11, 1423, 2003.
63. Kondo, J., Adachi, W., Umeda, S., Sunami, T. and Takénaka, A., Crystal structures of a DNA octaplex with I-motif of G-quartets and its splitting into two quadruplexes suggest a folding mechanism of eight tandem repeats, *Nucl. Acids Res.*, 32, 2541, 2004.
64. Cáceres, C., Wright, G., Gouyette, C., Parkinson, G. and Subirana, J.A., A thymine tetrad in d(TGGGGT) quadruplexes stabilized with Tl⁺/Na⁺ ions, *Nucl. Acids Res.*, 32, 1097, 2004.
65. Wu, G. and Wong, A., Direct detection of the bound sodium ions in self-assembled 5′-GMP gels: a solid-state ²³Na NMR approach, *Chem. Commun.*, 2658, 2001.
66. Rovnyak, D., Baldus, M., Wu, G., Hud, N.V., Feigon, J. and Griffin, R.G., Localization of ²³Na⁺ in a DNA quadruplex by high-field solid-state NMR, *J. Am. Chem. Soc.*, 122, 11423, 2000.
67. Arnott, S., Chandrasekaran, R. and Marttila, C.M., Structures of polyinosinic acid and polyguanylic acid, *Biochem. J.*, 141, 537, 1974.
68. Zimmerman, S., Cohen, G.H. and Davies, D.R., X-ray fiber diffraction and model-building study of polyguanylic acid and polyinosinic acid, *J. Mol. Biol.*, 92, 181, 1975.
69. Miles, H.T. and Frazier, J., Poly(I) helix formation. Dependence on size-specific complexing to alkali metal ions, *J. Am. Chem. Soc.*, 100, 8037, 1978.
70. Wu, G. and Wong, A. Solid-state ²³Na NMR determination of the number and coordination of sodium cations bound to *Oxytricha nova* telomere repeat d(G₄T₄G₄), *Biochem. Biophys. Res. Commun.*, 323, 1139, 2004.
71. Smith, F.W. and Feigon, J., Quadruplex structure of *Oxytricha* telomeric DNA oligonucleotides, *Nature (London)*, 356, 164, 1992.
72. Schultze, P., Hud, N.V., Smith, F.W. and Feigon, J., The effect of sodium, potassium and ammonium ions on the conformation of the dimeric quadruplex formed by the *Oxytricha nova* telomere repeat oligonucleotide d(G₄T₄G₄), *Nucl. Acids Res.*, 27, 3018, 1999.
73. Hud, N. V., Schultze, P., Sklená, V. and Feigon, J., Binding sites and dynamics of ammonium ions in a telomere repeat DNA quadruplex, *J. Mol. Biol.*, 285, 233, 1999.
74. Wu, G., Wong, A., Gan, Z. and Davis, J.T., Direct detection of potassium cations bound to G-quadruplex structures by solid-state ³⁹K NMR at 19.6 T, *J. Am. Chem. Soc.*, 125, 7182, 2003.
75. Tougard, P., Chantot, J.-F. and Guschlbauer, W., Nucleoside conformations. X. An x-ray fiber diffraction study of the gels of guanine nucleosides, *Biochim. Biophys. Acta*, 308, 9, 1973.
76. Doyle, D.A., Cabral, J.M., Pfuetzner, R.A., Kuo, A., Gulbis, J.M., Cohen, S.L., Chait, B.T. and MacKinnon, R., The structure of the potassium channel: molecular basis of K⁺ conduction and selectivity, *Science (Washington, DC)*, 280, 69, 1998.
77. Morals-Cabral, J.H., Zhou, Y. and MacKinnon, R., Energetic optimization of ion conduction rate by the K⁺ selectivity filter, *Nature (London)*, 414, 37, 2001.
78. Zhou, Y., Morals-Cabral, J.H., Kaufman, A. and MacKinnon, R., Chemistry of ion coordination and hydration revealed by a K⁺ channel-Fab complex at 2.0 Å resolution, *Nature (London)*, 414, 43, 2001.
79. Pilch, D.S., Plum, G.E. and Breslauer, K.J., The thermodynamics of DNA structures that contain lesions or guanine tetrads, *Curr. Opin. Struct. Biol.*, 5, 334, 1995.

80. Hardin, C.C., Perry, A.G. and White, K., Thermodynamic and kinetic characterization of the dissociation and assembly of quadruplex nucleic acids, *Biopolymers*, 56, 147, 2001.

81. Wong, A. and Wu, G., Selective binding of monovalent cations to the stacking G-quartet structure formed by guanosine 5'-monophosphate: a solid-state NMR study, *J. Am. Chem. Soc.*, 125, 13895, 2003.

82. Eisenman, G. and Horn, R., Ionic selectivity revisited: the role of kinetic and equilibrium processes in ion permeation through channels, *J. Membr. Biol.*, 76, 197, 1983.

83. Chantot, J.F. and Guschlbauer, W., Physicochemical properties of nucleosides 3. Gel formation by 8-bromoguanosine, *FEBS Lett.*, 4, 173, 1969.

84. Raghuraman, M.K. and Cech, T.R., Effect of monovalent cation-induced telomeric DNA structure on the binding of *Oxytricha* telomeric protein, *Nucleic Acids Res.*, 18, 4543, 1990.

85. Scaria, P.V., Shire, S.J. and Shafer, R.H., Quadruplex structure of d($G_3T_4G_3$) stabilized by K$^+$ or Na$^+$ is an asymmetric hairpin dimer, *Proc. Natl. Acad. Sci. USA*, 89, 10336, 1992.

86. Balagurumoorthy, P. and Brahmachari, S.K., Structure and stability of human telomeric sequence, *J. Biol. Chem.*, 269, 21858, 1994.

87. Ross, W.S. and Hardin, C.C., Ion-induced stabilization of the G-DNA quadruplex: free energy perturbation studies, *J. Am. Chem. Soc.*, 116, 6070, 1994.

88. Gu, J. and Leszczynski, J., Origin of Na$^+$/K$^+$ selectivity of the guanine tetraplexes in water: the theoretical rationale, *J. Phys. Chem. A*, 106, 529, 2002.

89. Meyer, M., Steinke, T., Brandl, M. and Sühnel, J., Density functional study of guanine and uracil quartets and of guanine quartet/metal ion complexes, *J. Comp. Chem.*, 22, 109, 2001.

90. Drew, H. and Dickerson, R.E., Structure of a B-DNA dodecamer. III. Geometry of hydration, *J. Mol. Biol.*, 151, 535, 1981.

91. Shui, X., Sines, C., McFail-Isom, L., VanDerveer, D. and William, L.D., Structure of the potassium form of CGCGAATTCGCG deformation by electrostatic collapse around inorganic cations, *Biochemistry*, 37, 16877, 1998.

92. Tereshko, V., Minasov, G. and Egli, M.A, "Hydrat-ion" spine in a B-DNA minor groove, *J. Am. Chem. Soc.*, 121, 3590, 1999.

93. Woods, K.K., McFail-Isom, L., Sines, C.C., Howerton, S.B., Stephens, R.K. and Williams, L.D., Monovalent cations sequester within the A-tract minor groove of [d(CGCGAATTCGCG)]$_2$, *J. Am. Chem. Soc.*, 122, 1546, 2000.

94. Howerton, S.B., Sines, C.C., VanDerveer, D. and Williams, L.D., Locating monovalent cations in the grooves of B-DNA, *Biochemistry*, 40, 10023, 2001.

95. Hud, N.V., Sklenar, V. and Feigon, J., Localization of ammonium ions in the minor groove of DNA duplexes in solution and the origin of DNA A-tract bending, *J. Mol. Biol.*, 285, 233, 1999.

96. Wong, A. and Wu. G., Solid-state ^{23}Na nuclear magnetic resonance of sodium complexes with crown ethers, cryptands, and naturally occurring antibiotic ionophores: a direct probe to the sodium-binding sites, *J. Phys. Chem. A*, 104, 11844, 2000.

97. Seeman, N.C., Rosenberg, J.M., Suddath, F.L., Park Kim, J.J. and Rich, A., RNA double-helical fragments at atomic resolution, *J. Mol. Biol.*, 104, 109, 1976.

Index

A

Afonin studies, 215–233
Agre and MacKinnon studies, 238
Alanine, *222*, 222–223
Alkali metal ions, nucleic acids and related
     systems
    basics, 317–318, 339
    bound Na⁺ and K⁺ ions, 328–329, *329–333*,
        331–334
    crystallographic data, *324,* 325, *326–327,*
        327–328
    double-stranded nucleic acids, 338
    G-quadruplexes, *324,* 324–338
    site-specific affinity, 334–338, *335–336*
    SSNMR techniques, 319–322, *320–324,* 324
Andrew and Lowe studies, 2
ϕ angle determination, 94–96, *95, 97–98,*
        114–118, *116–118*
ψ angle determination, 101–109, *102–103,*
        105–108, 110, 117–118, *117–118*
χ_1 angle determination, 114–116, *116–117*
Antimicrobial peptides
    gramicidin S, *217, 227–228,* 227–229
    K3, *230,* 231–232, *232*
    PGLa, *221,* 229–231, *230*
ANTIOPE computer program, 274
Antzutkin and Tycko studies, 90
Applications
    sensitivity enhancement, 164–166, 169, 172
    spectral assignment of proteins, 42–48
Arrairan, Vilma, 82

B

Bacillus brevis, 227
Backbone correlations, 66–67, *68,* 69
Balagurumoorthy and Brahmachari studies, 337
Baldus, Lange, Luca and, studies, 22
Baldus studies, 3, 39–49
Bax studies, 40
Benamira, Soraya, 233
Bennett studies, 12
Berditchevskaia studies, 215–233

Beveridge studies, 338
Bicelles, 178–185, *179, 181*
Bilayer environments, *see* Uniaxial motional
    averaging
Biological solids, *see* Torsion angle
    determination, biological solids
Biomaterial surfaces, *see* Peptide studies,
    biomaterial surfaces
Biomolecular applications, sensitivity
    enhancement, 166, 169, *170–171,*
    172
Blanco, Tycko and, studies, 130
BlochLib computer program, 274
Bökmann studies, 33, 44
Boltzmann polarization, 2
Bound Na⁺ and K⁺ ions, 328–329, *329–333,*
    331–334
Bower studies, 123–146
B18 peptide, 224–227, *226*
Brahmachari, Balagurumooorthy and, studies,
    337

C

C7, 6, 280, 282
Campbell studies, 132
Case, Dejaegere and, studies, 310, 313
Case, Sitkoff and, studies, 310
Castellani studies, 28
Castner, David, 146
Cataldi, Marcela, 82
Cations, 327–328
Cavity cations, 327–328
C'-C' distances, 114, *115*
Cech, Raghuraman and, studies, 337
2-CF₃-alanine, *222,* 223
4-CF₃-phenylglycine, *218, 221,* 221–222
Chan and Tycko studies, 109
Chantot and Guschlbauer studies, 336
Chemical labeling strategies, 219–223
Chemical shielding tensors (CST), 306
Chemical shift anisotropy (CSA), 109–114,
    111–113, see also Uniaxial
    motional averaging